Encyclopedia of Climate Models

Encyclopedia of Climate Models

Edited by **Andrew Hyman**

R CALLISTO REFERENCE

New York

Published by Callisto Reference,
106 Park Avenue, Suite 200,
New York, NY 10016, USA
www.callistoreference.com

Encyclopedia of Climate Models
Edited by Andrew Hyman

International Standard Book Number: 978-1-63239-224-4 (Hardback)

Printed in the United States of America.

Contents

Preface

I am honored to present to you this unique book which encompasses the most up-to-date data in the field. I was extremely pleased to get this opportunity of editing the work of experts from across the globe. I have also written papers in this field and researched the various aspects revolving around the progress of the discipline. I have tried to unify my knowledge along with that of stalwarts from every corner of the world, to produce a text which not only benefits the readers but also facilitates the growth of the field.

This book provides a collection of research studies done by various international scientists. These research works target our comprehension regarding physical environment for life on Earth. The studies within the book describe the recent developments in techniques and procedures for computer based simulation of environment variability. Planetary scale phenomena, regional ecology, effects of air pollution, floods and heat waves are some of the topics that it discusses. This book provides an insight to the modern methodologies and techniques used for a better understanding of the environment.

Finally, I would like to thank all the contributing authors for their valuable time and contributions. This book would not have been possible without their efforts. I would also like to thank my friends and family for their constant support.

Editor

Part 1

Atmospheric Models

Tropical Channel Model

Pallav Ray[1], Chidong Zhang[2], Jim Dudhia[3],
Tim Li[1] and Mitchell W. Moncrieff[3]
[1]International Pacific Research Center (IPRC), University of Hawaii,
[2]Rosenstiel School of Marine and Atmospheric Science (RSMAS), University of Miami,
[3]National Center for Atmospheric Research (NCAR),
USA

1. Introduction

Numerical models are primary tools used for weather forecast and climate projections. Richardson (1922) made the first effort to predict weather numerically. Subsequently, a number of simple numerical and analytical models were proposed to explain the general circulation of the atmosphere (e.g., Charney, 1948; Eady, 1949; Philips, 1956; Matsuno, 1966; Gill, 1980). The present state-of-the-art models are constantly being evaluated and refined with the help of observations and theory to better understand the earth's weather and climate. Our gradual progress in developing and utilizing complex numerical models have lead to a hierarchy of models: i) atmospheric global climate models (AGCMs, Smagornisky et al., 1965; Benwell and Bretherton, 1968; Phillips and Shukla, 1973; Manabe, 1975; Simmons and Bengtsson, 1984); ii) coupled atmosphere-ocean GCMs (Manabe and Bryan, 1969); iii) hydrostatic and non-hydrostatic regional models (Wang and Halpern, 1970; Dudhia, 1993); (iv) cloud-system resolving models with regional domains (Grabowski et al. 1998; Grabowski and Moncrieff, 2001) and global domains (Miura et al., 2007; Satoh et al., 2008). A comprehensive review of the present day numerical models can be found in Tao and Moncrieff (2009).

Each category of model has advantages and disadvantages in terms of area coverage, spatial resolution, computational efficiency, and the representations of the physical processes; i.e., convection, microphysics, radiation, and surface exchange. Typically, regional models have higher resolution albeit a limited computational domain. On the other hand, GCMs with lower resolution cut the advantage of global coverage. A recent development is the introduction of tropical channel models (TCMs) which are defined as models that are global in the zonal direction but bounded in the meridional direction. A TCM has the following considerable advantages over the aforementioned modeling approaches: (1) A standard regional model needs boundary conditions in the zonal and the meridional directions, whereas a TCM is continuous in the east-west direction and thereby isolates the influences that arrive solely from the meridional boundaries (i.e., extratropics). It also allows free circumnavigation of tropical modes in the zonal direction. (2) The use of lateral boundary conditions only in the meridional direction enables a controlled quantification of the effects of extratropical disturbances on the tropics. 3) TCMs with nonhydrostatic dynamical cores can have higher resolution and more sophisticated physics compared to GCMs. This is essential to capture the multi-scale organized convection in the tropics and its influence on the general circulation (Arakawa and Schubert, 1974; Chen et al., 1996; Houze, 2004; Moncrieff, 2010).

The objective of this chapter is to describe several TCMs that have been developed recently with an emphasis on their applications for the simulation and understanding of the tropical mean state and variabilty including the Madden-Julian oscillation (MJO), tropical cyclones (TCs) and double intertropical convergence zone (ITCZ).

Section 2 describes the models, data, and the design of the numerical simulations. Section 3 includes diagnoses of the simulations from three different TCMs. Section 4 summarizes the results along with the implications and limitations of the tropical channel model approach.

2. Model

We describe three different TCMs; two of them are constructed based on two different non-hydrostatic mesoscale models, and the third one is based on a hydrostatic GCM.

2.1 TCM based on MM5

A TCM was developed based on the fifth-generation Pennsylvania State University-National Center for Atmospheric Research (NCAR) Mesoscale Model (MM5; Dudhia, 1993; Grell et al., 1995). This atmosphere-only channel model employs a Mercator projection centered at the equator with open boundaries in the North-South direction. The model domain covers the entire tropics, with overlapping east-west boundaries. Tests ensured that simulated perturbations propagate zonally through the overlapping grids without distortion. Especially, two test runs are made in which the overlapping zone is located over the western Pacific and Atlantic Oceans, respectively. Results from these two simulations are the same over these regions with or without the overlapping zone. Hereafter, we refer to this tropical channel model as tropical MM5 (TMM5).

The dynamics and physics packages of TMM5 are same as those of the regular regional MM5, based on equations for a fully compressible, non-hydrostatic atmosphere. This TCM retains the two-way nesting capabilities. The spatial differencing is centered and of second order. There are 28 unevenly spaced full-sigma levels, with the maximum resolution in the boundary layer and the model top at 50 hPa. All nested domains are activated at the initial time of the simulation. The output is taken every 3 hours.

Several simulations are performed spanning between one to several weeks and are used to evaluate the skill of simulations by TMM5 against observations and reanalyses. Based on these tests and the work of Gustafson and Weare (2004a,b), the selected parameterizations are: (i) Betts-Miller convective scheme (Betts and Miller, 1986); (ii) explicit moisture calculations using a simple ice scheme (Dudhia, 1989); (iii) planetary boundary layer (PBL) scheme of the NCEP Eta model (Janjic, 1994), and (iv) Rapid Radiative Transfer Model longwave (RRTM, Mlïawer et al., 1997) and Dudhia (1989) shortwave radiation. The success of the Betts-Miller scheme, designed for the coarse resolution climate models, is perhaps because the simulations are in the hydrostatic regime at the resolutions employed (grid spacing of 111 and 37 km). Over the land, a 5-layer soil model option of the MM5 is used.

The model domains and the simulations are shown in Fig.1 and Table 1, respectively. The simulation 1DOM, with a model domain of 21°S-21°N, has a horizontal resolution of 111 km (D1 in Fig. 1). This is an ideal set up to study the intraseasonal varibility like the MJO which is of global scale in the zonal direction (Madden and Julian, 1971, 1972; Li and Zhou, 2009), and most of its variance is confined within the 20° latitude zones (Zhang and Dong, 2004). The purpose of the option of a two-way nested inner domain of 37 km over the Indian and western Pacific Oceans (D2 in Fig. 1) is to assess the effect of increasing horizontal resolution. This is our simulation 2DOM. Both simulations were integrated for four months from 1 March to 30 June, 2002, with sea surface temperature (SST) prescribed from observations (see section 2.4).

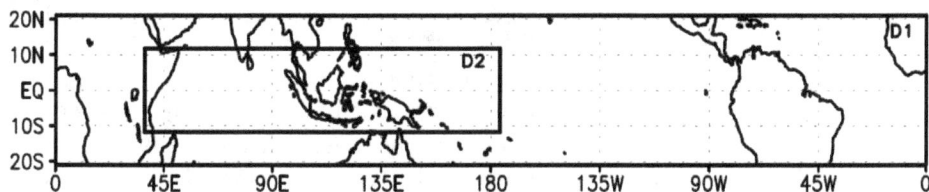

Fig. 1. Outer domain for the TMM5 (D1, 0-360°, 21°S-21°N) and the nested domain (D2, 37-183°E, 11°S-11°N). Domains D1 and D2 have resolutions of 111 km and 37 km, respectively.

Experiment	Integration Time	Remarks
1DOM	1 March- 30 June, 2002	Single domain simulation to test and validate the model
2DOM	1 March- 30 June, 2002	Nested domain (111 km / 37 km)

Table 1. The description of the simulations using the tropical channel MM5.

2.2 TCM based on WRF

Another tropical channel model is based on the Weather Research and Forecasting (WRF) model developed at the NCAR. We refer this as tropical WRF or TWRF (http://www.nrcm.ucar.edu). Conceptually, the configuration of TWRF is similar to TMM5. The horizontal grid-spacing of the TWRF is 36 km, and the meridional boundaries are placed at 30°S and 45°N. The model top is at 50 hPa, and 35 vertical levels are used. Output is archived every 3 hours. The configuration of the TWRF is shown in Fig. 2. The inner domains have grid-spacings of 12 km (domain D2 in Fig. 2) and 4 km (D3 in Fig. 2) respectively, and they are located over the warm pool region of the Indian and west Pacific oceans. Two-way interactions occur between the domains in the nested simulations. No cumulus parameterization is applied in the 4 km (cloud-system resolving) domain.

Preliminary simulations over the maritime continent evaluated the skill of the TWRF simulations. Based on these tests, the suite of parameterizations selected for this present study are: Kain-Fritsch cumulus parameterization (KF, Kain, 2004), WSM6 cloud microphysics (Hong et al., 2004), CAM 3.0 radiation scheme (Collins et al., 2006), YSU boundary layer scheme (Hong et al., 2006), and Noah land surface model (Chen and Dudhia, 2001). Note that the Betts-Miller scheme (BM, Betts and Miller, 1986) available within the MM5 differs from the Betts-Miller scheme (BMJ, Janjic, 1994) available within the WRF.

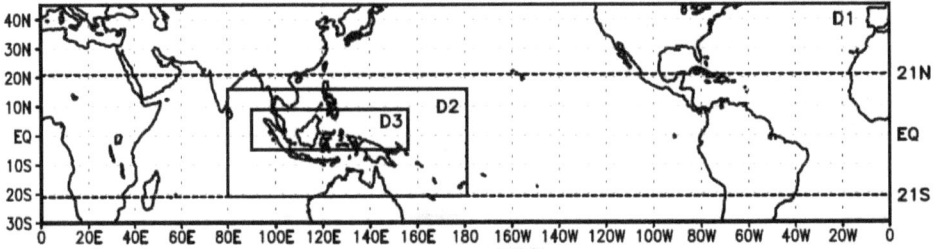

Fig. 2. Model domains for the tropical channel WRF (D1, 0°-360°, 30°S-45°N) and the nested domains (D2, 79°-183°E, 21°S-16°N and D3, 90°-157°E, 6°S-10°N). Domains D1, D2, and D3 have resolutions of 36 km, 12 km and 4 km, respectively. The simulation from 1996 to 2000 includes only the outer domain. The southern boundary was further moved to 45°S for the simulation from 1 December 1999 to 1 January 2006. The domain for the TMM5 is also marked here by the dashed lines (0°-360°, 21°S-21°N). See the text and Table 2 for further details.

The simulations using the TWRF are listed in Table 2. The simulation 1DOM in Table 2 is used to document the mean state. A similar multi-year simulation with a larger domain (1DOM_2 in Table 2) is also considered to evaluate its skill in capturing double ITCZ and tropical cyclones. Two other experiments (2DOM and 3DOM in Table 2) document the effect of increased horizontal resolution over the Indo-Pacific warm pool region.

Experiment	Integration Time	Description
1DOM	1 January 1996- 1 January 2001	1-way nested from NCEP-NCAR reanalysis with lateral boundaries at 30°S and 45°N
1DOM_2	1 December 1999-1 January 2006	Same as 1DOM, but with lateral boundaries at 45°S and 45°N
2DOM	1 January 1996-12 February 1998	2-way nested domains (36 km / 12 km)
3DOM	1 January 1997-1 July 1997	2-way nested domains (36 km / 12 km /4 km), 4 km domain is cloud resolving.

Table 2. The description of TWRF simulations. The horizontal resolution is 36 km for all the simulations except the nested runs (2DOM and 3DOM). See Fig. 2 for domain definitions and text for further details. [DOM : Domain].

2.3 TCM based on ECHAM4

TCMs based on regional models like MM5 and WRF, require boundary conditions from different datasets (other models or reanalysis). One disadvantage is that the variabilty in the prescribed boundary conditions may be different from the intrinsic variabilty produced by the model. To overcome this problem, a GCM-based framework, in which the boundary conditions come from the parallel simulation ('control') of the same model can be used. Such a GCM-based framework also presents a global view of the atmospheric variabilty and can be used for forecasting.

The GCM used is the atmosphere-only ECHAM4, which captures the tropical variability reasonably well (Sperber et al., 2005; Lin et al., 2006; Zhang et al., 2006). The model is integrated for 20 years using the prescribed monthly SST. This is our control simulation ('control' in Table 3). In the other experiment ('NS' in Table 3), the model prognostic variables are nudged toward the 'controlled' annual cycle state over the 20°-30° latitudinal zone (red in Fig. 3) to remove the extratropical influences without interfering with the influences from the zonal direction. By comparing the simulations in the control and the NS experiments, the influences of the extratropics on the tropics can be estimated.

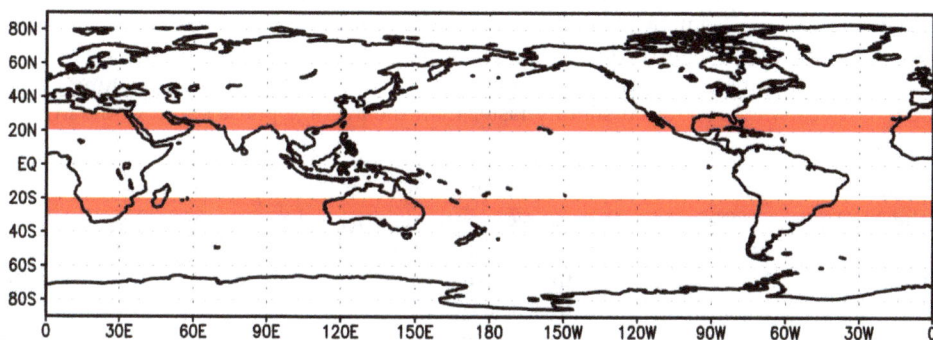

Fig. 3. Schematic diagram of the numerical experiments in which the prognostic variables in 20°-30° latitude zones (red) are relaxed toward the controlled climatological annual cycle. See text and Table 3 for further details.

Experiment	Description	Purpose
Control	ECHAM4 atmosphere only	To provide lateral boundary conditions for other simulations
NS	Relaxed to the annual cycle derived from the control simulation over 20°-30° latitudes	To evaluate the role of the extratropical influence on the tropics

Table 3. The description of the simulations using ECHAM4. (NS: North South).

2.4 Data

Model validation uses a number of observations and reanalyses data. They include: the NCEP-NCAR Re-analysis (Kalnay et al., 1996) winds; The European Centre for Medium-Range Weather Forecasts (ECMWF) 40-years Re-analysis (ERA40) winds (Uppala et al., 2005); Surface winds from the European Remote Sensing (ERS) satellites (Bentamy et al., 1999) and the NCEP-DOE Reanalysis (NCEP2, Kanamitsu et al., 2002); And two precipitation datasets including the merged analysis of precipitation (CMAP; Xie and Arkin, 1997), and the global precipitation climatological project (GPCP, Huffman et al., 1997) combined precipitation dataset.

The NCEP global tropospheric analyses (final or 'FNL' data, 1°x1°, 6 hourly) provide initial and lateral boundary conditions for the TMM5. The SSTs for TMM5 are also from this reanalysis, which contain intraseasonal fluctuations. The initial and boundary conditions of the TWRF are from the NCEP-NCAR reanalysis. The SSTs for TWRF are from the Atmospheric Model Intercomparison Project (AMIP; 1°x1°, 6-hourly; Taylor et al., 2000). For brevity, both reanalysis and CMAP/GPCP precipitation will be referred to as "observations".

3. Results

3.1 Tropical MM5 (TMM5)

The TMM5 simulations show considerable ability to capture an MJO event in comparison to most GCMs (Slingo et al., 1996; Lin et al., 2006; Zhang et al. 2006; Kim et al., 2009). The MJO event appeared in April-May 2002 (Fig. 4a), with the eastward propagating zonal wind anomalies switching from easterlies to westerlies on intraseasonal timescales. This event occured during a season in which the MJO is closest to the equator but, on average, is weaker than in other seasons (Zhang and Dong, 2004). After initiation, however, it propagated eastward at a speed slightly faster than the average phase speed of the MJO (5 m s⁻¹, marked by a straight line in Fig. 4a).

The start time (1 March) of the TMM5 simulations is about two months before the initiation of the MJO phase with active deep convection and low-level westerlies in May over the Indian Ocean. This choice of start time assesses the model's capability to reproduce the initiation of the MJO event, namely, the intraseasonal transition from low-level easterlies to westerlies (or from convectively inactive to active periods). In numerical models, forecast of future development of the MJO tends to have greater predictability when the MJO is already present at the initial time (Jones et al., 2000). Nevertheless, numerical experiments suggested predictability limit is about 10-15 days for rainfall, and about 25-30 days for upper-level winds (Waliser et al., 2003).

The results from the simulations 1DOM and 2DOM (Table 1) are shown in Fig. 4b and Fig. 4c, respectively. Simulated zonal winds at 850 hPa (hereafter U850) in D1 (middle panel) exhibits the same intraseasonal switch between easterly and westerly anomalies and eastward propagation over the Indian Ocean. Over the western Pacific, however, it moves faster than in reanalysis. This problem appears to be partially remedied by including the higher resolution nested domain D2 (Fig. 4c). In both simulations, the amplitudes of the anomalies are larger than that in reanalysis. Notice that the westward propagating synoptic-scale westerly anomalies embedded in the MJO envelope over the Indian Ocean are captured by the simulation with the nested domain (2DOM in Table 1). These detailed structures in simulated anomalies are perhaps due to the higher resolution of the model.

The most interesting result from this simulation is that the initiation of the MJO event over the Indian Ocean is reproduced by the model at about the same time as shown by reanalysis two months after the model initial time. The MJO is thought to be unpredictable beyond two to three weeks (e.g., Waliser et al., 2003). If this is correct, then the reproduction of the U850 anomalies by TMM5 cannot be attributed to the initial conditions.

The above results lead to a hypothesis that this MJO event is generated by the influences from the lateral boundaries. This hypothesis is supported by a series of sensitivity tests that demonstrate that the simulated MJO initiation is critically dependent on the time-varying lateral boundary conditions from the reanalysis (Ray et al., 2009). When such lateral boundary conditions are replaced by time-independent conditions, the model fails to reproduce the MJO initiation. In particular, the diagnoses of the zonal momentum budget for the MJO initiation region reveals that the advection by meridional winds is important prior to the initiation of this MJO (Ray and Zhang, 2010).

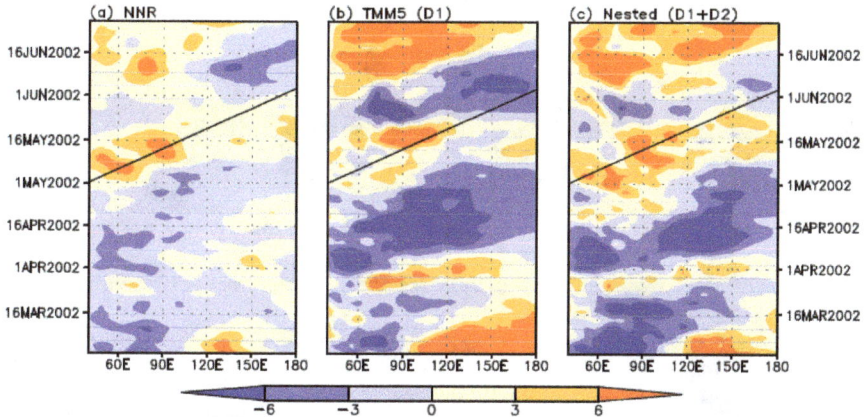

Fig. 4. Time-longitude diagrams of daily U850 anomalies (m s^{-1}) averaged over 10°S-10°N from the (a) NCEP-NCAR reanalysis (NNR), (b) TMM5 single domain simulation (1DOM in Table 1), and (c) TMM5 nested domain simulation (2DOM in Table 1). A 3-day running mean is applied.

3.2 Tropical WRF (TWRF)

We describe the performance of the TWRF in terms of its ability to capture the mean precipitation, double ITCZ, and hurricane statistics.

3.2.1 Mean state

Fig. 5 shows the mean precipitation from two different datasets and TWRF simulation (1DOM in Table 2). The model lacks precipitation over the equatorial Indian Ocean and the west African monsoonal region. However the model overestimates precipitation in the west Pacific, particularly in the region north of maritime continent and in the south Pacific convergence zone (SPCZ). The error in the mean state is found to be a primary reason for the poor simulation of the MJO in this model (Ray et al., 2011), even when higher resolution nested domains are included (2DOM and 3DOM in Table 2). However, such error does not seem to affect the simulation of convectively coupled Kelvin waves in the model (Tulich et al., 2011). Overall, precipitation is overestimated in the southern hemishere, and underestimated close to the equator (5°S-5°N, Murthi et al., 2011). The large bias in the precipitation over the southern Indian Ocean was thought to be due to the interactions between tropical cyclones and the southern boundaries. To rectify this problem, southern boundaries are further moved to 45°S in another experiment (1DOM_2 in Table 2). However, this does not improve the result significantly, indicating potential problems with the model physics (Tulich et al., 2011; Murthi et al., 2011). Nevertheless, the model reasonably captures the initiation of certain MJO events that are influenced by the extratropics (Ray et al., 2011), and genesis of tropical cyclones from easterly waves (see section 3.2.3).

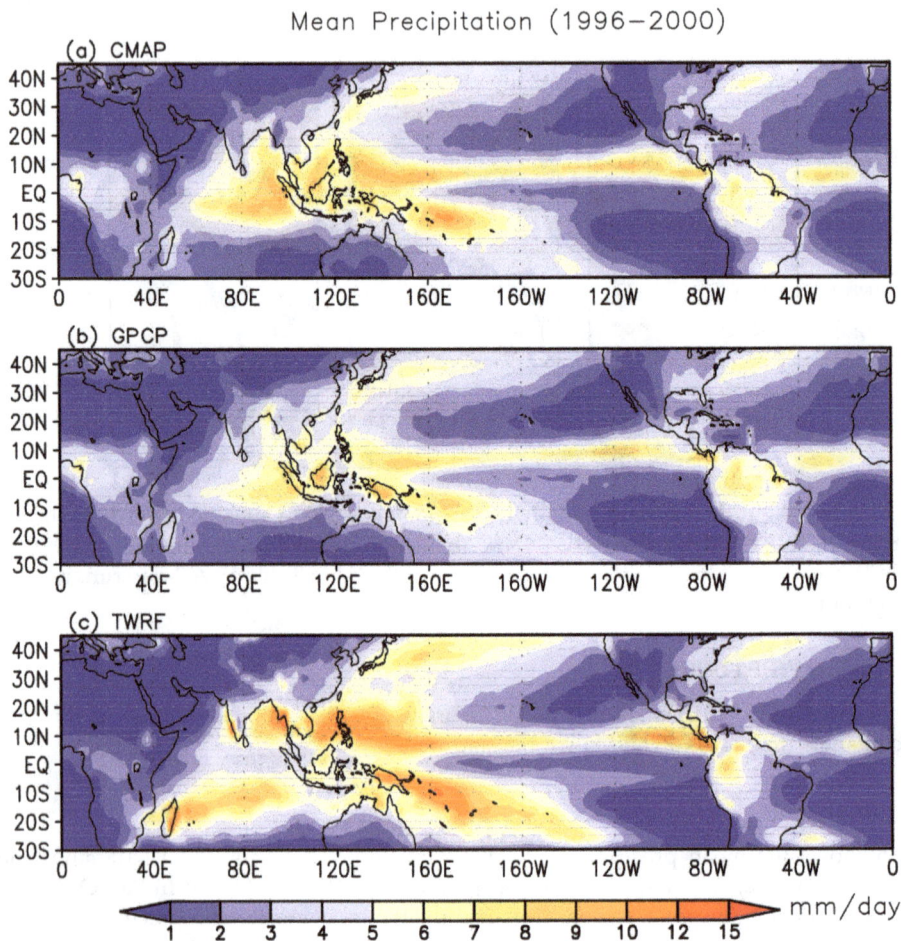

Fig. 5. Annual mean rainfall (mm day^{-1}) during 1996-2000 from the (a) CMAP (b) GPCP and (c) TWRF simulation 1DOM in Table 2.

3.2.2 Double ITCZ

It is known that the presence of a too-strong double ITCZ in the Pacific is a common bias in AGCMs (e.g., Meehl and Arblaster, 1998), as well as in the coupled GCMs (Mechoso et al., 1995). We examine the double ITCZ with respect to surface wind convergence and precipitation over the eastern Pacific, since the double ITCZ is most prominent over the eastern Pacific during March and April (Zhang, 2001).

Figure 6 shows the seasonal cycles in SST (shaded) and surface wind divergence (contoured) over the equatorial eastern Pacific (100°W-140°W). Note that our simulation period includes the strong ENSO event of 1997-98 (McPhaden, 1999). During March-April, the convergence south and north of the equator is captured well by the model. Over the equator, the observations show weak convergence, but the simulation indicates divergence. During boreal summer, the model shows much stronger convergence south of the equator compared to the observations. Overall, the TWRF has a distinct double ITCZ and no significant improvement occurred when nested domains were employed.

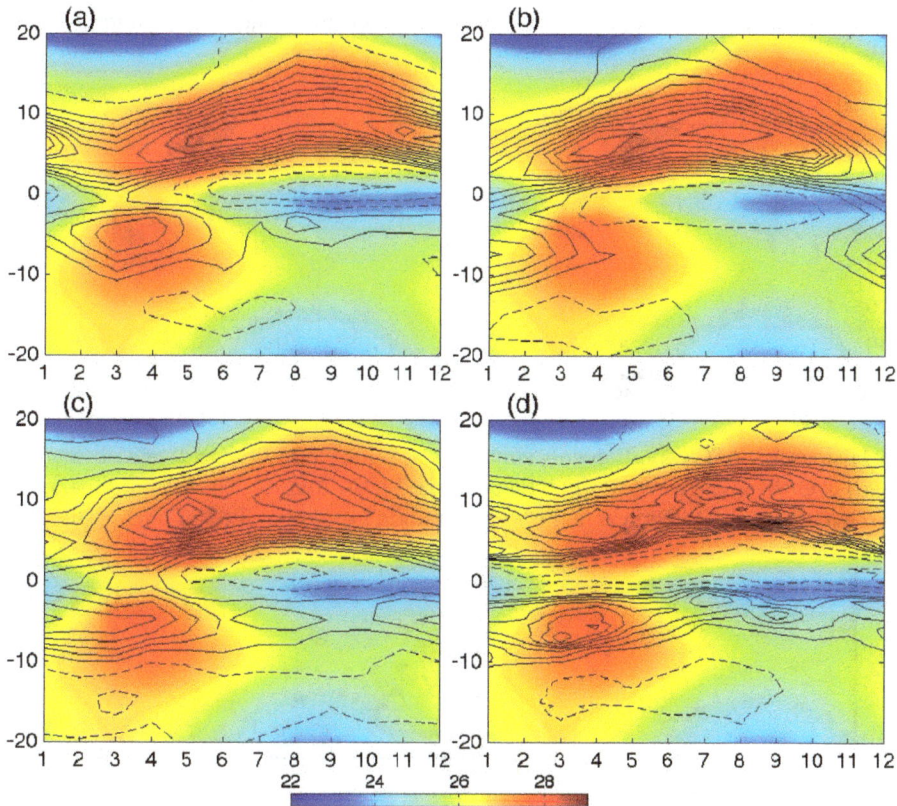

Fig. 6. Seasonal cycles of SST (shaded, °C) and surface wind divergence (contours, 10^6 s^{-1}) from the (a) ERS, (b) ERA40, (c) NCEP2, and (d) TWRF. All are averaged over the eastern Pacific (100°W-140°W) during the period 1996-2000. Solid (dashed) contours represent convergence (divergence).

3.2.3 Tropical cyclones

Using an objective tracking algoritm similar to Walsh et al. (2004), Tulich et al. (2011) estimated the genesis locations and tracks of tropical cyclones from the model simulation 1DOM_2 (Table 2). Fig. 7 shows the latitudinal distribution of TC genesis events over three different ocean basins. The error in the distribution is presumably due to the error in the model's easterly wave climatology that is intimately linked to the genesis of TCs (Landsea, 1993; Frank and Roundy, 2006). The model greatly overestimates the number of storms over the western Pacific (Fig. 7a), but underestimates over the north Atlantic (Fig. 7c). The model also overestimates over the eastern Pacific (Fig. 7b). Overall, while the TWRF simulation captures the global and seasonal distribution of tropical cyclones, their numbers are overestimated.

However, the simulation with a nested domain of 12 km over the Atlantic during the 2005 hurricane season produces resonable cyclone statistics (Fig. 8, Done et al., 2011). Compared to 27 observed tropical storms, the model produces 29 storms, indicating great improvement with the increase in the horizontal resolution.

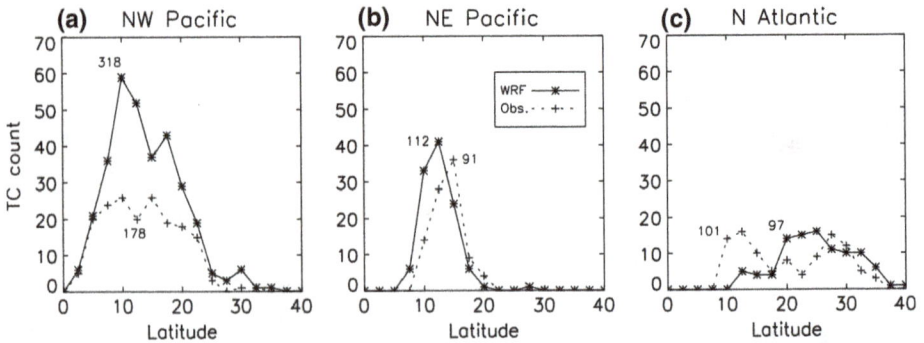

Fig. 7. Histogram of TC genesis events versus latitude for the TWRF (sold) and observations (dashed) over the (a) NW Pacific, (b) NE Pacific, and (c) N Atlantic. The total number of genesis events are indicated next to the corresponding histogram. (From Tulich et al., 2011).

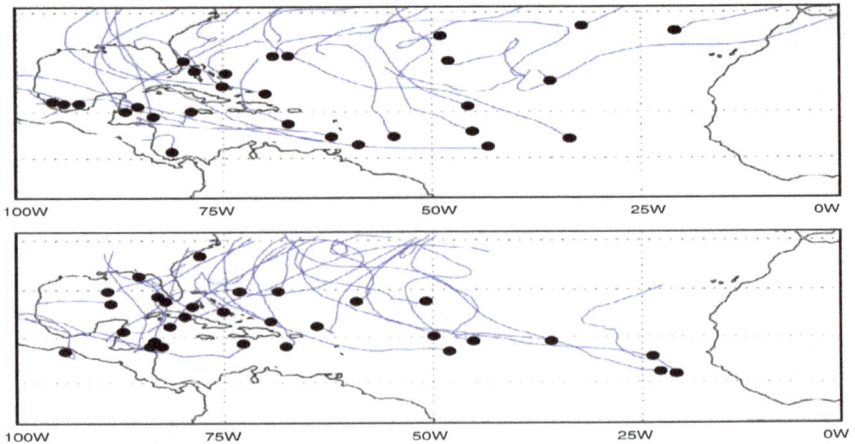

Fig. 8. Initial locations (black circles) and tracks (blue lines) of tropical cyclones in the Atlantic Basin during 2005 in (top) the IBTrACS dataset and (bottom) TWRF simulation. (From Done et al., 2011).

3.3 Tropical ECHAM4

Fig. 9 compares the simulated mean state from the observations and the model over the Indo-Pacific region. Lower tropospheric winds are westerly over the Indian Ocean (Fig. 9a, shaded) but easterlies prevail in the upper troposphere (Fig. 9a, contoured). This is captured well by the control simulation (Fig. 9b). However, in the experiment NS (Table 3), U850 is weaker over the equatorial western Indian Ocean, and U200 is of opposite sign over much of the Indian Ocean (Fig. 9c), and is underestimated over the west Pacific. The precipitation and OLR in the control (Fig. 9e) are similar to those in the observations (Fig. 9d), although precipitation (OLR) is everestimated (underestimated). Note that unlike TWRF, the control simulation does not lack precipitation over the equatorial Indian Ocean (Fig. 5c). However, in the NS, precipitation is significantly reduced over the equatorial Indian Ocean, and is increased over the SPCZ region (Fig. 9f). The ITCZ is shifted further from the equator in NS, with very little precipitation north of the equatorial Pacific. The MJO variance in the NS is also reduced substantially compared to the control indicating possible influences from the extratropics. To what extent the mean state is responsible for the lack of variability remains to be investigated more thoroughly (Ray and Li, 2011).

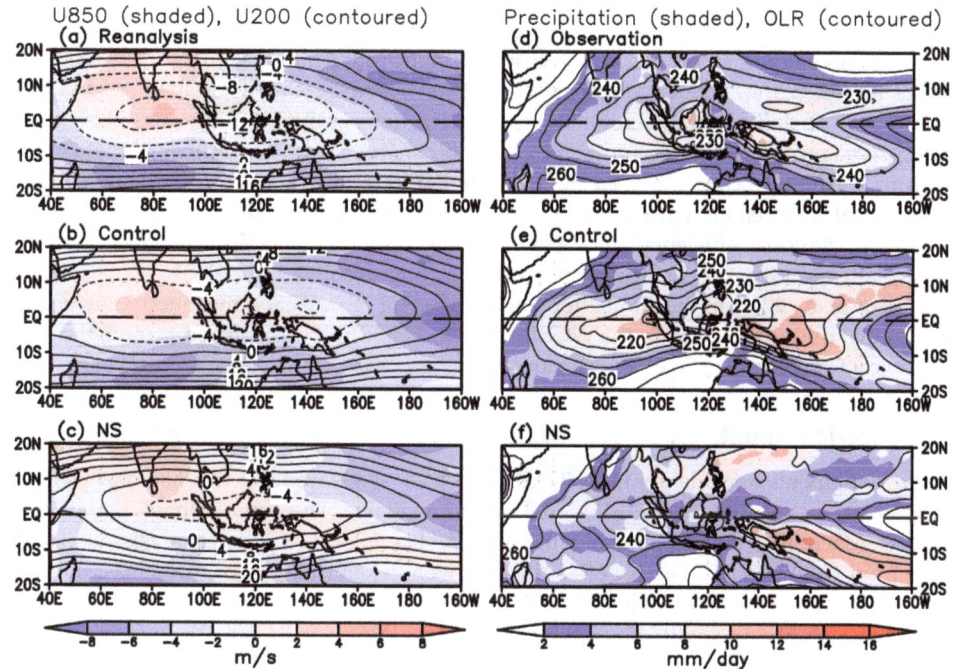

Fig. 9. (left) Mean zonal winds at 850 hPa (U850, m s⁻¹, shaded) and at 200 hPa (U200, m s⁻¹, contoured). (right) Mean precipitation (mm day⁻¹, shaded) and OLR (W m⁻², contoured). All are averaged over 20 years. Contour intervals are 4 m s⁻¹ for U200, and 10 W m⁻² for OLR. For U200, solid (dashed) lines represent westerlies (easterlies).

4. Conclusion

Tropical channel models (TCMs) are defined as models that are global in the zonal direction but bounded in the meridional direction. Such a model can be constructed using existing regional models or even the GCMs. Two TCMs based on regional models MM5 and WRF, and the third one based on the ECHAM4 GCM, were described. Although all three TCMs are atmosphere only, they could be coupled to an ocean model. Therefore the TCM is a unique tool to study the tropics and its interactions with the extratropics.

The simulations using TCMs are far from perfect. There are biases in the mean state, ITCZ, MJO, and other tropical modes. The gross fetures of an MJO event are reproduced after two months from the start of simulation using the TCM based on MM5 (TMM5). This is well beyond the usual MJO predictability limit of 15-20 days associated with global models (Waliser et al., 2003). The TCM simulations are forced by presecribed boundary conditions, and are not initial value problems. Longer integration of another TCM based on WRF (TWRF) does not lead to better tropical variabilty compared to that in GCMs. This is unexpected since, compared to a GCM, a TCM simulation has the added constraint of specified meridional boundary conditions. The error in the mean state is a possible reason for the poor representation of tropical variabilty in TWRF. It is not known to what extent the error in the mean state inhibits tropical variabilty, although it is likely to be model dependent.

These studies suggest a new practice in tropical prediction: a high-resolution domain of the tropics nested within a relatively coarse-resolution global model. The latter is known to suffer less from deficiencies in cumulus parameterizations in the extratropics because of the strong dynamic large-scale control of convection there. This approach of global nested domains, currently being explored by some modeling groups, permits two-way tropical-extratropical interactions with a more precise treatment of tropical convection and much more computational efficiency than a global high-resolution model. The results call for further attention to the untapped potential of high-resolution nonhydrostatic models (Moncrieff et al., 2007) in the simulation and forecasting of tropical atmospheric convection, such as that undertaken within the WCRP-WWRP/THORPEX coordinated project, the Year of Tropical Convection (YOTC, Waliser and Moncrieff, 2008; see www.ucar.edu/yotc).

5. Acknowledgment

Acknowledgment is made to the NCAR, which is sponsored by the National Science Foundation, for making the TWRF model output available. The authors thank Shuyi Chen and Joe Tenerelli, who helped build the TMM5. The NCEP-NCAR reanalysis data were taken from NOAA/CDC. The NCEP 'FNL' data were taken from the NCAR's mass storage. This work was supported by the Japan Agency for Marine-Earth Science and Technology (JAMSTEC), NASA (NNX07AG53G), and NOAA (NA09OAR4320075), which sponsor research at the International Pacific Research Center (IPRC). This is IPRC contribution 823.

6. References

Arakawa, A., & Schubert, W.H. (1974). Interaction of a cumulus cloud ensemble with the large-scale circulation. Part I, *J. Atmos. Sci.*, Vol.31, pp. 674-701

Bentamy, A.; Queffeulou, Y.; Quilfen, Y. & Katasaros, K. (1999). Ocean surface wind fields estimated from satellite active and passive microwave instruments. *IEEE Trans. Geosci. Sensing*, Vol.37, pp. 2469-2486

Benwell, G., & Bretherton, F. (1968). A pressure oscillation in a ten-level atmospheric model. *Quart. J. Roy. Meteor. Soc.*, Vol.94, pp. 123-131

Betts, A.K. & Miller, M.J. (1986). A new convective adjustment scheme. II. Single column testing using GATE wave, BOMEX, ATEX, and arctic air-mass data sets. *Quart. J. Roy. Meteor. Soc.*, Vol.112, pp. 693-709

Charney, J.G. (1948). On the scale of atmospheric motions. *Geophys. Publ.*, Vol.17, No.2, pp. 1-17

Chen, F. & Dudhia, J. (2001). Coupling an advanced land surface-hydrology model with the Penn State-NCAR MM5 modeling system, Part I: Model implementation and sensitivity. *Mon. Wea. Rev.*, Vol.129, pp. 569-585

Chen, S.S.; Houze, R.A. & Mapes, B.E. (1996). Mult-scale variabilty of deep convection in relation to large-scale circulation in TOGA-COARE. *J. Atmos. Sci.*, Vol.53, pp. 1380-1409

Collins, W.D.;Bitz, C.M.; Blackmon, M.L.; Bonan, G.B.; Bretherton, C.S.; Carton, J.A.; Chang, P.; Doney, S.C.; Hack, J.J.; Henderson, T.B.; Kiehl, J.T.; Large, W.G.; McKenna, D.S.; Santer, B.D. & Smith, R.D. (2006). The Community Climate System Model: CCSM3. *J. Clim.*, Vol.19, pp. 2122-2143

Done, J.M.; Holland, G.J. & Webster, P.J. (2011). The role of wave energy accumulation in tropical cyclogenesis over the tropical north Atlantic. *Clim. Dyn.*, Vol.36, pp. 753-767, doi 10.1007/s00382-010-0880-5

Dudhia, J. (1989). Numerical study of convection observed during the Winter Monsoon Experiment using a mesoscale two-dimensional model. *J. Atmos. Sci.*, Vol.46, pp. 3077-3107

Dudhia, J. (1993). A nonhydrostatic version of the Penn State-NCAR Mesoscale Model: Validation tests and simulation of an Atlantic cyclone and cold front. *Mon. Wea. Rev.*, Vol.121, pp. 1493-1513

Eady, E.T. (1949). Long waves and cyclone waves. *Tellus*, Vol.1, No.3, pp. 33-52

Frank, W.M. & Roundy, P.E. (2006). The role of tropical waves in tropical cyclogenesis. *Mon. Wea. Rev.*, Vol.134, pp. 2397-2417

Gill, A. (1980). Some simple solutions for heat-induced tropical circulation. *Quart. J. Roy. Meteor. Soc.*, Vol.106, pp. 447-462

Grabowski, W.W.; Wu, X., Moncrieff, M.W. & Hall, W.D. (1998). Cloud-resolving modeling of cloud systems during Phase III of GATE. Part II: Effects of resolution and the third spatial dimension. *J. Atmos. Sci.*, Vol.55, pp. 3264-3282

Grabowski, W.W. & Moncrieff, M.W. (2001). Large-scale organization of tropical convection in two-dimensional explicit numerical simulations. *Quart. J. Roy. Meteor. Soc.* Vol.127, pp. 445-468

Grell, G.A.; Dudhia, J. & Stauffer, D.R. (1995). A description of the fifth-generation Penn-State/NCAR Mesoscale Model (MM5). NCAR/TN-398.

Gustafson, W.I. & Weare, B.C. (2004a). MM5 modeling of the Madden-Julian oscillation in the Indian and west Pacific Oceans: Model description and control run results. *J. Clim.*, Vol.17, pp. 1320-1337

Gustafson, W.I. & Weare, B.C. (2004b). MM5 modeling of the Madden-Julian oscillation in the Indian and west Pacific Oceans: Implications of 30-70 day boundary effects on MJO development. *J. Clim.*, Vol.17, pp. 338-1351

Hong, S.-Y.; Dudhia, J. & Chen, S.-H. (2004). A revised approach to ice microphysical processes for the bulk parameterization of clouds and precipitation. *Mon. Wea. Rev.*, Vol.132, pp. 103-120

Hong, S.-Y.; Noh, Y. & Dudhia, J. (2006). A new diffusion package with an explicit treatment of entrainment processes. *Mon. Wea. Rev.*, Vol.134, pp. 2318-2341

Houze, R.A. (2004). Mesoscale convective systems. *Rev. Geophys.*, Vol.42, RG4003, doi:10.1029/2004RG000150

Huffman, G.J. & Coauthors (1997). The global precipitation climatology project (GPCP) combined precipitation dataset. *Bull. Amer. Meteor. Soc.*, Vol.78, pp. 5-20

Janjic, Z.I. (1994). The step-mountain coordinate model: Further development of the convection, viscous sublayer, and turbulence closure schemes. *Mon. Wea. Rev.*, Vol.122, pp. 927-945

Jones, C.; Waliser, D.E.; Schemm, J.K.E. & Lau, W.K.M. (2000). Prediction skill of the Madden-Julian oscillation in dynamical extended range forecasts. *Clim. Dyn.*, Vol.16, pp. 273-289

Kain, J.S. (2004). The Kain-Fritsch convective parameterization: An update. *J. Appl. Meteorol.*, Vol.43, pp. 170-181

Kalnay, E. & Coauthors (1996). The NCEP-NCAR 40-year reanalysis project. *Bull. Amer. Meteor. Soc.*, Vol.77, pp. 437-471

Landsea, C.W. (1993). A climatology of intense (or major) Atlantic hurricanes. *Mon. Wea. Rev.*, Vol.121, pp. 1703-1713

Kanamitsu, M.; Ebisuzaki, W.; Woollen, J.; Yang, S.-K.; Hnilo, J.J.; Fiorino, M. & Potter, G.L. (2002). NCEP-DOE AMIP-II Reanalysis (R-2). *Bull. Amer. Meteor. Soc.*, Vol.83, pp. 1631-1643.

Kim, D. & Coauthors (2009). Application of MJO simulation diagnostics to climate models. *J. Clim.*, Vol.22, 6413-6436

Li, T. & Zhou, C. (2009). Planetary scale selection of the Madden-Julian oscillation. *J. Atmos. Sci.*, Vol.66, pp. 2429-2443

Lin, J.-L., & Coauthors (2006). Tropical intraseasonal variability in 14 IPCC AR4 climate models. Part I: Convective signals. *J. Clim.*, Vol.19, 2665-2690

Madden, R.A. & Julian, P.R. (1971). Detection of a 40-50 day oscillation in the zonal wind in the tropical Pacific. *J. Atmos. Sci.*, Vol.28, pp. 702-708

Madden, R.A. & Julian, P.R. (1972). Description of global-scale circulation cells in the tropics with a 40-50 day period. *J. Atmos. Sci.*, Vol.29, pp. 1109-1123

Manabe, S., & Bryan, K. (1969). Climate calculations with a combined ocean-atmosphere model. *J. Atmos. Sci.*, Vol.26, No.4, pp. 786-789

Manabe, S. (1975). The use of comprehensive general circulation modeling for studies of the climate and climate variation, the physical basis of climate and climate modeling. Report of the International Study Conference in Stockholm, *GARP Publication Series*, Vol.16, pp. 148-162

Matsuno, T. (1966). Quasi-geostrophic motions in the equatorial area. *J. Meteor. Soc. Jpn.*, Vol.44, pp. 25-43

McPhaden, M.J. (1999). Genesis and evolution of the 1997-98 El Nino. *Science*, Vol.283, pp. 950-954

Mechoso, C.R. & Coauthors (1995). The seasonal cycle over the tropical Pacific in coupled ocean-atmosphere general circulation models. *Mon. Wea. Rev.*, Vol.123, pp. 2825-2838

Meehl, G.A. & Arblaster, J.M. (1998). The Asian-Australian monsoon and El Niño Southern Oscillation in the NCAR Climate System Model. *J. Clim.*, Vol.11, pp. 1356-1385

Miura, H.; Satoh, M.; Nasuno, T., Noda, A.T. & Oouchi, K. (2007). A Madden-Julian oscillation event realistically simulated by a global cloud-resolving model. *Science*, Vol.318, pp. 1763-1765

Mlawer, E.J.; Taubman, S.J.; Brown, P.D.; Jacono, M.J. & Clough, S.A. (1997). Radiative transfer for inhomogeneous atmosphere: RRTM, a validated correlated-k model for the longwave. *J. Geophys. Res.*, Vol.102, pp. 16663-16682

Moncrieff, M.W. (2010). The multi-scale organization of moist convection and the interaction of weather and climate, In D.-Z. Sun and F. Bryan (Eds.), Climate Dynamics: Why Does Climate Vary? *Geophysical monograph series*, Vol.189, American Geophysical Union, Washington DC, 3-26, doi:10.1029/2008GM000838

Moncrieff, M.W.; Shapiro, M.A.; Slingo, J.M. & Molteni, F. (2007). Collaborative research at the intersection of weather and climate. *WMO Bulletin*, Vol. 56, pp. 1-9

Murthi, A.; Bowmann, K.P. & Leung, L.R. (2011). Simulations of precipitation using NRCM and comparisons with satellite observations and CAM: annual cycle. *Clim. Dyn.* (in press)

Phillips, N.A. (1956). The general circulation of the atmosphere: A numerical experiment. *Quart. J. Roy. Meteorol. Soc.*, Vol.82, pp. 123-164

Phillips, N.A. & Shukla, J. (1973). On the strategy of combining coarse and fine grid meshes in the numerical weather predition models. *J. Appl. Meteorol.*, Vol.12, pp. 736-770

Ray, P.; Zhang, C.; Dudhia, J. & Chen S.S. (2009). A numerical case study on the initiation of the Madden-Julian oscillation. *J. Atmos. Sci.*, Vol.66, pp. 310-331

Ray, P. & Zhang, C. (2010). A case study of the mechanics of extratropical influence on the initiation of the Madden-Julian oscillation. *J. Atmos. Sci.*, Vol.67, pp. 515-528

Ray, P.; Zhang, C.; Moncrieff, M.W.; Dudhia, J., Caron, J., Leung, L.R. & Bruyere, C. (2011). Role of the atmospheric mean state on the initiation of the Madden-Julian oscillation in a tropical channel model. *Clim. Dyn.*, Vol.36, pp. 161-184, doi 10.1007/s00382-010-0859-2

Ray, P. & Li, T. (2011). On the relative roles of the circumnavigating waves and the extratropics on the Madden-Julian oscillation (MJO), (in preparation).

Richardson, L.F. (1922). Weather prediction by numerical process. Cambridge University press (reprt. Dover).

Satoh, M.; Matsuno, T.; Tomita, H.; Miura, H.; Nasuno, T. & Iga, S. (2008). Nonhydrostatic icosahedral atmospheric model (NICAM) for global cloud resolving simulations. *J. Comp. Phys.*, Vol.227, pp. 3486-3514

Simmons, A.J., & Bengtsson, L. (1984). Atmospheric general circulation models: Their design and use for climate studies. In The Global Climate, ed. J. T. Hougton. Cambridge University Press., Cambridge, pp. 37-62

Slingo J.M., and Coauthors (1996) Intraseasonal oscillations in 15 atmospheric general circulation models: Results from an AMIP diagnostic subproject. *Clim. Dyn.*, Vol.12, pp. 325-357

Smagorinsky, J.; Manabe, S. & Holloway, Jr. J.L. (1965). Numerical results from a nine-level general circulation model of the atmosphere. *Mon. Wea. Rev.*, Vol.93, No.12, pp. 727-768

Sperber, K.P.; Gualdi, S.; Legutke, S. & Gayler, V. (2005). The Madden–Julian oscillation in ECHAM4 coupled and uncoupled general circulation models. *Clim. Dyn.*, Vol.25, pp. 117–140

Tao, W.-K. & Moncrieff, M.W. (2009). Multiscale cloud system modeling. *Rev. Geophys.*, Vol.47, RG40002, doi:10.1029/2008RG000276

Taylor, K.E.; Williamson, D. & Zwiers, F. (2000). The sea surface temperature and sea-ice concentration boundary conditions for AMIP II simulations. PCMDI Report No. 60 and UCRL-MI-125597, 25 pp

Tulich, S.N.; Kiladis, G.N. & Suzuki-Parker, A. (2011). Convectively coupled Kelvin and easterly waves in a regional climate simulation of the tropics. *Clim. Dyn.*, Vol.36, pp. 185-203, doi 10.1007/s00382-009-0697-2

Uppala, S.M. & Coauthors (2005). The ERA-40 reanalysis. *Quart. J. Roy. Meteorol. Soc.*, Vol.131, pp. 2961-3012

Waliser, D.E.; Lau, K.M.; Stern, W. & Jones, C. (2003). Potential predictability of the Madden-Julian oscillation. *Bull. Amer. Meteor. Soc.*, Vol.84, pp. 33-50

Waliser, D.E. & Moncrieff, M.W. (2008). Year of Tropical Convection (YOTC) Science Plan, WMO/TD-No. 1452, WCRP -130, WWRP/THORPEX, No.9, 26 pp

Wang, H.-H., & Halpern, P. (1970). Experiments with a regional fine-mesh prediction model. *J. Appl. Meteor.*, Vol.9, pp. 545-553

Walsh, K.J.E.; Nguyen, K.-C. and McGregor, J.L. (2004). Fine resolution regional climate model simulations of the impact of climate change on tropical cyclons near Australia. *Clim. Dyn.*, Vol.22, pp. 47-56

Xie, P. & Arkin, P.A. (1997). Global precipitation: A 17-year monthly analysis based on gauge observations, satellite estimates, and numerical model outputs. *Bull. Amer. Meteor. Soc.*, Vol.78, pp. 2539-2558

Zhang, C. (2001). Double ITCZs. *J. Geophys. Res.*, Vol.106, No.D11, pp. 11785-11792

Zhang, C. & Dong, M. (2004). Seasonality of the Madden-Julian oscillation. J. *Clim.*, Vol.17, pp. 3169-3180

Zhang, C. (2005). Madden-Julian oscillation. *Rev. Geophys.*, Vol.43, RG2003, doi:10.1029/2004RG000158

Zhang, C.; Dong, M.; Gualdi, S.; Hendon, H.H.; Maloney, E.D.; Marshall, A.; Sperber, K.R. & Wang, W. (2006). Simulations of the Madden-Julian oscillation by four pairs of coupled and uncoupled global models. *Clim. Dyn.*, DOI: 10.1007/s00382-006-0148-2

2

Regional Climate Model Applications for West Africa and the Tropical Eastern Atlantic

Leonard M. Druyan and Matthew Fulakeza
Center for Climate Systems Research, Columbia University,
NASA/Goddard Institute for Space Studies, New York,
USA

1. Introduction

The chapter reviews applications of a regional climate model for studies of climate variability over West Africa and the adjacent eastern tropical Atlantic Ocean. Fig. 1a shows the region and identifies some of the geographic features mentioned in the chapter. The savannah region south of the Sahara Desert, known as the Sahel (situated between 10-15°N), is of particular interest to climate scientists because of its vulnerability to recurring drought. The West African monsoon (WAM) is characterized by the advance of precipitation northward to the Sahel region during late June or early July (onset) and its southward retreat during September. Below normal Sahel rainfall during some summers is associated with an abbreviated northward advance of this rain belt (Hastenrath and Polzin, 2010). A shortfall in seasonal rainfall has an especially severe negative impact on the pastoral and agricultural economies of the region, causing famine and widespread social upheaval in worst case scenarios. The socio-economic repercussions of drought or even floods lend importance to the ultimate goals of climate model research relevant to this area: improving seasonal forecasts, predicting monsoon onset dates, determining the role of sea-surface temperature anomalies and land surface characteristics on the variability of the developing monsoon and creating reliable projections of the future Sahel climate under a range of greenhouse warming scenarios. Early warning of imminent drought, for example, can allow mitigation strategies to blunt some of the negative impacts. In addition, research is also devoted to better understanding African easterly wave disturbances (AEWs). AEWs enter the tropical Atlantic from West Africa during the summer and can develop into tropical cyclones, and occasionally into hurricanes.

Regional climate models (RCMs) are integrated over limited area domains. Accordingly, RCMs can afford higher horizontal resolution than global domain models for the same dedication of computer resources. Computation over the limited area domains alternatively facilitates multiple and longer simulations. The usually higher horizontal resolution of RCMs (compared to global models) offers the potential for better definition of spatial gradients of topography, land surface characteristics and the atmospheric variables involved in the models' integration and output. More accurate gradients, in turn, improve the simulations and provide better detail in the modeled results of interest to users of climate data.

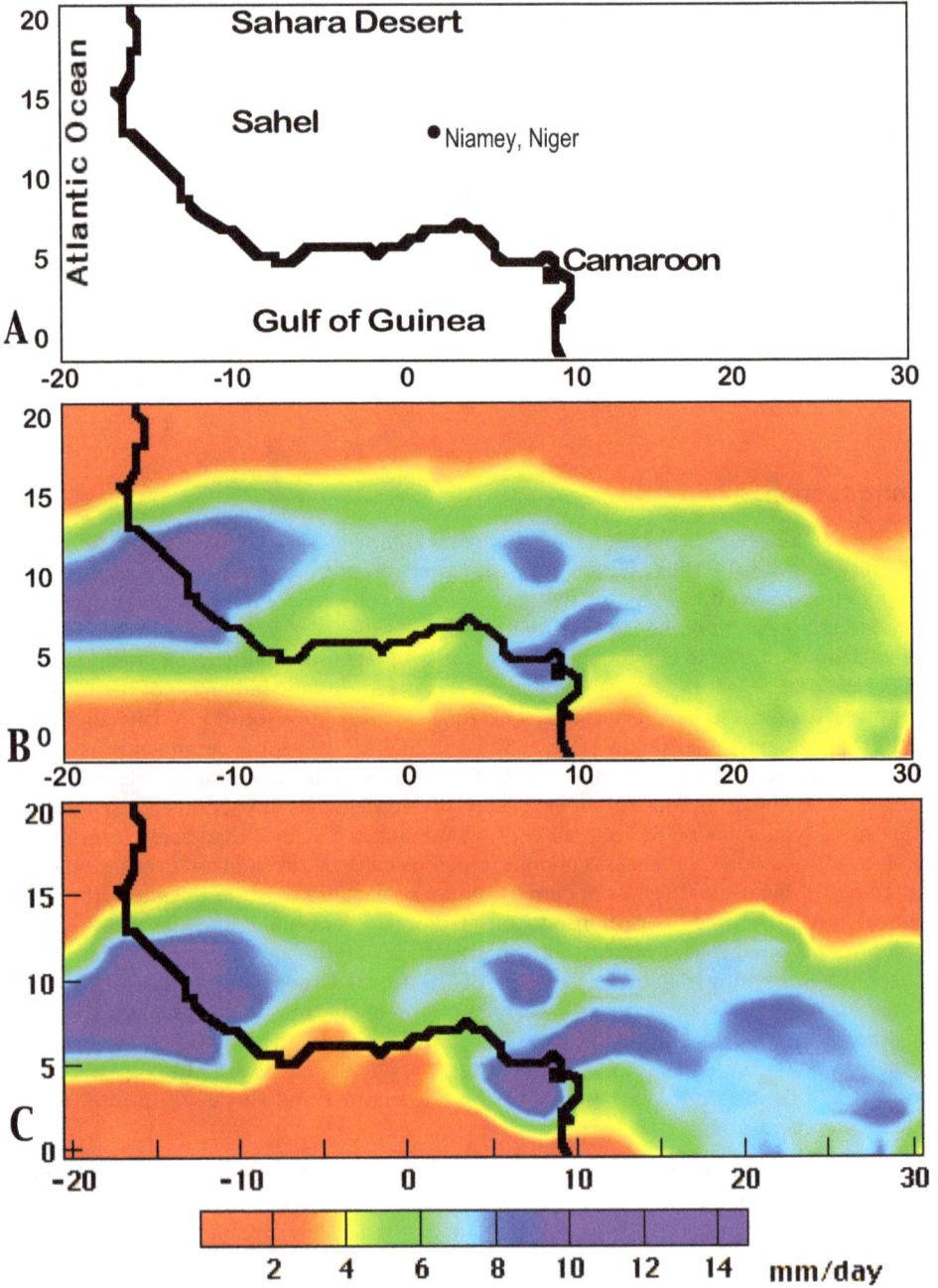

Fig. 1. a. Geography of West Africa. b. Mean RM3 16L precipitation rates (mm day^{-1}) for June-September 1998-2003, c. corresponding TRMM means (courtesy of NASA's Goddard Earth Sciences [GES] Data and Information Services Center).

For example, better resolution of the exceptionally strong and dynamically important summertime meridional temperature and moisture gradients south of the Sahara should contribute to more skillful simulations than are possible using coarser gridded global models. In addition, projected rainfall distributions mapped on a 0.5° grid are much more useful for agricultural applications than corresponding data on a 2.5° grid. The chapter gives a detailed description of one such RCM applied to WAM studies.

Although some model-based climate studies focus only on monthly or seasonally mean fields, confidence in results improves if the models realistically capture the characteristics of relevant daily weather phenomena. For example, precipitation triggered and modulated by transient AEWs plays a crucial role in West African summer monsoon hydrology. Accordingly, modeling the behavior of AEWs is a fruitful application of RCMs, made more meaningful by validation against corresponding observational evidence. The chapter describes RCM simulations of AEWs and their associated precipitation patterns.

2. The regional climate model of the Center for Climate Systems Research (CCSR, Columbia University) and the NASA/Goddard Institute for Space Studies in New York City (GISS)

Druyan et al. (2006) and Druyan et al. (2008) describe the regional climate model (3rd generation version), the *RM3*, which runs exclusively at CCSR/GISS. The authors have published some 13 papers since 2000 describing various characteristics of RM3 climate simulations. The earlier version integrated the equations of motion at 16 vertical sigma levels, while the current version of the RM3 has been expanded to 28 vertical sigma levels. The RM3 was improved over the years by incorporating land surface hydrology and moist convection modules originally developed for the global climate model of NASA/GISS.

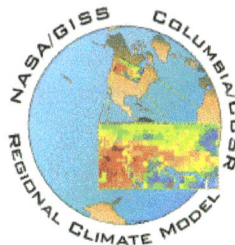

Model computations are made on a horizontal grid with either 0.5° latitude/longitude spacing, or, for the currently underway CORDEX (Coordinated Regional Climate Downscaling Experiment) simulations, 0.44° grid spacing. Given typical AEW wavelengths of 2800 km (25° longitude), these grid resolutions resolve AEWs quite well and can also represent mesoscale convective complexes imbedded within AEWs. Integration of the RM3 simulates the 4-D evolution of atmospheric temperature, humidity and circulation as well as the location, rate and timing of precipitation.

The RM3 incorporates a land surface (LS) process model, originally developed by Rosenzweig and Abramopoulos (1997) (See also Hansen et al. 2002), that computes ground temperatures, soil moisture variability and evapotranspiration. The LS model consists of two integrated parts, the soil and the canopy, and it conserves water and heat while simulating their vertical fluxes. The RM3 modeled soil is divided into six layers to a depth of 3.5 m, and the model distinguishes between five textures of soil. The canopy, modeled as a separate layer located above the soil, is responsible for the interception of precipitation, evaporation of accumulated water and removal of soil water through transpiration. Ocean temperatures are specified boundary data that are updated daily, usually from the same data set that provides the lateral boundary conditions.

The DelGenio and Yao (1993) moist convection parameterization and the Del Genio et al. (1996) scheme for the effects of cloud liquid water are important RM3 modules. The convection scheme incorporates entraining and non-entraining plumes, downdrafts, and subsidence. Vertical cumulus mass fluxes are proportional to the moist static stability and are constrained to relax the atmosphere to a neutrally stable state at the cloud base. The convective plume and subsiding environment transport grid-scale horizontal momentum. Convective cloud cover is assigned as proportional to the mean pressure thickness of all model layers up to the cloud top. The cloud liquid water scheme (DelGenio et al., 1996) allows for life cycle effects in stratiform clouds and permits cloud optical properties to be determined interactively. Cloud optical thickness is calculated from the predicted liquid/ice water path and a variable droplet effective radius estimated by assuming constant droplet number concentration. Microphysical and radiative properties are assumed to be different for liquid and ice clouds, and for liquid clouds over land and ocean. Mixed phase processes can change the phase if ice falls into a lower layer containing supercooled liquid water. The scheme parameterizes Bergeron-Findeisen diffusional growth of the ice phase at the expense of the liquid phase via the "seeder-feeder" process by allowing a layer with supercooled water to glaciate if sufficient ice falls into it from above.

Three RM3 domains for simulations are used. For the Druyan and Fulakeza (2011) study of systems in the tropical eastern Atlantic, the domain for the 0.5° grid is 20°S-35°N, 50°W-20°E, representing a westward shift of 15° of longitude relative to the domain used by Druyan et al. (2008), which focuses more on West Africa. CORDEX simulations are made over a 0.44° grid domain expanded to include all of Africa and adjacent coastal oceans: 50°S-50°N, 35°W-64°E.

Atmospheric boundary conditions and SST to drive the RM3, for most of the published studies, are taken from the US National Center for Environmental Prediction and Department of Energy reanalysis (NCPR, 1 and 2), archived at 2.5° latitude/longitude grid spacing (Kanamitsu et al. 2002). Driving the RCM with NCPR is not appropriate for any of several important applications: not for daily weather predictions, nor seasonal climate outlooks, nor for investigating the implications of decadal climate change on a regional scale. However, the analysis of downscaling NCPR does enable an assessment of whether the RCM can successfully provide more realistic spatial detail of climate variables than the driving data set. Over West Africa in August, this could mean generating and propagating a squall line that is not resolved at 2.5° grid spacing. Optimization of the RCM downscaling of NCPR then provides a better RCM tool, which can subsequently be driven by boundary conditions from global models' weather or climate forecasts or decadal climate projections, for the stated applications. Moreover, the RM3/NCPR system can also be used to test the sensitivity of the local climate to prescribed changes in SST and/or land surface conditions or to modeling changes.

3. Results of selected RM3 experiments

Druyan et al. (2006) describe the results of RM3 16L (16 level version) simulations spanning six summers (June-September), 1998-2003. Initial conditions for each seasonal run are NCPR atmospheric, soil moisture and SST data for May 15. RM3 simulations on the 0.5° grid are then driven by synchronous NCPR lateral boundary data and sea-surface temperatures, four times daily, interpolated to the RM3 grid.

Six-year mean June-September RM3 precipitation is validated against gridded observational data from the Climate Research Unit (CRU) of East Anglia University (New et al. 2002) as well as against data from the Tropical Rainfall Measuring Mission satellite- TRMM (3B43 V6) (Kummerow et al., 2000). All data sets were gridded at 0.5° latitude/longitude. Figs. 1b and 1c show that the RM3 reproduces the TRMM observed orographic precipitation maxima along the southwest (Guinea) coast and over coastal Cameroon, albeit with some underestimates. It also features the West African rain belt along 10°N, although too much precipitation is simulated along the Gulf of Guinea coast, which climatologically experiences a dry period in August.

3.1 RM3 daily precipitation time series compared to rain gauges and TRMM

Daily precipitation variability across the Sahel in summer is to a large measure regulated by the transit of AEWs. Time series of RM3 and TRMM (3B42 V6) simulated daily precipitation accumulations are compared to a time series of co-located rain gauge measurements near Niamey, Niger (from Thorncroft et al. 2003), in Fig. 2. Niamey is identified on the map in Fig. 1a.

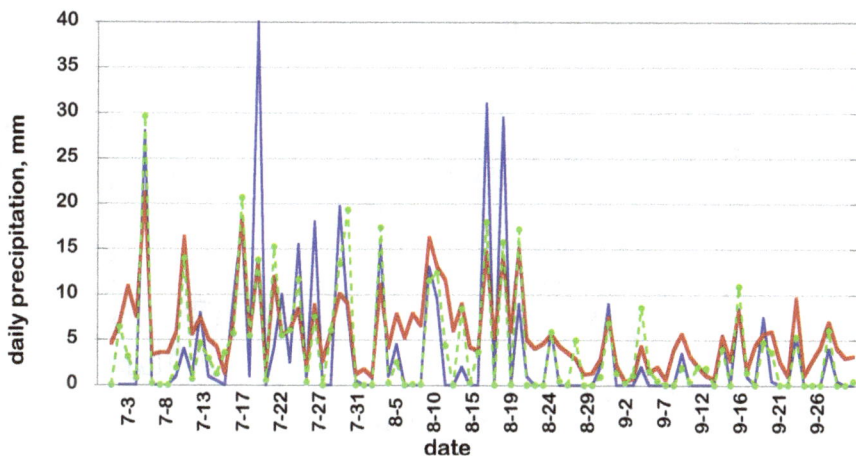

Fig. 2. Time series of the mean of 34 rain gauge observations (blue) within the area bounded by 13-13.9°N, 1.7-3.1°E (Thorncroft et al. 2003) versus RM3 daily values (red) for 15 co-located grid elements and TRMM daily estimates (green) for four co-located 1° squares, July-September 2000.

Note that the RM3 and TRMM represent area averages, compared to the average of 34 single point observations of the rain gauges. One model deficiency is that, even on days observed to be rainless, RM3 daily precipitation is non-zero. However, the model simulation does indicate a peak on every rainy day, and many of the peaks are comparable to observations. There are three days on which much larger accumulations were measured by rain gauges than indicated by the RM3 or by TRMM. The correlation of the 3-month time series of the RM3 simulation versus TRMM is 0.86. The correlation between the rain gauge time series and TRMM is 0.79, and between the RM3 and the rain gauges, 0.73. Thus, both TRMM and the RM3 produce a daily variability that matches rain gauge measurements quite well. Note that the RM3/NCPR system does not share any data source with the TRMM system so that all three sources of precipitation data are completely and mutually independent.

3.2 The spectral signature of AEWs

AEW-related vorticity is particularly strong at 700 mb, just below the level of the African Easterly Jet. Transient AEWs cause perturbations in the 700 mb meridional wind (v7), related to their cyclonic circulation. Time series of v7 at a given West African location therefore exhibit a periodicity determined by the frequency of AEW traversal. Spectra of v7 time series at such locations show peaks for periods ranging between 3-6 days, which are the spectral signature of AEWs. The example shown in Fig. 3, for the location 15°N, 12°W, is from a RM3 16L simulation forced by NCPR data for June-September 2002. This particular example shows spectral peaks at 3.9 and 5.1 days. Spatial mapping of the spectral amplitudes between 3-6 days indicates preferred AEW tracks.

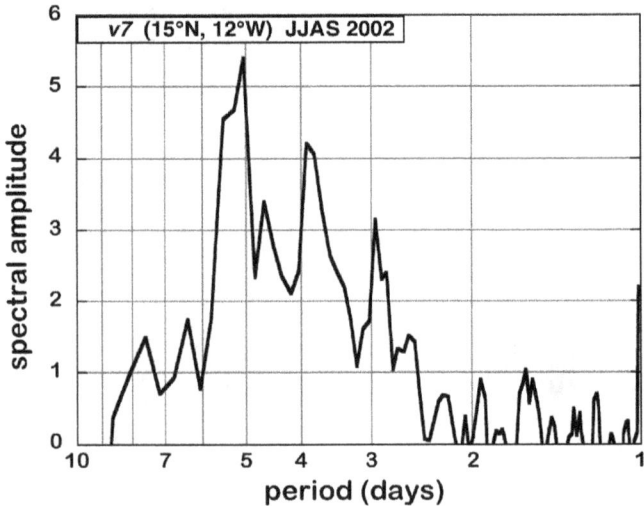

Fig. 3. Spectrum for June-September 2002 time series of RM3 16L simulated v7 at 15°N, 12°W. The red noise spectrum has been subtracted. Peaks at 3.9 days and 5.1 days reflect the periodicity of v7 due to the traversal of AEWs.

RM3 700 mb circulation shows evidence of considerable interannual variability in spectral properties that relate to AEWs. Spectral amplitudes for 700 mb meridional wind time series peak most often in the range of 3-6- day periods over swaths traversed by AEWs. RM3 results favor AEW tracks near 17°N and 5°N.

3.3 Validation of RM3 daily circulation variability

AEWs are also manifest by westward propagating bands of alternating southerlies and northerlies in NCPR, European Center for Medium- range Weather Forecasting 40- year reanalysis (ERA-40) and RM3 700 mb circulations. Fig. 4 shows time series of once daily v7, averaged over 5-15°N along 7.5°W, for each of six Augusts from the three data sets.

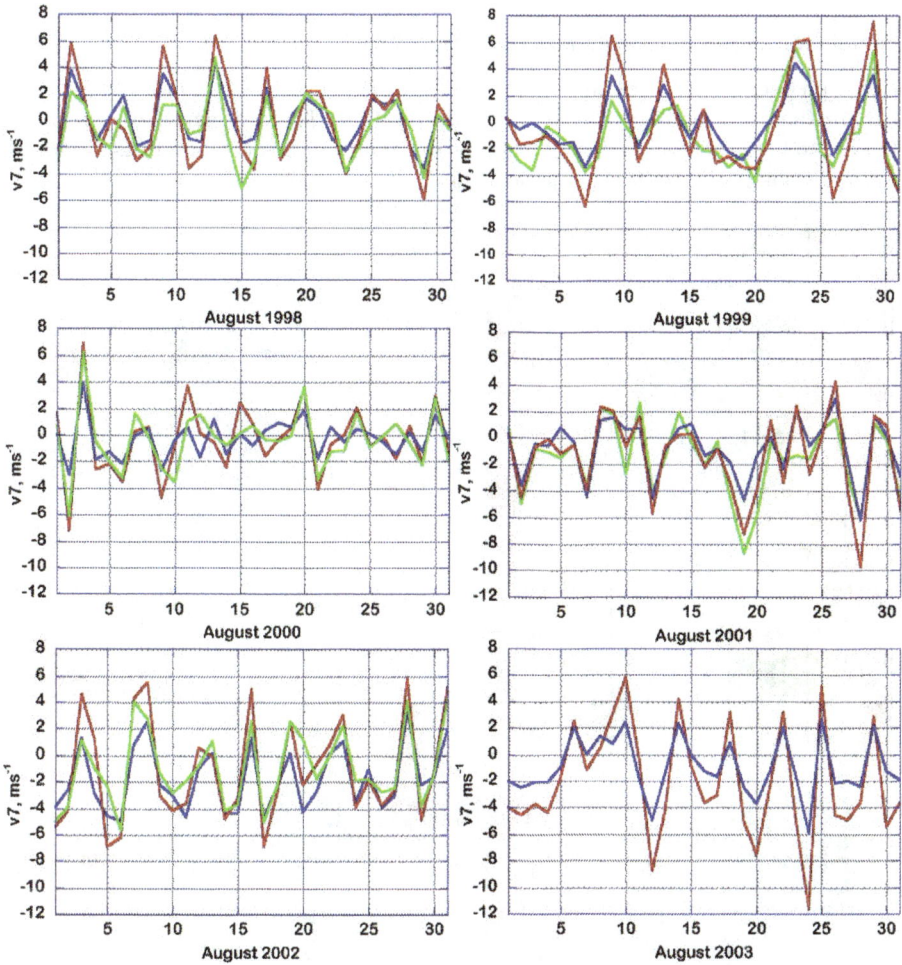

Fig. 4. Time series of v7 along 7°W, averaged over 5-15°N, for each of six Augusts. Simulated RM3 16L v7 (blue) are compared to NCPR v7 (red) from the same data set as the lateral boundary conditions and to ERA-40 v7 (green) (except for 2003).

Periodicities in v7 associated with AEWs are evident. RM3 v7 daily variability is highly correlated with corresponding NCPR (correlations ranging from 0.86 to 0.97) and ERA-40 values (correlations ranging from 0.76 to 0.86), but the RM3 v7 amplitudes in Fig. 4 are the lowest of the three. Although the RM3 simulations are driven by NCPR data at the lateral boundaries (LBC), circulation within the domain is entirely the product of the RM3 integration. The implication is that, based on the LBC, the RM3 generates and propagates the actual AEWs through the domain. Druyan et al. (2008) show that the spectral amplitudes and vorticity variances associated with RM3 AEWs double in amplitude between the eastern boundary and the center of the domain, so the model is not merely transporting an existing signal from NCPR westward.

3.4 Simulation of the seasonal march of precipitation

Precipitation over the Gulf of Guinea coast (about 6°N) typically wanes near the end of June and thereafter reforms over the Sahel closer to 10°N. Rapid advance of precipitation northward to the Sahel region during late June or early July, termed "onset", is discussed by Hagos and Cook (2007), Sultan and Janicot (2000) and Ramel et al. (2006). Onset is related to

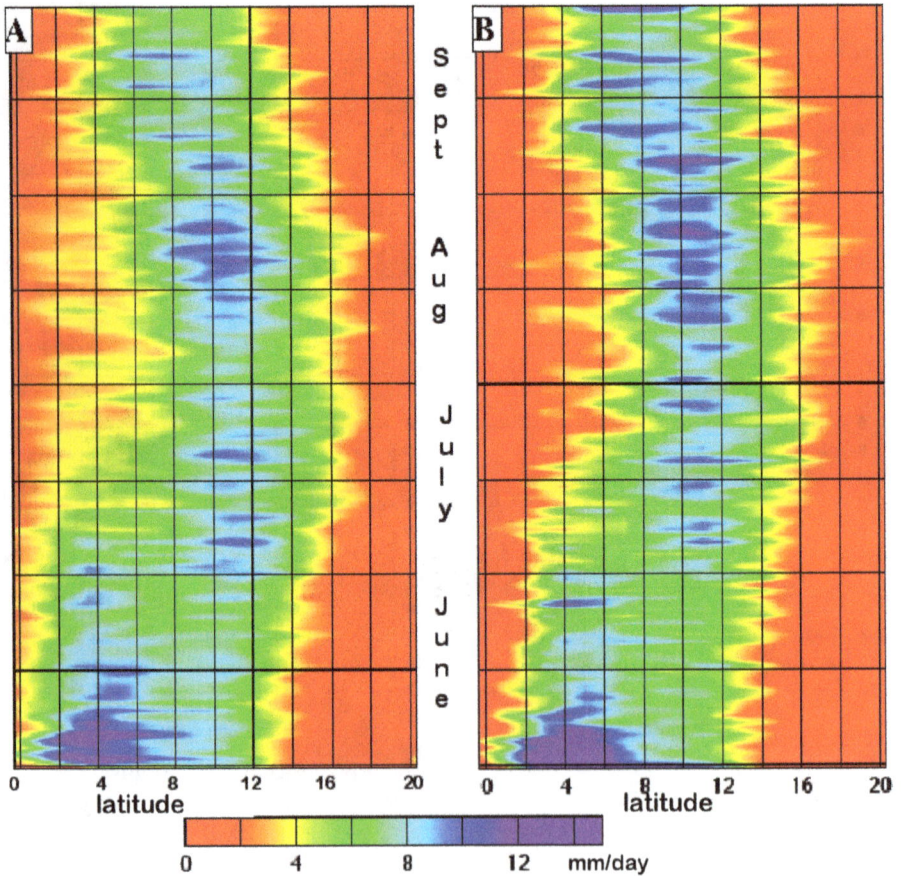

Fig. 5. Hovmöller time-latitude distributions of daily precipitation rates (mm day^{-1}) averaged over 15°W-10°E for June-September 1998-2003. a. RM3, b. TRMM (courtesy of NASA's Goddard Earth Sciences [GES] Data and Information Services Center).

increased continental sensible heating which, in turn, initiates a surge in moisture convergence over the Sahel. The ability of RCMs to reproduce the northward migration of the rain belt and its subsequent retreat is fundamentally important to both their usefulness in forecasting and climate sensitivity studies. Skillful simulation of the northward jump is especially desirable since it represents the onset of monsoon rains over the semi-arid Sahel region. Fig. 5a shows the latitude versus time progression of RM3 16L precipitation averaged over 15°W-10°E and over six seasons and Fig. 5b shows the corresponding graphic for TRMM data. The abrupt shift of heavy rainfall from 4°N to 10°N at the end of June is prominent on both panels. Moreover, the RM3 16L simulates the onset and other meridional shifts in the West African rain band in close agreement with TRMM observations.

3.5 The impact of increasing RM3 vertical resolution

Six June-September seasons simulated by the RM3 16L are compared to the corresponding simulations using the RM3 28L in Druyan et al. (2008). In general, the increased vertical resolution creates stronger circulation, manifest by a stronger African Easterly Jet (AEJ), stronger near-surface monsoon westerlies, stronger v7 spectra and stronger vorticity centers. Fig. 6 compares the 5-season mean zonal wind (m s-1) at 700 mb for the two model versions, where the AEJ is represented by the diagonal swath of minimum zonal wind over West Africa. Corresponding NCPR zonal wind speeds (on the 2.5° grid) are intermediary between the 16L and 28L values, but the higher resolution RM3 28L may be a better representation of the actual circulation. Figs. 7a and 7b show the spatial distributions of vorticity of the 700 mb circulation by the 16L and 28L resolutions, respectively, for a composite of seven events that feature an AEW trough near 10°W. Each of these events occurs in both the 28L and 16L simulations. The positive and negative vorticity centers are about 50% more extreme for the 28L version, mostly because of stronger overall circulation. Higher vertical resolution does not have an impact on simulating the timing of monsoon onset.

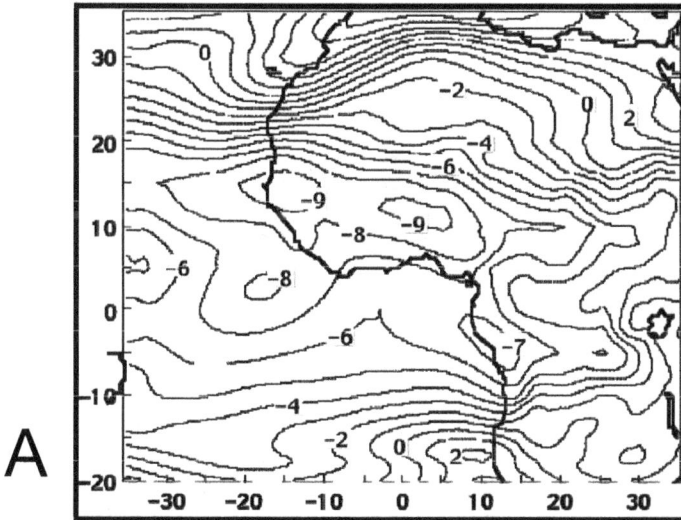

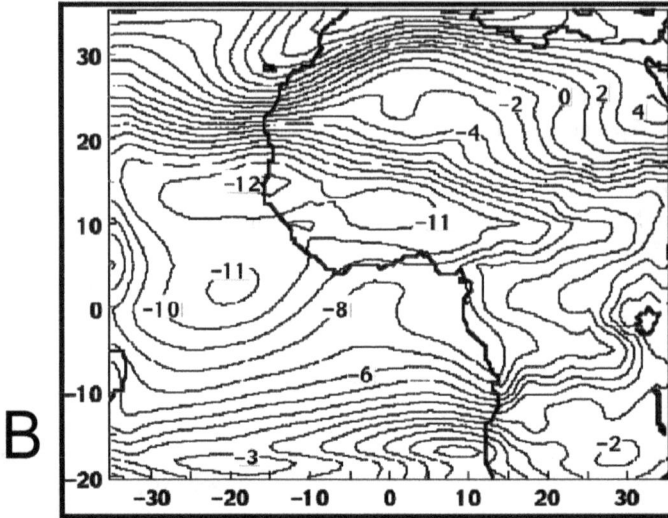

Fig. 6. Horizontal distributions of RM3 zonal winds (m s⁻¹) averaged over June September 1998-2003. a. 16L, b. 28L.

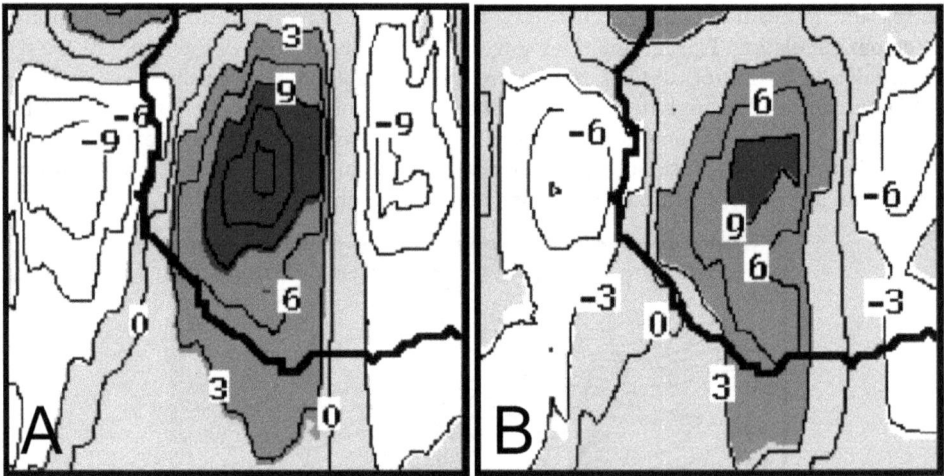

Fig. 7. Vorticity of the RM3 simulated 700 mb circulation for composites of seven events featuring an AEW near 10°W. a. RM3 16L, b. RM3 28L. Units: x10⁻⁶ s⁻¹.

Druyan and Fulakeza (2011) used the RM3 28L to investigate the influence of the eastern tropical Atlantic SST maximum on the WAM climate, and in particular on transient AEWs. A control simulation was forced with May-October 2006 SST from NCPR2 and a second simulation experiment with the same forcing, except for -3°K SST anomalies between 0-15°N. Subtracting 3°K from the SST boundary conditions at every time step eliminated the SST maximum in the eastern tropical Atlantic. Both simulations were driven by the same synchronous NCPR2 atmospheric lateral boundary data and both reproduced realistic ITCZ precipitation maximums and the actual transient AEWs with their associated rain shields. Fig. 8 shows time series of daily precipitation from each of the two experiments over the eastern tropical Atlantic (5-15°N, 25°W). The precipitation peaks in Fig. 8 represent the passage of convective complexes, presumably associated with AEWs. Results show that the absence of the SST maximum does not affect the timing of the precipitation systems, but rather the amplitude of each rain event. Fig. 8 shows that in the absence of the SST maximum, each precipitation peak is diminished, for example, on September 9th by some 83% and on September 12th by only about 10%.

4. WAMME

RM3 28L simulations are included in the West African Monsoon Modeling and Evaluation (WAMME) project, a recognized sub-group within the African Monsoon Multidisciplinary Analysis (AMMA) (Redelsperger et al., 2006). Druyan et al. (2010) summarized WAMME results from five regional models simulating the West African summer monsoon with both reanalysis and GCM forcing.

WAMME makes an intercomparison of the results from five RCMs including the RM3, each driven by synchronous NCEP reanalysis II and C20C SST data over four May- October seasons. The second part of the study analyzes results from two of the RCMs, driven with GCM (HadAM3) forcing, but the same SST lower boundary conditions.

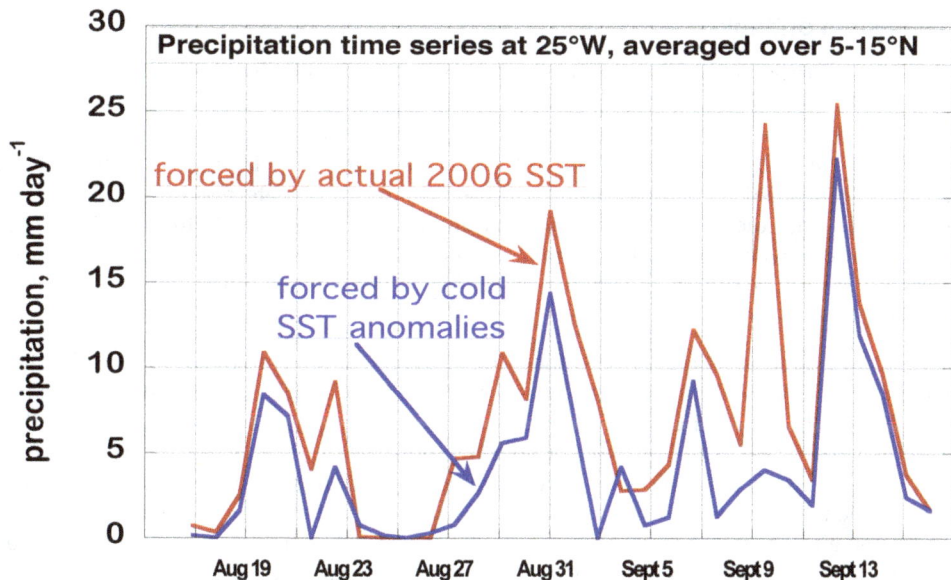

Fig. 8. Daily precipitation time series for two RM3 28L simulations at 25°W, averaged over 5-15°N. The red curve represents the August 16-September 16 interval from a May-September 2006 simulation forced with NCPR2 SST. The blue curve represents the same interval from a May-September 2006 simulation forced with NCPR2 SST minus 3°K between 0°-15°N.

The RCMs in the WAMME study simulate the northward jump of the precipitation band that represents monsoon onset over Sahelian Africa. However, the jumps in three of the four models are 2-5 weeks earlier than observed, suggesting serious model limitations for predicting monsoon onset. The RCMs show positive precipitation biases over much of West Africa, consistent with positive precipitation biases within the West African rain belt in NCEP reanalysis II and HadAM3 (Xue et al., 2010), the two data sets used for LBC. A five-model average precipitation bias for June-September mean rates over West Africa (5-20°N, 15°W-20°E) was about 1.7 mm day^{-1} and the average spatial correlation between modeled and observed mean rainfall rates over the same area was 0.82, which means that the simulated pattern accounts for 67% of the observed spatial variance. Much of that success comes from correctly locating orographic precipitation maxima, the latitude of the main west to east rain band and the transition to the desert regime to the north. RCM performance in simulating the seasonal mean precipitation distribution compared to observations was generally better north of 10°N. Spatial correlation coefficients against the observed pattern are near 0.90 for all RCMs within the Sahel belt centered on 15°N, but are in the range of 0.30 to 0.65 for more southerly sub-regions. The four- model average surface air temperature bias for June-September means over West Africa (5-20°N, 15°W-20°E) is -2.6°K and the average spatial correlation between modeled and observed mean surface air temperature is 0.88. RM3 convective rainfall rates are excessive even with reasonable meridional moisture advection, while they are closer to observations with diminished moisture advection, suggesting the need to better optimize the convective parameterization.

5. Research outlook

WRF: We are creating and analyzing simulations of the West African summer monsoon using the Weather, Research and Forecasting (WRF) regional model and comparing results to RM3 performance. WRF is the product of considerable research and development (coordinated at the National Center for Atmospheric Research) and is considered to be state-of-the-art in sophistication. WRF is widely used for many climate research applications. However, to date there have been few applications of WRF in WAM studies. Moreover, there is no consensus yet on which combination of the many available alternative physical parameterizations for WRF is best suited for WAM studies. Erik Noble is heading this research effort.

ACMAD: We have collaborated with the African Center of Meteorological Application for Development (ACMAD), in Niamey, Niger, since 2005. ACMAD is a consortium of more than 50 African countries charged with providing meteorological and climate data to the constituent societies. The RM3 and supporting programs are installed at the ACMAD computing facility, producing daily weather forecasts that are posted on the ACMAD web site. The model is updated as needed via remote access. All RM3 simulations require initial conditions and boundary conditions, and those needed for the daily forecasts are downloaded from the Internet site of the US National Oceanic and Atmospheric Administration (NOAA). The system uses NOAA Global Forecast System data sets for initial conditions and boundary conditions, although special initialization techniques are required. Collaboration between CCSR/GISS and ACMAD compares RM3 forecasts of daily precipitation and surface temperature to co-located station observations. Results of this validation are not yet available.

Dr. Patrick Lonergan installed a program to automatically initiate the download of boundary condition data at designated appropriate times within the 24- hr cycle. Weather simulations are rather time consuming, but need to be timely if they are to be useful. The automation, using "crontabs," is particularly valuable in that it allows the simulation jobs to get set up and completed overnight as the data becomes available, so that the forecasts can be finished early during the day shift. Posting of RM3 forecasts on the ACMAD website is interrupted by frequent down time of the internet connection and even the electric power supply.

CORDEX: RM3 simulations are underway for planned participation in the Africa subproject of the Coordinated Regional Climate Downscaling Experiment (CORDEX). CORDEX is an initiative of the World Climate Research Project (WCRP) of the UN. It aims to improve coordination of international efforts in regional climate downscaling, both by dynamical and statistical methods. Results for Africa are intended to provide support for the Intergovernmental Panel on Climate Change (IPCC) 5th Assessment Report (AR5), due out by 2013-2014. In the first phase, lateral boundary conditions (LBC) for driving the RM3 are taken from a 0.75°x0.75° gridded version of the European Center for Medium Range Weather Forecasts (ECMWF) interim reanalysis (ERA-I), available for the base period 1989-2008. Subsequently, LBC will be derived from atmosphere-ocean global model (AOGCM) climate change projections (from the Coupled Model Intercomparison Project [CMIP5]), for the future climate and for the base period. RM3 simulations covering all of Africa for the complete annual cycle will be provided to the CORDEX archive, which will facilitate model intercomparisons and multi-model ensemble/consensus results. Preliminary results (Jones and Nikulin 2011; Jones et al., 2011) show that 4 of the 5 RCMs (not including RM3) simulate too rapid an onset of summer monsoon rains near 10°N, based on validation of 1998-2008 simulations.

Fig. 9a shows some preliminary results of the RM3 simulation over the CORDEX domain (which encompasses all of Africa) driven by ERA-I and initialized on November 30, 1990. No adverse lateral boundary effects are evident in this precipitation rate difference field, nor in the monthly mean fields on which Fig. 9a is based (not shown). August 1991 minus January 1991 differences in precipitation rate reflect the migration of the rain belt from southern Africa in January, northward to northern hemisphere Africa in August. Fig. 9a, accordingly, features large gains over the Sahel latitudes centered at 10°N, and large deficits over southern Africa. In addition, negative differences along 5°S-3°N over the Atlantic and Gulf of Guinea mark the vacated position of the January rain band during northern hemisphere summer. Fig. 9b shows the corresponding difference distribution based on CMAP (Climate Prediction Center Merged Analysis of Precipitation) data (Xie and Arkin, 1997). The CMAP observations validate the latitudes and orientation of the major RM3 difference extremes, evidence that the RM3 performs the seasonal transitions well. Excess positive RM3 differences in the Sahel reflect RM3 overestimates of August rainfall, although some of the higher RM3 values may be a consequence of the higher horizontal resolution that resolves orographic maxima better than CMAP, which is represented on a 2.5° grid. RM3 projections of climate change to be made in future research will be evaluated in the context of the performance for recent climate history.

Decadal climate change will generally warm near-surface air temperatures. Impacts on regional hydrology are less certain. Druyan (2011) reviews 10 climate model investigations of 21st century trends of Sahel rainfall. There is no consensus among the studies as to whether climate change during this century will favor more frequent Sahel droughts or rainier summers. In order to properly simulate decadal precipitation trends for the Sahel a number of competing physical influences must be accurately considered by climate models. Perhaps the most challenging influence is to account for the future pattern of SST in the Atlantic Ocean. During the 20th century, cold (warm) SST anomalies in the tropical North Atlantic in juxtaposition with warm (cold) anomalies south of the Equator favored Sahel drought (positive rainfall anomalies) (Hastenrath and Polzin, 2010). If this teleconnection continues over the following decades, models will have to accurately project the sign and magnitude of the cross-equatorial SST gradient in the eastern Atlantic, which is a daunting challenge. Atmospheric and SST predictions from atmospheric-ocean global climate models (AOGCMs) forced by greenhouse gas trend scenarios will be used as the boundary conditions for regional model simulations. Results will therefore be derived from the latest generation of CMIP5 AOGCMs' projections of SST and atmospheric conditions, downscaled by regional models to provide climate evolution scenarios with high spatial detail over West Africa. The regional models, in turn, will then simulate high resolution projections of the West African climate that account for other forcings, among them projected changes in land use and vegetation cover. This line of research will accordingly test the sensitivity of climate change over West Africa to desertification, deforestation, changes in cropland, urbanization, etc.

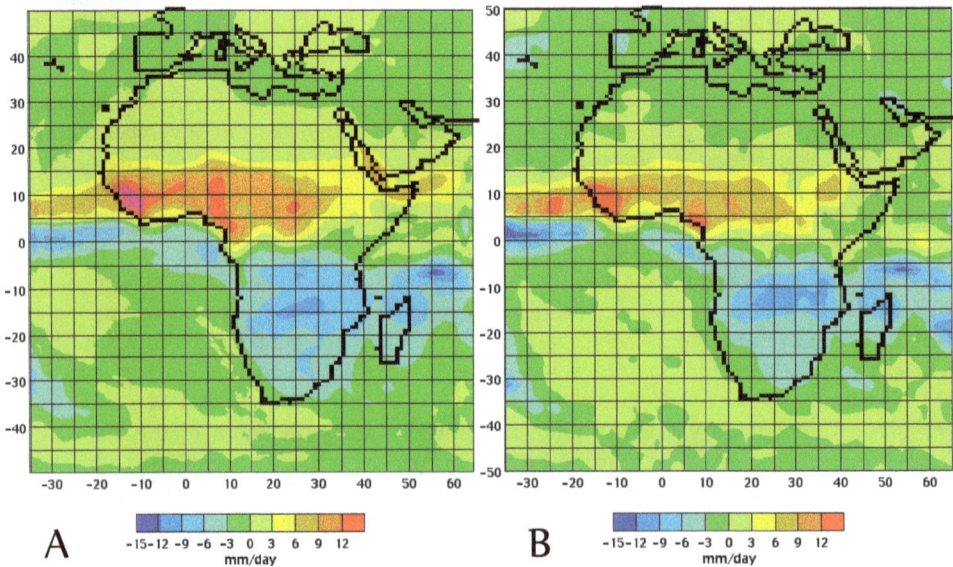

Fig. 9. August 1991 minus January 1991 precipitation rate differences (mm day^{-1}).
a. RM3 driven by ERA-I data; b. CMAP observations (gridded at 2.5°, data courtesy of the Earth System Research Laboratory, Physical Sciences Division, NOAA).

6. Acknowledgments

Research reported here was supported by National Science Foundation grants AGS-0652518 and AGS-1000874, NASA grant NNX07A193G and NASA cooperative agreement NNG04GN76A. TRMM images and data are acquired using the GES-DISC Interactive Online Visualization ANd aNalysis Infrastructure (Giovanni) as part of the NASA's Goddard Earth Sciences (GES) Data and Information Services Center (DISC). Reanalysis and CMAP data are provided by the NOAA-ESRL Physical Sciences Division, Boulder Colorado from their Web site at http://www.esrl.noaa.gov/psd/.

7. References

Druyan L (2011) Studies of 21st century precipitation trends over West Africa. *Int. J. Climato*logy 31: 1415-1424. DOI: 10.1002/joc.2180.

Druyan, L., M. Fulakeza, and P. Lonergan (2006) Mesoscale analyses of West African summer climate: Focus on wave disturbances. *Clim Dyn* 27: 459-481. doi:10.1007/s00382-006-0141-9.

Druyan, L., M. Fulakeza, and P. Lonergan (2008) The impact of vertical resolution on regional model simulation of the West African summer monsoon. *Int. J. Climatology*, 28: 1293-1314 doi:10.1002/joc.1636.

Druyan L, Fulakeza M (2011) The sensitivity of African easterly waves to eastern tropical Atlantic sea-surface temperatures. *Meteorol. Atmos. Phys.* 113: 39-53. DOI 10.1007/s00703-011-0145-9

Druyan L.M.; J. Feng; K.H. Cook; Y. Xue; M. Fulakeza; S.M. Hagos; A. Konare; W. Moufouma-Okia; D.P. Rowell; E.K. Vizy (2010) The WAMME regional model intercomparison study. *Clim Dyn* 35, 175-192. DOI 10.1007/s00382-009-0676-7.

Hansen J, co-authors (2002) Climate forcings in Goddard Institute for Space Studies SI2000 simulations. *JGR* DOI: 10.1029/200IJD001143

Hagos S M, Cook KH (2007) Dynamics of the West African Monsoon Jump. *J Clim* 20: 5264–5284.

Hastenrath S, Polzin D (2010) Long-term variations of circulation in the tropical Atlantic sector and Sahel rainfall. *Int. J. Climatol.* DOI: 10.1002/joc.2116.

Jones C, Nikulin G (2011) Evaluating the first CORDEX simulations over Africa. http://www.smhi.se/forskning/ forskningsomraden/klimat forskning/1.11299.

Jones C, Giorgi F, Asrar G (2011) The Coordinated Regional Downscaling Experiment: CORDEX, an international downscaling link to CMIP5. *CLIVAR Exchanges* 16: 34-39.

Kanamitsu M, Ebisuzaki W, Woollen J, Yang S K, Hnilo J J, Fiorino M, Potter G L (2002) NCEP-DOE AMIP-II reanalysis (R-2). *Bull Am Meteor Soc* 83: 1631-1643.

Kummerow C, co-authors (2000) The status of the Tropical Rainfall Measuring Mission (TRMM) after two years in orbit. *J. Appl. Meteor.* 39: 1965–1982.

New M, Lister D, Hulme M, Makin I, 2002: A high-resolution data set of surface climate over global land areas. *Clim Res* 21: 1-25.

Ramel R, Gallée H, Messager C (2006) On the northward shift of the West African monsoon. *Clim Dyn* 26: 429–440. DOI 10.1007/s00382-005-0093-5

Redelsperger JL, C.D. Thorncroft, A. Diedhiou, T. Lebel, D.J. Parker, J. Polcher (2006) African monsoon multidisciplinary analysis: an international research project and field campaign. *Bulletin of AMS* 87: 1739-1746.

Rosenzweig C, Abramopoulos F (1997) Land-surface model development for the GISS GCM. *J Climate* 10: 2040-2054.

Sultan B, Janicot S (2000) Abrupt shift of the ICTZ over West Africa and intra-seasonal variability. *Geophys Res Lett* 27: 3353–3356.

Thorncroft and co-authors (2003) The JET2000 Project: Aircraft observations of the African Easterly Jet and African easterly waves. Bull Am Met Soc 84: 337-351.

Xie, P., and P.A. Arkin, 1997: Global precipitation: A 17-year monthly analysis based on gauge observations, satellite estimates, and numerical model outputs. *Bull. Amer. Meteor. Soc.*, 78, 2539 - 2558.

Xue Y et al. (2010) Intercomparison of West African monsoon and its variability in WAMME: First GCM experiment. *Clim Dyn* 35: 3-27.

3

Seasonal Climate Prediction and Predictability of Atmospheric Circulation

June-Yi Lee and Bin Wang
Department of Meteorology and International Pacific Research Center,
University of Hawaii, Honolulu, HI,
USA

1. Introduction

While the detailed evolution of weather events may not be predictable beyond the span of a few days to two weeks due to the chaotic internal dynamics of atmospheric motion (Lorenz 1965), the statistical behavior of weather, that is the time or space averages, may be predictable over timescales of a season or longer due to the interaction between the atmosphere and the slowly varying lower boundary including ocean and land surface properties (Shukla 1998; Wang et al. 2009). In the past few decades, climate scientists have made tremendous advances in understanding and modeling the variability and predictability of the climate system. As a result, prediction of seasonal-to-interannual climate variations and the associated uncertainties using multiple dynamical models has become operational (Palmer et al. 2004; NOAA Climate Test Bed 2006; Lee et al. 2009). This chapter reviews and discusses the current status of seasonal climate prediction for the upper-tropospheric atmospheric circulation over the Northern Hemisphere (NH), in particular, using eight coupled, state-of-the art models that have participated in the Asia-Pacific Economic Cooperation Climate Center/Climate Prediction and its Application to Society (APCC/CliPAS) (Wang et al. 2009; Lee et al. 2010) and ENSEMBLE-based predictions of climate changes and their impactS (ENSEMBLES) project (Weisheimer et al. 2009; Alessandri et al. 2011). This is an extension of work from Lee et al. (2011), which addressed prediction and predictability of the NH summer upper-tropospheric circulation.

The multi-model ensemble (MME) approach was designed to quantifying forecast uncertainties due to model formulation near the turn of this century (Krishnamurti et al. 1999, 2000; Doblas-Reyes et al. 2000; Shukla et al. 2000; Palmer et al. 2000). The idea behind the MME is that if the model parameterization schemes are independent of each other, the model errors associated with the model parameterization schemes may be random in nature; thus, an averaging may cancel out the model errors contained in individual models. In general, the MME prediction is superior to the predictions made by any single-model component for both two-tier systems (Krishnamurti et al. 1999, 2000; Palmer et al. 2000; Shukla et al. 2000; Barnston et al. 2003) and one-tier systems (Hagedorn et al. 2005, Doblas-Reyes et al. 2005).

Climate forecast skill and predictability depend on the spatial location and season. Using 14 climate prediction models, Wang et al. (2009) showed that the one-month lead seasonal MME prediction is more (less) skillful in predicting wintertime atmospheric variability in the Pacific North America (East Asia) region compared to its summer counterpart. Jia et al. (2011) demonstrated that the current climate models have a season-dependent forecast skill regarding the dominant atmospheric circulation pattern in the NH extratropics. While the climate models have a significant skill for the leading atmospheric pattern and the associated time variation in MAM, JJA, and DJF, they have a difficulty in capturing the leading mode in SON.

The source of predictability of seasonal atmospheric anomalies over the NH extratropics is mainly attributable to teleconnection patterns, often linked to tropical boundary forcing. In boreal winter, the Pacific-North American (PNA) pattern and North Atlantic Oscillation (NAO) tend to significantly influence surface climate conditions and they explain a significant part of the interannual variance of the atmospheric anomalies over the NH extratropics (Wallace and Guztler 1981; Barnston and Livezey 1987; Trenberth et al. 1998; Hoerling et al. 2001). In boreal summer, the circumglobal teleconection (CGT) pattern is dominant over the NH extratropics (Ding and Wang 2005), acting as a significant source of climate variability and predictability over the region (Lee et al. 2011; Ding et al. 2011).

How to determine the signal variance (the predictable part of total variance) and predictability of atmospheric variability on seasonal time scale is still an open issue. According to the conventional signal-to-noise ratio approach determined by ensemble simulations of a stand-alone atmospheric model (Charney and Shukla 1981; Shukla 1998; Rowell 1998; Kang and Shukla 2006), the summertime atmospheric variability in the NH extratropics is less predictable than its winter counterpart and far less than that in the tropics. However, predictability obtained from the AGCM-alone approach is highly model-dependent. To better estimate predictability of seasonal-to-interannual climate variations, Wang et al. (2007) and Lee et al. (2011) suggested a "predictable mode analysis (PMA)" approach, which relies on identification of the predictable leading modes of the interannual variations in observations and retrospective MME forecast. The predictability is estimated by the fractional variance accounted for by the predictable leading modes. This chapter uses the PMA approach to estimate predictability of upper-level atmospheric circulation in each season in comparison with Mean Square Error Method suggested by Kumar et al. (2007).

Section 2 introduces the observational and prediction data and analysis methods used in this study. In Section 3, the current status of dynamical prediction of the seasonal atmospheric circulation is investigated using eight coupled models' hindcast data. Section 4 is devoted to estimate potential predictability for seasonal-mean atmospheric circulation anomalies. The summary is given in section 5.

2. Data and analysis method

2.1 Retrospective forecast data

This study uses one-month lead seasonal hindcast products of eight fully coupled ocean-land atmosphere models. Table 1 lists the acronyms of the institutions and models mentioned in the text. The eight coupled models are from CAWCR, NCEP, and GFDL in the CliPAS project (Lee et al. 2011) and CMCC-INGV, ECMWF, IFM-GEOMAR, MF, and UKMO in the ENSEMBLES project (Weisheimer et al. 2009). None of the coupled models has flux adjustments. A brief summary of the coupled models and their retrospective forecasts is presented in Table 2. For more models' descriptions, refer to Lee et al. (2011) for the three CliPAS models and Weisheimer et al. (2009) for the five ENSEMBLES models.

The common retrospective forecast period of the models covers the 25 years of 1981-2005. All models were integrated from around February 1 to at least May 31 for the boreal spring season (hereafter MAM), from around May 1 to at least August 31 for the boreal summer season (hereafter JJA), from around August 1 to at least November 30 for boreal fall season (hereafter SON), and from around November 1 to at least February 28 for boreal winter season (hereafter DJF). Each model has a different ensemble size (Table 2). The one-month lead MME prediction was made by simply averaging the eight coupled models' ensemble means.

2.2 Validation data

The observed data for validating models' performance are as follows. Data for 200-hPa zonal wind and 200-hPa GPH are from the NCEP/department of Energy (DOE) reanalysis II data (Kanamitsu et al. 2002). SST data is from the improved Extended Reconstructed Sea Surface Temperature Version 2 (ERSST V2) data (Smith and Reynolds 2004).

Acronym	Full names
AMIP	Atmospheric general circulation model intercomparison project
APCC	Asia-Pacific Economic Cooperation Climate Center
CAWCR	Centre for Australia Weather and Climate Research
CFS	Climate Forecast System
CliPAS	Climate Prediction and its Application to Society
CMCC-INGV	Euro-Mediterranean Centre for Climate Change
DEMETER	Development of a European Multimodel Ensemble System for Seasonal to Interannual Prediction
ECMWF	European Centre for Medium-Range Weather Forecast
ENSEMBLES	ENSEMBLE-based predictions of climate changes and their impactS
IFM-GEOMAR	Leibniz Institute of Marine Sciences at Kiel University
GFDL	Geophysical Fluid Dynamic Lab
MF	Météo France
NCEP	National Center for Environmental Prediction
POAMA	Predictive Ocean Atmosphere Model for Australia
UKMO	UK Met Office

Table 1. Acronym names of institutions and models used in the text.

Institute	Model Name	AGCM	OGCM	Ensemble member	Reference
CAWCR	POAMA1.5	BAM 3.0d T47 L17	ACOM3 0.5-1.5o lat x 2.0o lon L31	10	Zhong et al (2005)
GFDL	CM2.1	AM2.1 2olat x 2.5olon L24	MOM4 1/3olat x 1olon L50	10	Delworth et al (2006)
NCEP	CFS	GFS T62 L64	MOM3 1/3olat x 5/8olon L27	15	Saha et al (2006)
CMCC-INGV	CMCC	ECHAM5 T63 L19	OPA 8.2 2.0o x2.0o L31	9	Alessandri et al (2011) Pietro and Masina (2009)
ECMWF	ECMWF	IFS CY31R1 T159 L62	HOPE-E 1.4o x 0.3o-1.4o L29	9	Stockdale et al. (2011) Balmaseda et al. (2008)
IFM-GEOMAR	IFM	ECHAM4 T42 L19	OPA 8.2 2.0o lat x 2.0o lon L31	9	Keenlyside et al. (2005) Jungclaus et al. (2006)
MF	MF	IFS T95 L40	OPA 8.0 182GPx152GP L31	9	Daget et al. (2009) Salas Melia (2002)
UKMO	UKMO	ECHAM5 T42 L19	MPI-OM1 2.5o lat x 0.5o-2.5o lon L23	9	Collins et al. (2008)

Table 2. Description of the coupled models and their retrospective forecast used in this study.

2.3 Forecast quality measures

The measure of prediction skill includes the temporal correlation coefficient (TCC) skill, evaluating the interannual variability, and anomaly pattern correlation coefficient (PCC) skill for spatial similarity. We also calculated the area-averaged TCC skill over the NH tropics (Eq-30°N) and extratropics (30°-80°N), respectively, taking latitudinal weight into account.

We define a skill score as representing the coupled models' capability in predicting empirical orthogonal function (EOF) modes in terms of the PCC score for eigenvector and TCC score for the principal component (PC) time series for each mode. The skill score for each mode (i) is calculated by

$$Skill\ Score(i) = \sqrt{PCC(i) \times TCC(i)}\ .$$

The skill score ranges from 0 (no skill at all) to 1 (for perfect forecast). It should be mentioned that we reordered the EOF modes of the MME prediction according to the skill score because the order of the predicted EOF mode is not necessarily the same as its observed counterpart. In order to reorder the predicted EOF modes, the skill score for the first observed mode was calculated against all of the predicted modes and then the predicted mode that had the best skill score with the first observed mode was taken as the first predicted mode. Repeating the above process, other predicted modes were similarly determined. In the case of 200-hPa geopotential height (Z200) in all season, there was no change of mode order until the 3rd predicted mode.

3. Dynamical prediction of seasonal-mean atmospheric circulation

3.1 Variance

The upper tropospheric circulation is represented by geopotential height at 200-hPa (Z200 hereafter). Figure 1 compares the observed and hindcast interannual variance of the upper-level circulation along with the climatological jet stream in each season. Individual ensemble hindcasts of each model are used to estimate standard deviation of the Z200 anomalies for individual models in each season. The composite of individual models' total standard deviation is used for the MME. Note that the MME is able to capture the location of jet streams in both hemispheres and the atmospheric variability centers over the Tropics in all seasons. However, it has difficulty in capturing the observed variability centers particularly over the NH extratropics, except over the PNA region for most seasons. Models tend to overestimate the tropical variability and underestimate the high latitude variability.

3.2 Season-dependence

Season-dependent MME forecast skills are investigated for the period 1981-2005. Figure 2 shows the TCC skill of one-month lead MME predictions for Z200 in each season. The skillful forecast for Z200 is primarily confined in the tropics (Fig. 2a) in all seasons that are consistent with earlier results (Peng et al. 2000). Nonetheless, the MME has significant skills over some specific geographical locations in the NH extratropics depending on season, suggesting that predictable patterns (or modes) of the seasonal upper-level circulation in the region may exist. For example, atmospheric variability over the PNA region is more predictable during MAM and DJF while that over the Asian region is more predictable during JJA and SON.

In the tropics, the area-averaged TCC skill is the lowest in JJA and SON. On the other hand, in the entire NH extratropics (0-360°E, 30-80°N) TCC is the highest in JJA, although over the PNA region it is higher in DJF than in JJA. The TCC skill over the Southern Hemisphere (SH) shows a season-dependent pattern as well. The skill tends to increase in southern hemisphere subtropics and midlatitude during DJF and SON. The South Pacific convergence region tends to have higher skill than other southern hemisphere regions.

We also investigate the TCC skill for 500-hPa geopotential height (Z500) which is often used for representing mid-tropospheric atmospheric circulation. Comparison between Fig. 2 and 3 indicates that Z200 is more predictable than Z500 although the spatial pattern of the TCC for Z200 is very similar to that for Z500.

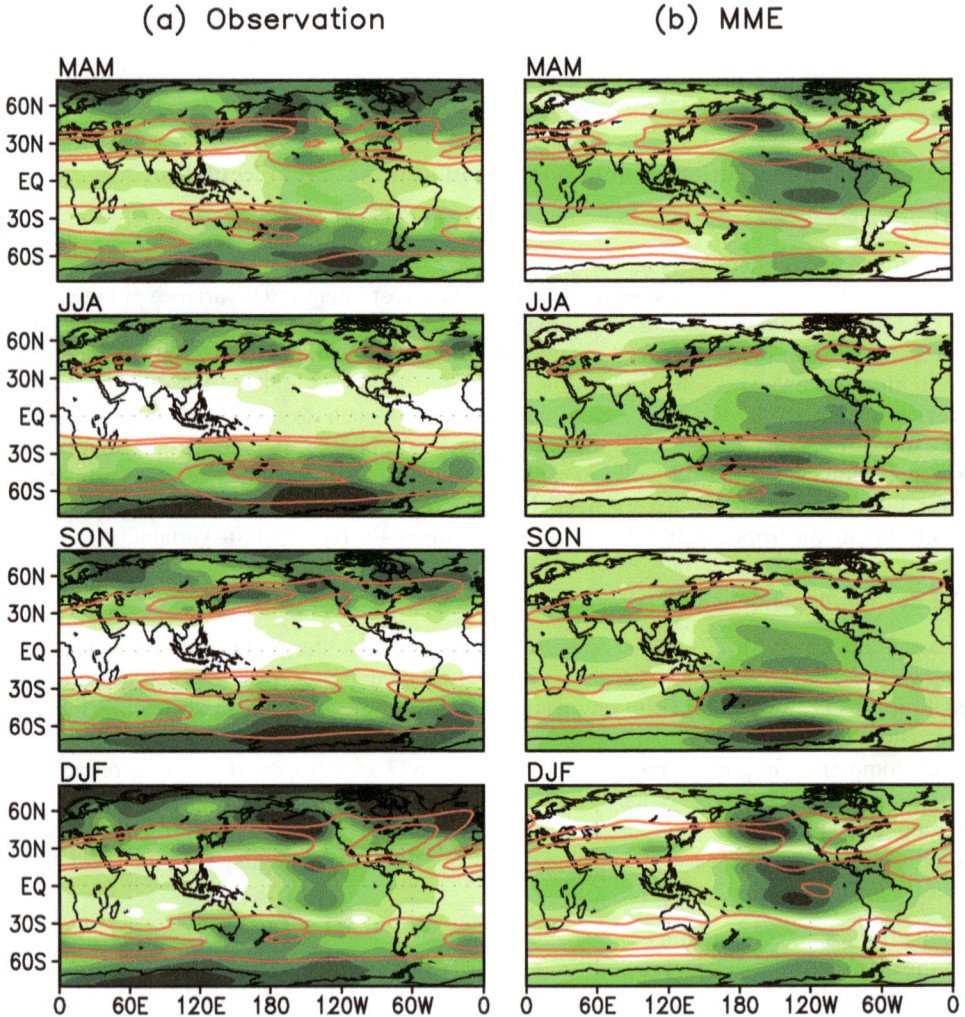

Fig. 1. Standard deviation of geopotential height (shading) anomalies and climatological mean of zonal wind (contour) at 200 hPa in each season obtained from (a) observations and (b) the one-month lead MME seasonal prediction. The units are *m* for geopotential height and *m* s^{-1} for zonal wind. 20 and 30 *m* s^{-1} zonal wind are contoured.

Temporal Correlation Skill for Z200

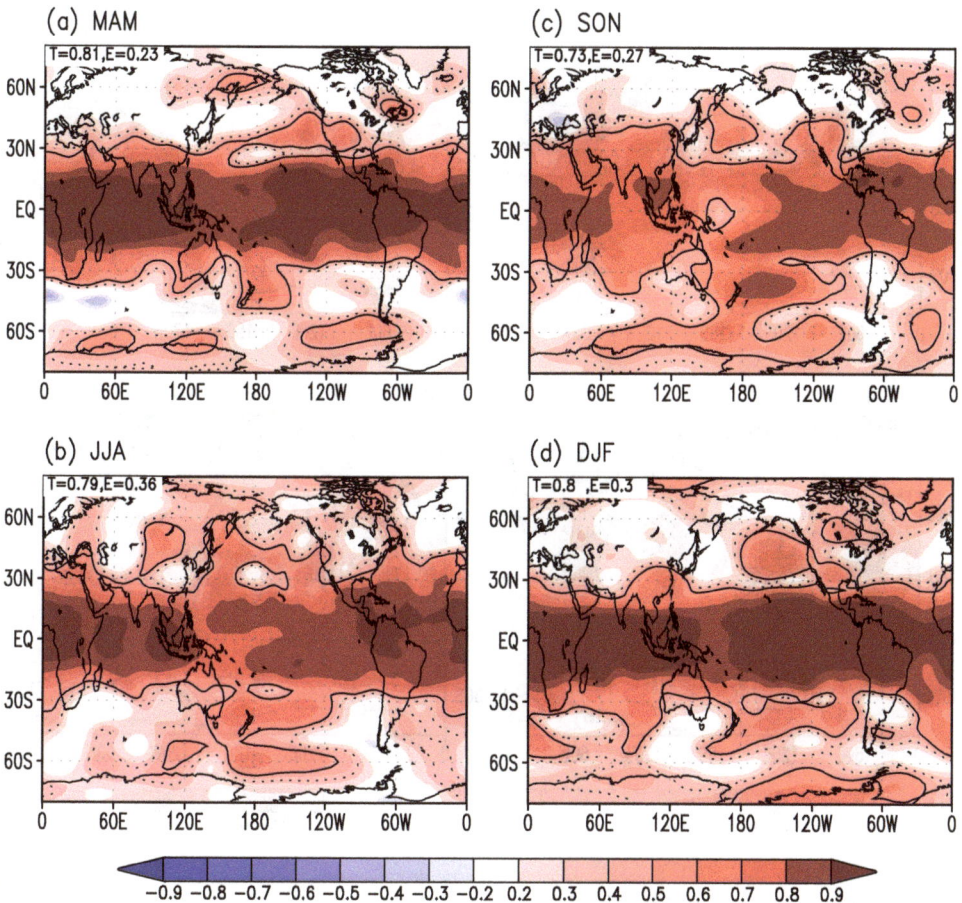

Fig. 2. The temporal correlation coefficient (TCC) skill for the one-month lead prediction of 200-hPa geopotential height (Z200) in (a) MAM, (b) JJA, (c) SON, and (d) DJF obtained from the eight coupled models' multi-model ensemble (MME) for the period of 1981-2005 in the Northern Hemisphere (NH). Solid (dashed) line represents statistical significance of the correlation coefficients at 95% (99%) confidence level. The numbers in the left upper corners indicate averaged correlation skill over the tropics (T: 0-360°E, Eq-30°N) and extratropics (E: 0-360°E, 30-80°N) in the NH.

Temporal Correlation Skill for Z500

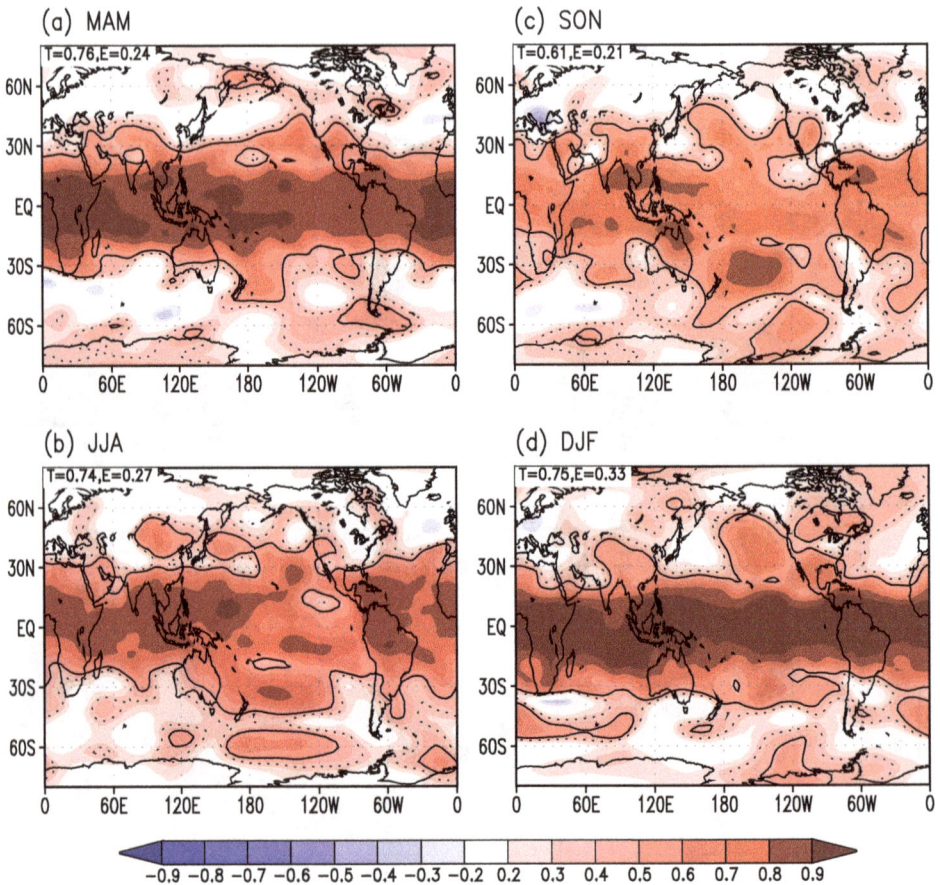

Fig. 3. Same as Fig. 2 except for 500-hPa geopotential height (Z500).

3.3 Predictable modes

It is important to understand the source of the prediction skill in the current coupled models and identify predictable modes of climate variability. To this end, we evaluate how well the MME hindcast captures the first two dominant modes of the Z200 variability in each season. To identify the major modes of upper-tropospheric atmospheric circulation, we first applied EOF analysis, using the correlation matrix, to the seasonal Z200 over the entire NH (0-360°E, Eq-80°) for observations and for the MME prediction. All data were interpolated to the same geographic grid to avoid the latitudinal weighting effect and data were normalized by their own standard deviation.

Figures 4-7 show the first and second EOF modes of the NH Z200 variability for observations and the one-month lead MME prediction in each season. Similar to the major modes of JJA Z200 discussed in Lee et al. (2011), the first two EOF modes of the NH Z200 variability in other seasons are also understood in terms of ENSO teleconnection dynamics. For all seasonal cases, a strong positive (negative) phase in the first mode tends to occur simultaneously or after the mature phase of La Niña (El Niño), and is thus driven by prolonged impacts of the ENSO from the preceding SON to following JJA. The second mode, on the other hand, is regulated by the developing phase of the ENSO starting from the preceding DJF to following SON on the interannual time scale and is correlated with the SST anomalies over the North Pacific, North Atlantic, and tropical Western Pacific on the interdecadal time scale.

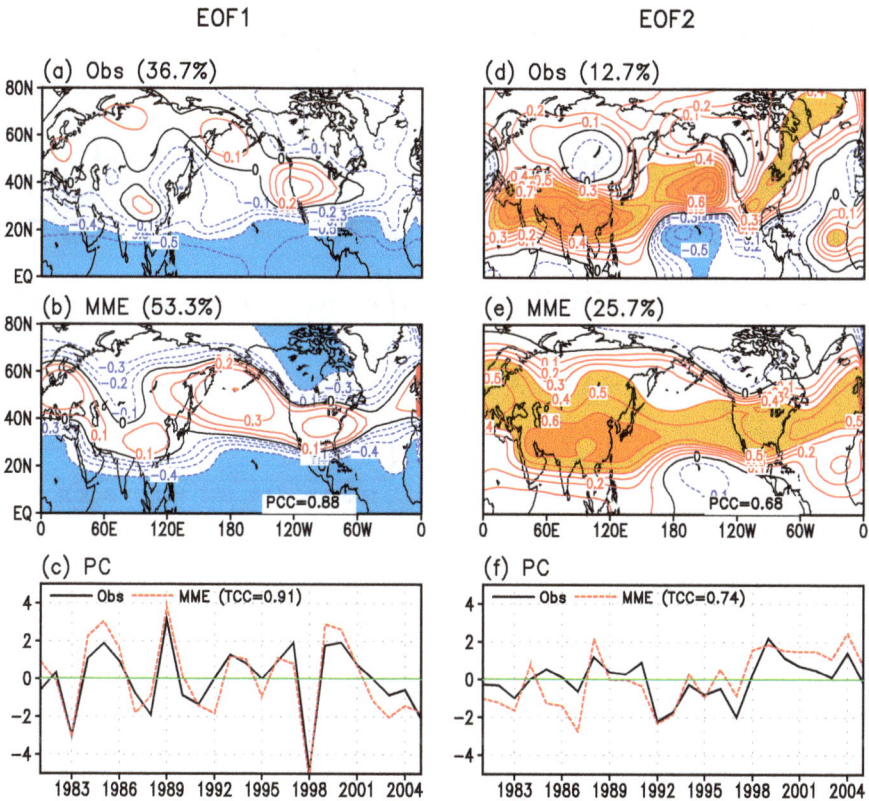

Fig. 4. Spatial patterns of the first (left panels) and second (right panels) eigenvectors of 200-hPa GPH anomalies in MAM obtained from observation (a, d) and 8-coupled modes' MME prediction (b, e), respectively. The numbers in the right lower corners in (d) and (e) indicate pattern correlation coefficient (PCC) between observation and the corresponding prediction. (c) and (f) principal components (PCs) of the first and the second EOF modes obtained from observation (black solid line) and MME (red dashed line). The numbers within the parenthesis in the figure legend in lower panels indicate the temporal correlation coefficients (TCC) between the observed and MME PC time series.

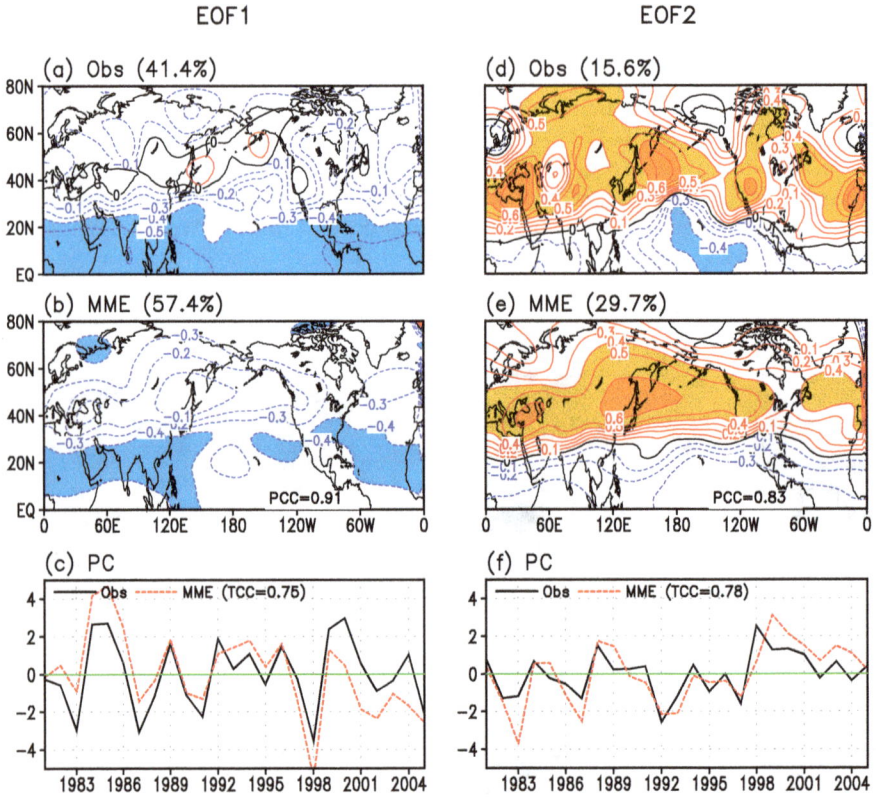

EOF1   EOF2

(a) Obs (41.4%)   (d) Obs (15.6%)

(b) MME (57.4%)   (e) MME (29.7%)

(c) PC   (f) PC

Fig. 5. Same as Fig. 4 except for JJA.

The predictability source of the first two modes can be more clearly depicted by the lead-lag correlation coefficients for seasonal SST against the first and second PC of Z200 shown Fig. 8. We just show the DJF case, since the other seasons have similar results. Figure 8 indicates that the first mode is associated with a typical or conventional eastern Pacific El Nino with complimentary same-sign SST anomalies occurring in Indian Ocean and Atlantic. On the other hand, the second mode seems to be associated with a central Pacific El Nino with pronounced opposite-sign SST anomalies in the western Pacific and Atlantic but little SST anomalies in the Indian Ocean. This suggest that the tropical precipitation anomalies associated with the locations of the tropical SST anomalies are critical important for determining the extratropical response.

The MME is capable of predicting the spatial and temporal structures of the two leading EOF modes one-month ahead with high fidelity in each season, except for the second EOF mode of DJF Z200. The MME's forecast skill for the EOF modes will be more discussed in section 4.2.

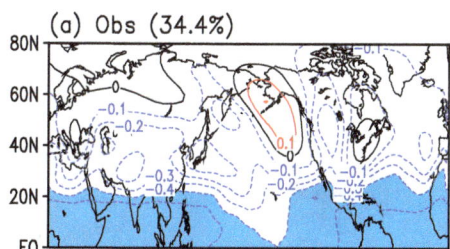

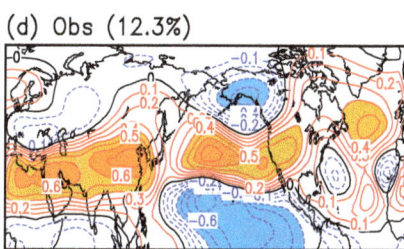

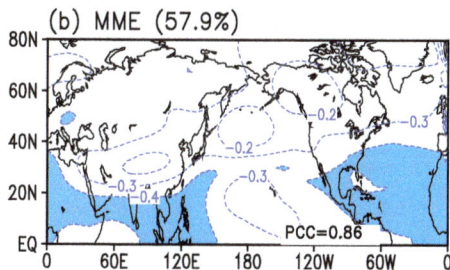

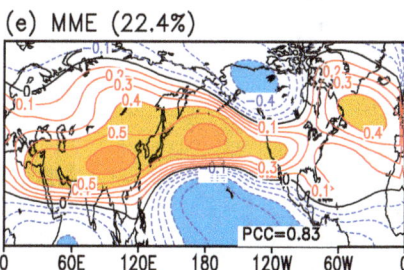

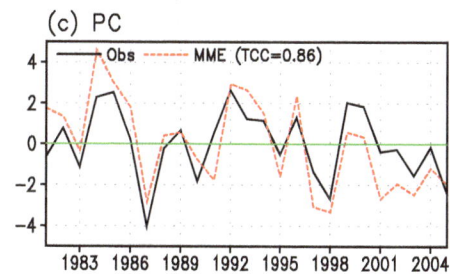

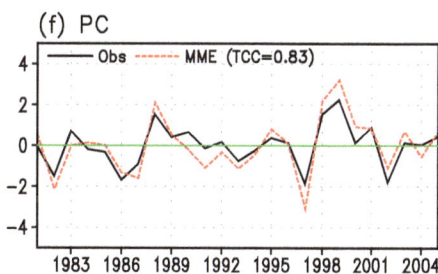

Fig. 6. Same as Fig. 4 except for SON.

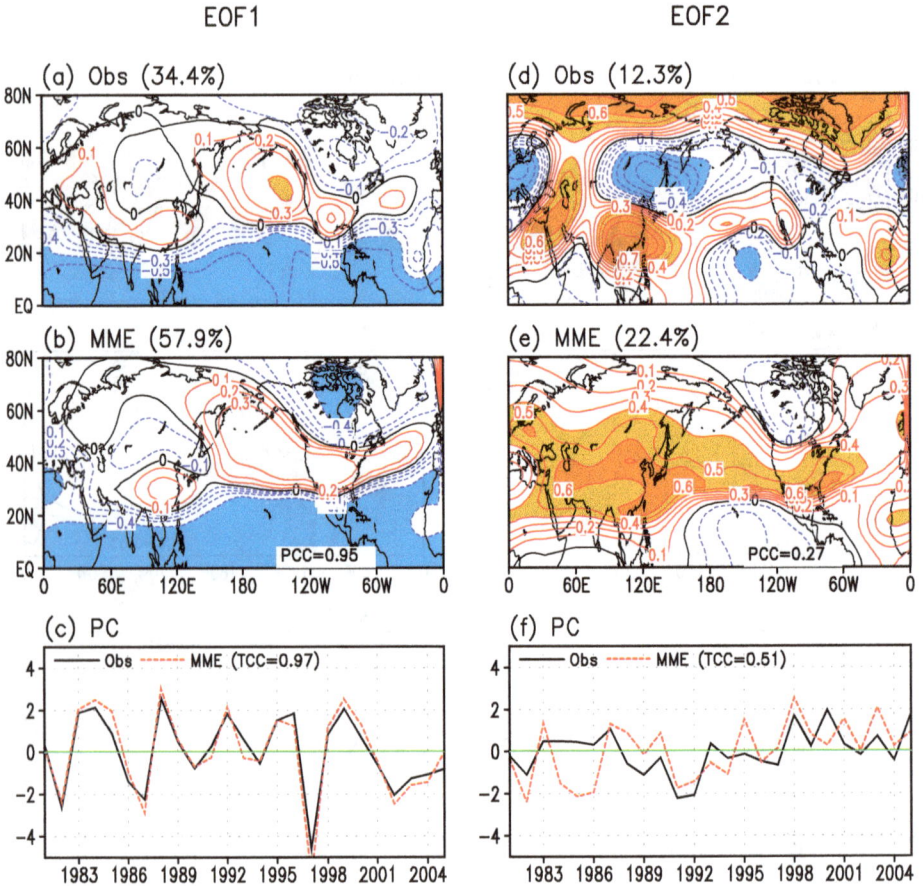

Fig. 7. Same as Fig. 4 except for DJF.

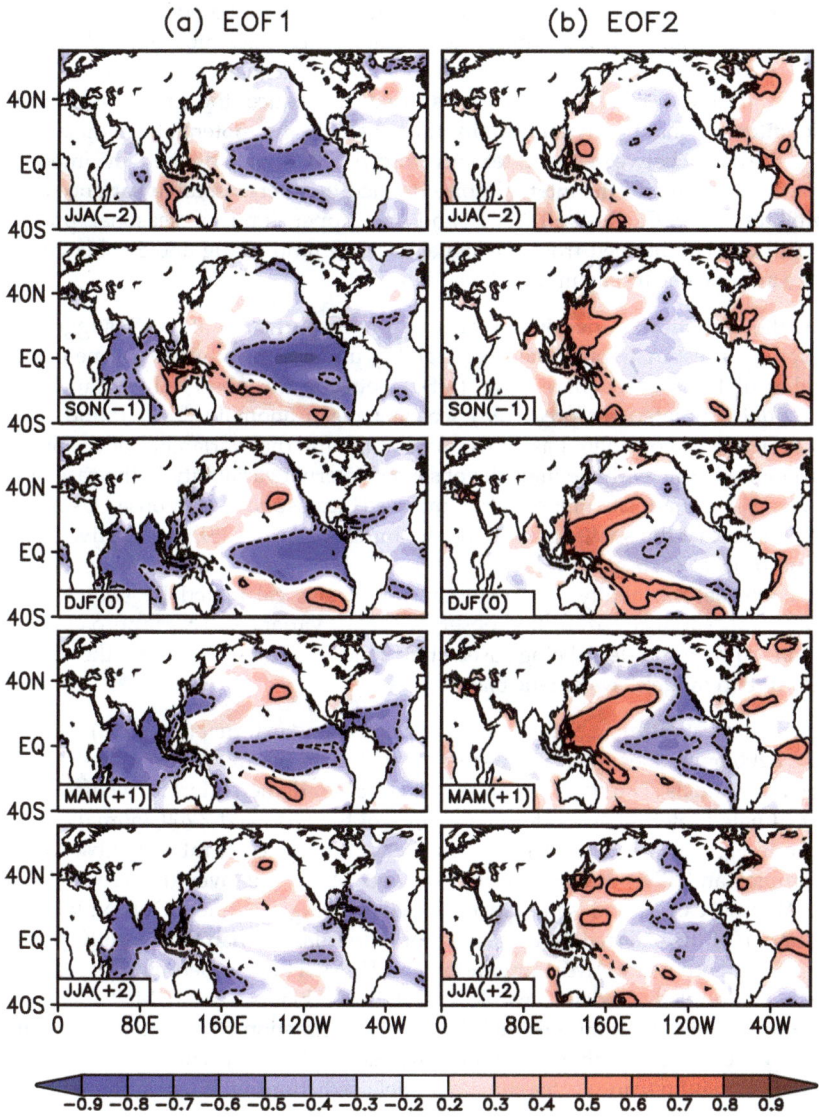

Fig. 8. Spatial patterns of the lead-lag correlation coefficients for seasonal mean SST against (a) the first PC and (b) the second PC, respectively, of DJF Z200 obtained from observations. Correlation coefficients which are statistically significant at 99% confidence level are contoured.

4. Predictability of seasonal-mean atmospheric circulation

4.1 Mean Square Error method

For a two-tier prediction, the physical basis for the seasonal predictability lies in slowly varying lower boundary forcing, especially the anomalous SST forcing (Charney and Shukla 1981; Palmer 1993; Shukla 1998). Thus, the predictability does not depend on initial atmospheric conditions. To better extract a predictable signal, an ensemble forecast approach with different initial conditions is used to reduce the errors arising from atmospheric internal chaotic dynamics. In this system, the potential predictability is measured as the ratio between the externally forced SST signal defined by interannual ensemble mean variance and the internal noise defined by inter-ensemble variance (Shukla 1981; Roswell et al. 1995; Roswell 1998) and the upper limit of seasonal prediction correlation skill can be obtained from the signal-to-noise ratio (Kang and Shukla 2006). By definition, the approach is highly model-dependent.

To overcome the limitation of the conventional approach, Kumar et al (2007) suggested the method to optimally estimate potential predictability under a multi-model frame using the expected value of the mean square error (MSE) between the observed and the general circulation model simulated (or predicted) seasonal mean anomaly. They found that the property of MSE can be considered as an estimate of the internal variability of the observed seasonal variability because MSE equals the observed internal variability in the case of large ensembles using a dynamical model with unbiased atmospheric response. Using multi models, they found that the spatial map of the minimum value of MSE irrespective of which model it came from at each geographical location can be regarded as the best estimate for the observed internal variability. The external variance is calculated by subtracting the estimated internal variance from the observed total variance. This approach is less dependent on climate models being used but tends to underestimate the observed predictability due to models' systematic bias.

Figure 9 shows the best estimate of the internal (or noise) and external (or signal) variance of the observed Z200 anomalies in each season obtained from the minimum value of MSE in the eight coupled models. The signal-to-noise ratio (ρ; SN ratio hereafter) is calculated by the ratio of the estimated internal and external variance of the observed Z200 anomalies in each season shown in Fig. 10. Significant signal variance and thus the SN ratio are observed over the entire Tropics and PNA region during DJF and MAM but over the Tropical Indian Ocean and Atlantic Ocean and some parts of Asia during JJA and SON. Note that the current estimation of the external variance and SN ratio is larger than that from Kumar et al. (2007)'s estimation both using AMIP simulation and DEMETER coupled hindcast indicating that the current coupled models have less systematic biases than AMIP models and DEMETER coupled models. It implies that the current estimation of potential predictability is closer to the observed true value than that of Kumar et al. (2007). There is always a room to improve the estimation using better climate model predictions with less systematic biases.

In order to compare the estimated SN ratio (ρ) with the MME TCC skill, a theoretical limit of correlation coefficient (R_{Limit}, Kang and Shukla 2006) is calculated using the following equation

$$R_{Limit} = \sqrt{\frac{\rho}{\rho+1}}$$

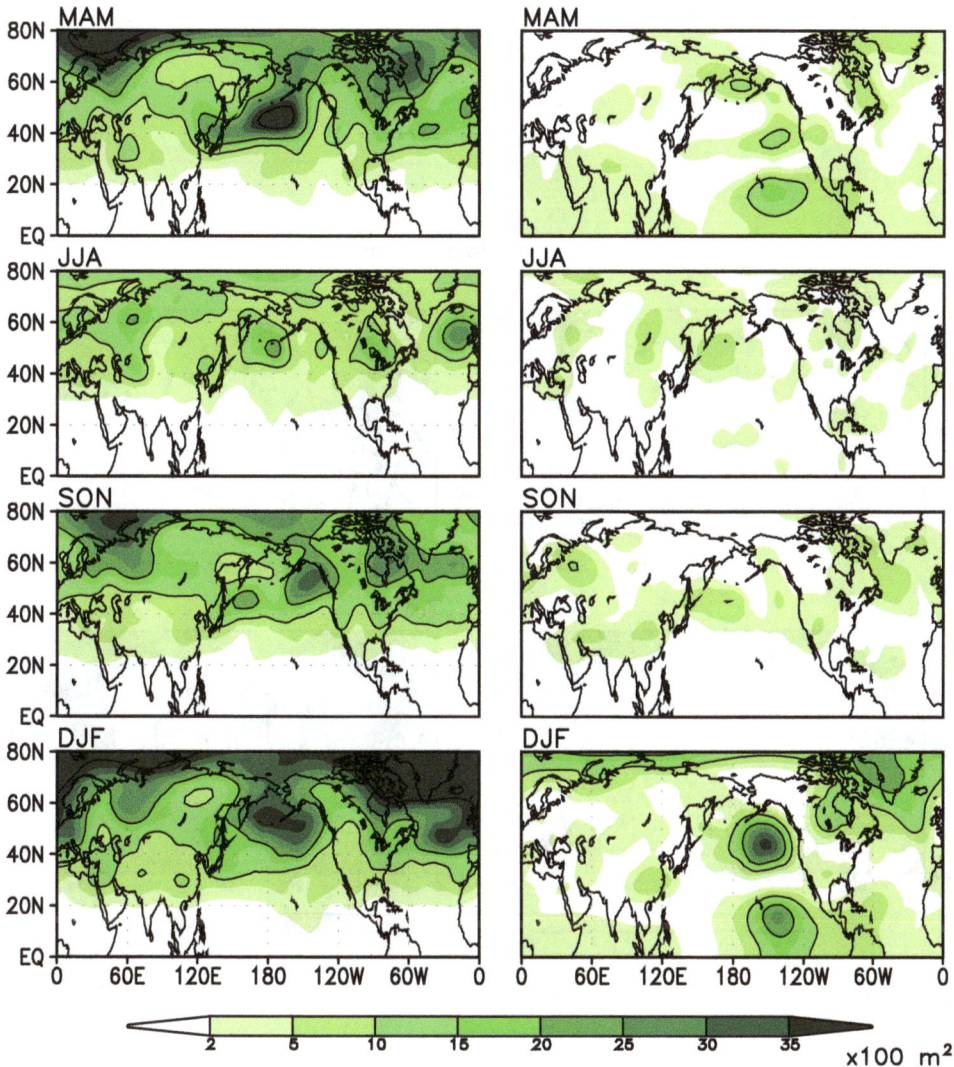

Fig. 9. The best estimate of the (a) internal and (b) external variance of observed Z200 anomalies in each season. The unit is m^2 and 1000, 2000, and 4000 m^2 are contoured.

The R_{Limit} measures the intrinsic limits of the predictability due to the internal dynamics of atmospheric variability. Figure 11 indicates that the MME skill (Fig. 2) reaches the theoretical limit of TCC over the Tropics during all seasons but is less over most of the extratropical region, particularly over continents. It is interesting to note that R_{Limit} is less than the MME skill over the Tropical Pacific during SON and JJA implying that climate models used in this study tend to have large systematic biases over the region during those seasons.

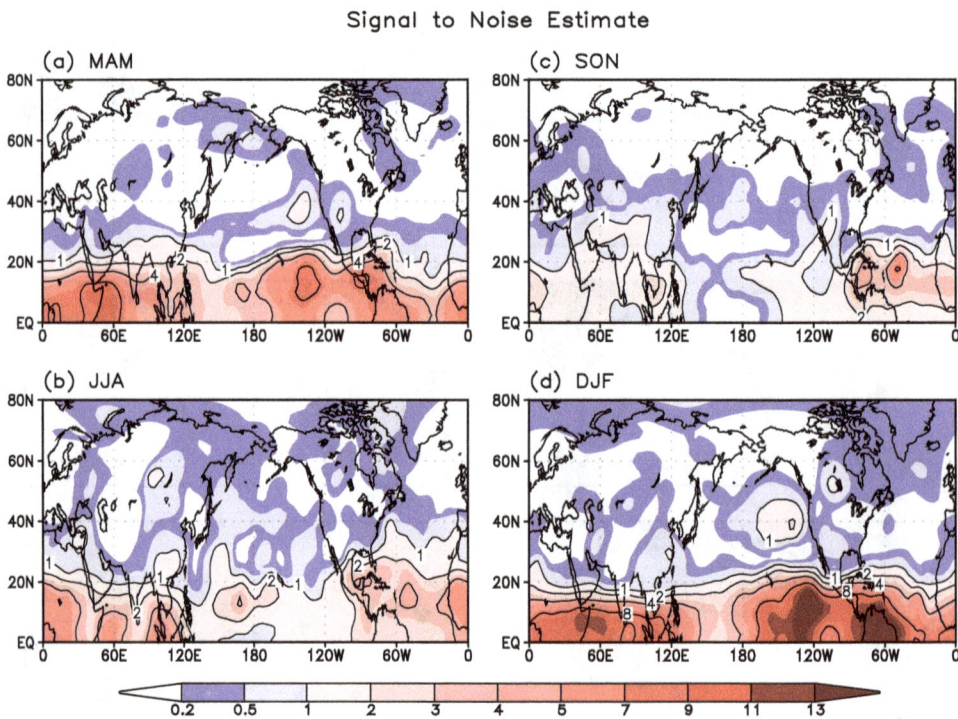

Fig. 10. The best estimate of signal-to-noise ratio for the observed Z200 anomalies in (a) MAM, (b) JJA, (c) SON, and (d) DJF computed as the ratio of external-to-internal variance. Higher ratios imply higher potential predictability.

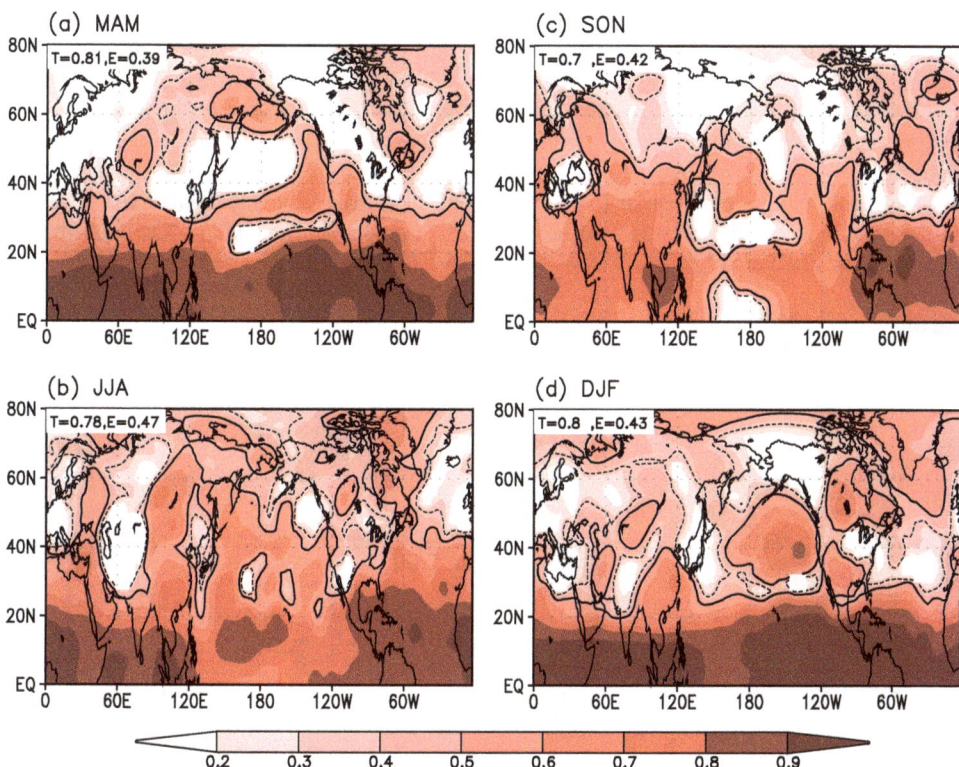

Fig. 11. Potential predictability in terms of the theoretical limit of correlation skill for Z200 in (a) MAM, (b) JJA, (c) SON, and (d) DJF using the best estimate of the internal and external variance based on MSE method. Solid (dashed) line represents statistical significance of the correlation coefficients at 95% (99%) confidence level. The numbers in the left upper corners indicate averaged correlation skill over the tropics (T, 0-360°E, Eq-30°N) and extratropics (E, 0-360°E, 30-80°N) in the NH.

4.2 Predictable mode analysis approach

Wang et al. (2007) and Lee et al. (2011) suggested a way to determine predictable modes and thus potential predictability using observations and the state-of-the-art climate models' predictions. There are two basic criteria for the determination. First, for observations, predictable modes should explain a large part of the total variability and be statistically separated from other higher modes. Second, the climate prediction models should be capable of predicting these major modes. According to these principles, the predictable modes are identified using percentage variance for each EOF mode in observation and the skill score of the MME prediction in terms of the combined spatial and temporal skill for each mode.

Figure 12 shows a scatter diagram between the observed percentage variance (ordinate) and the skill score for each EOF (abscissa) mode. The first two modes are not only well separated from higher modes statistically but also predicted with high fidelity by the current coupled models' MME in MAM, JJA, and SON similar as the result of Lee et al. (2011). However, the second EOF mode in DJF is much less predictable while its first mode is the most predictable mode with the skill score of 0.96. The low skill of the second mode is mainly attributable to the fact that the current MME has difficulty in capturing its spatial distribution over the high-latitude while it has a useful skill in capturing PC time variation (Fig. 7). Nonetheless, we consider the second mode as a predictable mode since the physical basis of the mode is understood as described in Section 3 and the MME's TCC skill for the second PC is relatively high.

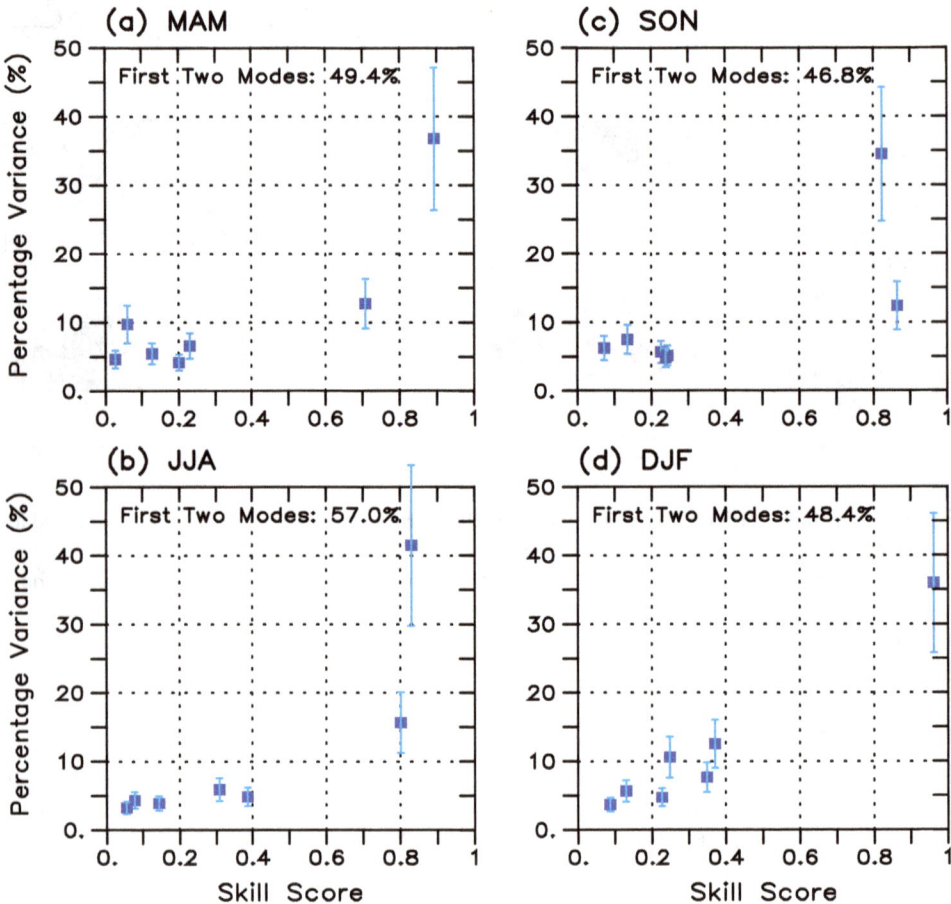

Fig. 12. The percentage variances that are accounted for by the observed first seven EOFs (ordinate) and the combined forecast skill score for the eigenvector and principal component for each mode (abscissa) for 200-hPa GPH in NH. The leading two modes of the observed 200-hPa GPH in JJA capture about 57% of the total interannual variability.

The first two modes capture about 49.4%, 57.0%, 46.8%, and 48.4% of the observed total variability in MAM, JJA, SON, and DJF, respectively, over the entire NH. Over the extratropics only, 18.8% (65.9%), 35.4% (82.7%), 18.9% (69.8%), and 23.2% (53.5%) of the observed (predicted) total variability can be explained by the first two modes in MAM, JJA, SON, and DJF, respectively (Table 3). The fractional variance can be interpreted as potential predictability from a conventional point of view. That means summer-time upper-level circulation is most predictable than the other seasons in the current approach. About 35.4 % of total variability over the NH extratropics in JJA is potentially predictable but only 18.8% is predictable in MAM.

Given the assumption that the first two leading modes are more predictable and the higher modes are less predictable (Fig. 12), the total predicted as well as observed fields are decomposed into the predictable and unpredictable parts. The predictable part is reconstructed by the linear combination of the first two EOF modes and the unpredictable part is then calculated by subtracting the predictable part from the total field. The realizable potential predictability is estimated by the TCC between the observed total field and the observed predictable part in order to facilitate comparison with the MME prediction skill. Thus, it represents the achievable forecast skill if climate models can perfectly predict the observed predictable modes.

Season		EOF1		EOF2	
		Tropics	Extratropics	Tropics	Extratropics
MAM	Obs	63.5%	6.1%	13.3%	12.7%
	MME	74.5%	28.2%	16.4%	37.7%
JJA	Obs	76.8%	13.4%	7.2%	22.0%
	MME	72.7%	41.8%	17.8%	40.9%
SON	Obs	58.7%	6.9%	13.0%	12.0%
	MME	68.6%	46.7%	21.6%	23.1%
DJF	Obs	60.0%	8.0%	9.8%	15.2%
	MME	73.7%	22.5%	15.0%	31.0%

Table 3. Area-averaged fractional variance for the first two EOF modes over the tropics and extratropics in NH obtained from observation and the one-month lead MME prediction.

Comparison between Figs. 13 and 11 indicates that potential predictability measured by the PMA method is higher over the NH tropics than that by the MSE method. Over the NH extratropics, the two methods have a comparable result in MAM and DJF but the PMA has higher (lower) predictability in JJA (SON) than the MSE. It is suggested that the two methods are complimentary to better estimate the true value of the observed predictability because they each have their own shortcoming. By definition, the two methods tend to underestimate the true value.

Potential Predictability /PMA Method

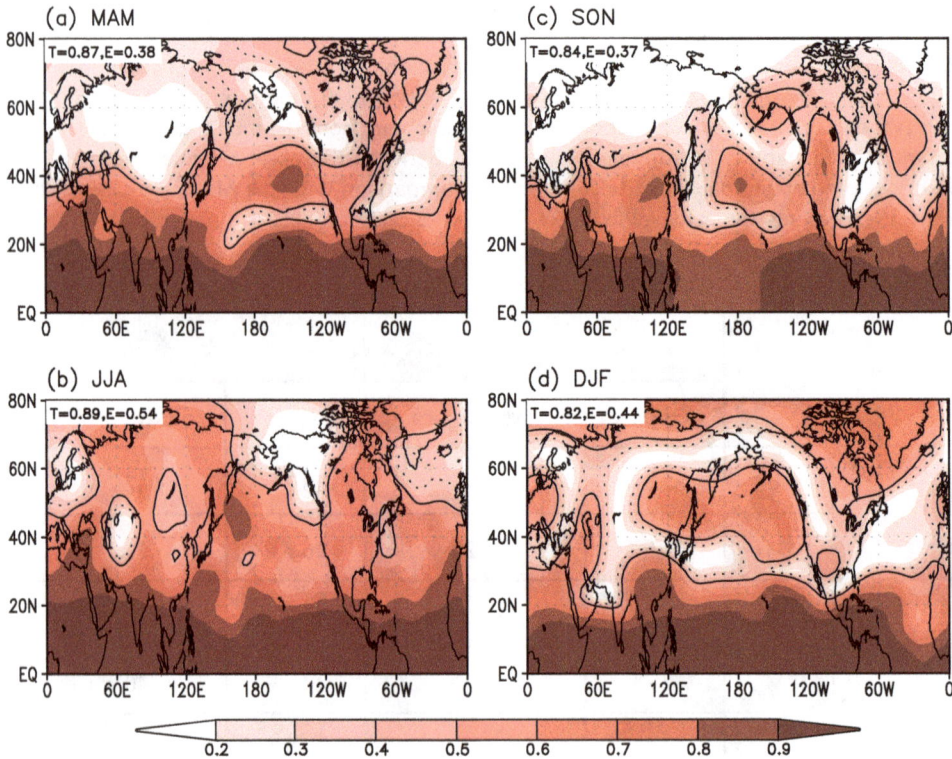

Fig. 13. Same as Fig. 12 except for potential predictability in terms of the TCC skill based on PMA method.

5. Summary and discussion

Using the eight fully coupled models in the CliPAS and ENSEMBELS project, prediction skill and predictability of seasonal mean upper-level atmospheric circulation are investigated over 25 years, 1981-2005. The eight coupled models' MME predicts the seasonal Z200 anomalies over the NH tropics with high fidelity, but has significantly lower skills in predicting those over the NH extratropics. Nonetheless, over specific geographic locations in the extratropics depending on season, the MME has considerable forecast skill, suggesting that predictable patterns may exist over the region of interest in each season. The MME has better skill for the Z200 anomalies over the PNA (Asian) region during MAM and DJF (JJA and SON) than the Asian (PNA) region. The area-averaged TCC skill over the NH tropics is the lowest in SON, while over the NH extratropics, it is highest in JJA.

The first two EOF modes of the Z200 anomalies in all seasons are identified as predictable leading modes for a number of reasons. First, these observed modes are statistically well separated from the higher modes and account for the large fraction of total variability; second, these modes have clear physical interpretations and their sources of variability are understood in terms of ENSO teleconnection dynamics and third, the current MME is capable of predicting the spatial structure and temporal variation of these modes with high fidelity except the second mode of the DJF Z200. Note that the low skill of the DJF second mode is mainly attributable to the fact that the current MME has difficulty in capturing its spatial distribution over high latitudes.

The first mode is associated with a typical or conventional eastern Pacific El Nino with complimentary same-sign SST anomalies occurring in Indian Ocean and Atlantic. On the other hand, the second mode seems to be associated with a central Pacific El Nino with pronounced opposite-sign SST anomalies in the western Pacific and Atlantic but little SST anomalies in the Indian Ocean. This suggest that the tropical precipitation anomalies associated with the locations of the tropical SST anomalies are critical important for determining the extratropical response.

Given the assumption that the first two leading modes can be perfectly predicted and the higher modes are noise patterns, we define the realizable potential predictability by the total fractional variances of the predictable modes. The first two modes capture about 49.4%, 57.0%, 46.8%, and 48.4% of the observed total variability in MAM, JJA, SON, and DJF, respectively, over the entire NH, which may represent an upper limit for the forecast skill potentially obtainable using the MME prediction; thus offering an estimate of attainable potential predictability. The realizable potential predictability is calculated by the TCC between the observed total field and the observed predictable component. Potential predictability of Z200 in each season reveals that the interannual variability over particular geographic locations in the NH extratropics is to a large degree predictable with the first two modes, although the extratropical upper-tropospheric circulation is generally much less predictable than over the tropics.

We compare the potential predictability measured by the PMA method with that estimated by the MSE method suggested by Kumar et al. (2007). Kumar et al. (2007) demonstrated that MSE can be considered as an estimate of the internal variability of the observed seasonal variability because MSE equals the observed internal variability in the case of large ensembles using a dynamical model with unbiased atmospheric response. Using multi models, they found that the spatial map of the minimum value of MSE, irrespective of which model it came from at each geographical location, can be regarded as the best estimate for the observed internal variability. The external variance is calculated by subtracting the estimated internal variance from the observed total variance. We demonstrate that the two estimates compliment each other to better measure the true value of the observed predictability, taking into account their own shortcomings. It is noted that potential predictability measured by the PMA method is higher over the NH tropics than by the MSE method. Over the NH extratropics, the two methods have a comparable result in MAM and DJF, but the PMA has higher (lower) predictability in JJA (SON) than the MSE. The estimations shown here are based on the state-of-the art climate model predictions. With further improvement of climate prediction models, estimates of potential predictability will become closer to their true values.

6. Acknowledgement

This research was supported by APEC Climate Center and IPRC (which is in part supported by JAMSTEC, NOAA (grant No. NNX07AG53G)), and NOAA (grant No. NA09OAR4320075). Authors would like to acknowledge contributions from the CliPAS and ENSEMBLES team members. This is IPRC publication 825 and SOEST publication 8513.

7. References

Alessandri A., A. Borrelli, A. Navarra, A. Arribas, M. Déqué, P. Rogel, and A. Weisheimer, 2011: Evaluation of probabilistic quality and value of the ENSEMBLES multi-model seasonal forecasts: comparison with DEMETER. Mon. Wea. Rev., 139, 581-607, doi:10.1175/2010MWR3417.1.

Barnston, A. and R. E. Livezey, 1987: Classification, seasonality and persistence of lowfrequency atmospheric circulation patterns. Mon. Wea. Rev., 82, 1083-1126.

Barnston, AG, Mason SJ, Goddard L, Dewitt DG, Zebiak SE, 2003: Multimodel ensembling in seasonal climate forecasting at IRI. Bull Am Meteor Soc 84:1783-1796

Charney, J. G., J. Shukla, 1981: Predictability of monsoons. Paper presented at the Monsoon Symposium in New Delhi, India, 1977. In: Lighthill J Sir, Pearce RP (eds) Monsoon dynamics, Cambridge University Press

DelSole, T., 2004: Predictability and information theory. Part I: Measures of predictability, J. Atmos. Sci., 61, 2425-2440.

Ding, Q., and B. Wang, 2005: Circumglobal teleconnection in the Northern Hemisphere summer. J Clim 18:3483-3505

Ding, Q., B. Wang, J. M. Wallace, and G. Branstator, 2011: Tropical-extratropical teleconnections in boreal summer: observed interannual variability. J. Climate, 24, 1878-1896.

Doblas-Reyes, F. J., Déqué, M. and Piedelievre, J.-P., 2000: Multi-model spread and probabilistic forecasts in PROVOST. Q. J. R. Meteorol. Soc. 126, 2069-2087.

Folland, C., J. Knight, H. Linderholm, D. Fereday, S. Ineson, and J. Hurrell, 2009: The summer North Atlantic Oscillation: Past, Present, and Future. J. Climate, 22, 1082-1103.

Hagedorn R, Doblas-Reyes FJ, Palmer TN, 2005: The rationale behind the success of multi-model ensembles in seasonal forecasting – I. Basic concept. Tellus 57A: 219-233

Hoerling, M., J. W. Hurrell, and T. Xu, 2001: Tropical origins for recent north atlantic climate change. Science, 292, 90-92.

Jia, XiaoJing, H. Lin, J.-Y. Lee, and B. Wang, 2011: Season-dependent forecast skill of the dominant atmospheric circulation patterns over the Pacific North-American region. Submitted to J. Climate.

Kang, I. S. and J. Shukla, 2006: Dynamic seasonal prediction and predictability of the monsoon. In: Wang B (ed) The Asian monsoon, Springer-Paraxis, Chichester, UK

Krishnamurti, T. N., Kishtawal, C. M., LaRow, T. E., Bachiochi, D. R., Zhang, Z. and co-authors. 1999. Improved weather and seasonal climate forecasts from multi-model superensemble. Science 285, 1548–1550.

Krishnamurti, T. N., Kishtawal, C. M., Shin, D. W. and Williford, C. E. 2000. Multi-model superensemble forecasts for weather and seasonal climate. J. Climate 13, 4196–4216.

Kumar A., B. Jha, Q. Zhang, and L. Bounoua, 2007: A new methodology for estimating the unpredictable component of seasonal atmospheric variability. J. Climate, 20, 3888-3901.

Lee J.-Y., B. Wang, I.-S. Kang, J. Shukla et al., 2010: How are seasonal prediction skills related to models' performance on mean state and annual cycle? Clim Dyn 35:267–283

Lee, J.-Y., Wang, B., Ding, Q., Ha, K.-J., Ahn, J.-B. et al. 2011:How predictable is the Northern Hemisphere summer upper-tropospheric circulation? Clim. Dyn. 37, 1189-1203.

Lee, W.-J. et al. 2009: APEC Climate Center for climate information services, APCC 2009 Final Report (Available at http://www.apcc21.net/activities/activities03_01.php)

Leung, L.-Y., and G. R. North, 1990: Information theory and climate prediction, J. Climate, 3: 5–14.

Lorenz, E. N. 1965: A study of the predictability of a 28-variable model. Tellus, 17, 321-333

NOAA Climate Test Bed, 2006: MME White Paper (Available at http://www.cpc.noaa.gov/products/ctb/ctb-publications.shtml)

Palmer TN, Brankovic C, Richardson DS, 2000: A probability and decision-model analysis of PROBOST seasonal multi-model ensemble integrations. Q J R Meteorol Soc 126: 2013-2034

Palmer, T. N., Alessandri, A., Andersen, U. et al., 2004: Development of a European multi-model ensemble system for seasonal to interannual prediction (DEMETER). Bull. Amer. Meteor. Soc., 85, 853-872

Rowell, D. P.,1998: Assessing potential seasonal predictability with an ensemble of multidecadal GCM simulations. J. Climate 11:109–120

Shukla J et al, 2000: Dynamical seasonal prediction. Bull Am Meteorol Soc 81:2493-2606

Shukla, J., 1998: Predictability in the midst of chaos: a scientific basis for climate forecasting. Science, 282, 728-731

Trenberth, K. E., G. W. Branstator, D. Karoly, A. Kumar, N. Lau, and C. Ropelewski, 1998: Progress during toga in understanding and modeling global teleconnections associated with tropical sea surface temperature. J. Geophys. Res., 103, 12,324-14,291.

Wallace, J. M. and D. Guztler, 1981: Teleconnections in the geopotential height field during the northern hemisphere winter. Mon. Wea. Rev., 109, 784-812.

Wang, B., J.-Y., I.-S. Kang IS, J. Shukla et al.,2007: Coupled predictability of seasonal tropical precipitation. CLIVAR Exchanges, 12:17–18

Wang, Bin, June-Yi Lee, J. Shukla, I.-S. Kang, C.-K. Park and et al., 2009: Advance and prospectus of seasonal prediction: Assessment of APCC/CliPAS 14-model ensemble retrospective seasonal prediction (1980-2004). Clim. Dyn. 33, 93-117

Weisheimer, A., F. J. Doblas-Reyes, T. N. Palmer et al., 2009: ENSEMBLES:A new multi-model ensemble for seasonal-to-annual predictions-Skill and progress beyond DEMETER in forecasting tropical Pacific SSTs. Geophys. Res Lett, 36, L21711.

Part 2

Hydrology and Extreme Climate Events

4

Impact of Global Climate Change on Regional Water Resources: A Case Study in the Huai River Basin

Ju Qin[1], Hao Zhen-chun[1], Ou Geng-xin[2],
Wang Lu[3] and Zhu Chang-jun[4]
[1]State Key Laboratory of Hydrology-Water Resources and Hydraulic Engineering,
Hohai University, Nanjing,
[2]School of Natural Resources,University of Nebraska–Lincoln, Lincoln,
[3]Delft University of Technology, Delft,
[4]College of Urban Construction,
Hebei University of Engineering, Handan,
[1,4]China
[2]USA
[3]The Netherlands

1. Introduction

As one of the most important environmental issues, climate change has attracted great attention of various governmental agencies and scientists around the world in the last decade. The Intergovernmental Panel on Climate Change (IPCC), which was co-established by the United Nations Environment Program and the World Meteorological Organization, regularly assesses the global impact of climate change. Climate change, with the main feature of global warming, has become an indisputable fact. The earth's surface temperature increased by 0.74°C in the recent 100 years (1906 ~ 2005). Temperature change of China over the recent 100 years is consistent with the global trends, with the average increase of 0.5~0.8°C (IPCC, 2007). Global warming will probably accelerate the global hydrological cycle, resulting in temporal and spatial changes of various meteorological and hydrological variables such as rainfall and evapotranspiration, which directly leads to frequent occurrence and constant increase of such extreme hydrological events as floods and droughts, thereby further affecting the regional ecology as well as the survival environment of the human being.

When hydrological models driven by climate scenarios are used to evaluate the impact of climate change on water resources, climate scenarios can be divided into two types. One is the hypothetical climate change scenario based on the climate change trend, which is mainly used to analyze the sensitivity of water resources to climate change. Nash and Gleick (1991) used the conceptual hydrological model to study the influence of climate change on the basin's annual runoff assuming temperature increased by 2°C and the precipitation increased or decreased by 10% to 20%. The improved distributed Xin'anjiang hydrological model was used to analyze the sensitivity of Huai River Basin runoff to climate change by assuming various temperature and precipitation changes (Hao et al., 2006). The hydrological impact of fictitious climate scenarios changes were reported in different basins (e.g., Schwarz, 1977; Beyene et al., 2010; Young et al., 2009). The other is the climate prediction of global circulation models (GCMs). With the continual improvement of global and regional climate models, researchers have made great progress in studying the attribution and prediction of spatial and temporal changes in hydrological processes. GCMs have become one of the most effective tools in constructing future climate change scenarios. Ozkul (2009) used the predicted results of climate change scenarios released publicly by the Fourth Assessment of IPCC in the Gediz and Buyuk Menderes River Basins, combined with water balance model to predict a 20% reduction of surface water resources under future climate change before 2030, and the percentage will increase to 35% and 50% by the years of 2050 and 2100, respectively. Nijssen et al. (2001) adopted the predicted scenario results of four climate models, coupled with the Variable Infiltration Capacity (VIC) model to predict the change of hydrological cycle change trends in nine large, continental river basins in 2025 and 2045. Other scholars have also conducted considerable research (Beyene et al., 2010; Minville et al., 2008; Ellis et al., 2008; Edwin, 2007; Christensen and Lettenmaier, 2007; Koutsoyiannis and Efstratiadis, 2007; Nemec and Schaake, 1982;).

Many scientists used the prediction results of GCMs to drive a hydrological model to evaluate the impact of climate change on water resources. The research, however, are mostly based on a single GCM output or a specified number of GCMs. Regional simulation ability of different models is quite different from each other (Zhou and Yu, 2006; Hao, et al., 2010). In addition, uncertainty is greater for a single GCM simulation. However, the differences between different GCMs can be considered when more than one GCM is selected. Moreover, prediction stability and reliability could be increased through the statistical estimations, Very few studies, however, offer sufficient reasons for selecting GCM. Therefore, studying principles and methods of selecting GCMs are extremely important in the coupling study of global climate models and hydrological models.

2. The evaluation of IPCC AR4 climate model's simulation performance

2.1 Introduction of global climate models

Climate scenarios simulation and prediction results of 24 GCMs were published by the fourth assessment of IPCC Data Distribution Centre (IPCC-DDC), including nine prediction results under different emission scenarios, three of which—high emission SRES A2 (Special Report on Emissions Scenarios A2), middle- emission SRES A1B and low-emission emissions SRES B1— are identified as the major future climate change scenarios. More detailed information about the models can be found at http://www-pcmdi.llnl.gov/ipcc/about-ipcc.Php. Basic information of the 22 global climate models provided by IPCC AR4, is listed in Table 1 (Both the BCC_CM1 of China and the CCCMA_CGCM3.1 T61 of Canada are not available respectively for non-reference period data and vague description of the data structure).

Model Research Center	Model	Abbreviations	Grid resolution (°)		Time range
			Latitude	longitude	
Bjerknes Center for Climate Research, Norway	BCM2.0	BCM2	2.8125	2.79	1850~2099
Canadian Center for Climate Modelling and Analysis, Canada	CGCM3 (T47 resolution)	CGMR	3.75	3.71	1850~2300
Centre National de Recherches Meteorologiques, France	CM3	CNCM3	2.8125	2.79	1860~2299
Australia's Commonwealth Scientific and Industrial Research Organisation, Australia	Mk3.0	CSMK3	1.875	1.8652	1871~2200
Max-Planck-Institut for Meteorology, Germany	ECHAM5-OM	MPEH5	1.875	1.865	1960~2200
Meteorological Institute, University of Bonn, Germany Meteorological Research Institute of KMA, Korea Model and Data Groupe at MPI-M, Germany	ECHO-G	ECHOG	3.75	3.711	1860~2100
Institude of Atmospheric Physics, China	FGOALS-g1.0	FGOAL	2.8125	2.79	1850~2199
Geophysical Fluid Dynamics Laboratory (US)	CM2.0	GFCM20	2.5	2	1861~2100
Geophysical Fluid Dynamics Laboratory (US)	CM2.1	GFCM21	2.5	2	1861~2300
Goddard Institute for Space Studies, USA	AOM	GIAOM	4	3	1850~2100
	E-H	GIEH	5	4	1880~2099
	E-R	GIER	5	4	1880~2300
Institute for Numerical Mathematics, Russia	CM3.0	INCM3	5	4	1871~2200
Institut Pierre Simon Laplace, France	CM4	IPCM4	3.75	2.535	1860~2230
National Institute for Environmental Studies, Japan	MIROC3.2 hires	MIHR	1.125	1.1214	1900~2100
	MIROC3.2 medres	MIMR	2.8125	2.79	1850~2300
Meteorological Research Institute, Japan	CGCM2.3.2	MRCGCM	2.8125	2.79	1851~2300
National Centre for Atmospheric Research, USA	PCM	NCPCM	2.8125	2.79	1870~2099
	CCSM3	NCCCSM	1.40625	1.400763	1890~2099
UK Met. Office, UK	HadCM3	HADCM3	3.75	2.5	1860~2199
	HadGEM1	HADGEM	1.875	1.25	1860~2100
National Institute of Geophysics and Volcanology, Italy	SXG 2005	INGSXG	1.125	1.1215	1870~2100

Table 1. Information of 22 climate models of IPCC AR4.

2.2 Study area and data

2.2.1 Study area

Located in eastern China, the Huai River originates in Tongbai mountain, south Henan Province, flowing through Henan and Anhui Provinces from west to east, and finally entering the Hongze Lake of Jiangsu Province. The Huai River Basin is between the Yangtze River and Yellow River Basins, stretching from 111°55'E to 121°25'E and from 30°55'N to 36°36'N. The basin is in a north to south climatic transition zone, thus having a temperate monsoon climate with hot and rainy summer, as well as cold and dry winter. Rainfall is of uneven distribution during the year. About 60% of the annual precipitation concentrates between June and August. From mid-June to early July, the basin turns into rainy season, with a large amount of rainfall across a wide range, easily causing severe basin-wide floods. In wet years, the total precipitation is 300-500% of dry years. According to historical data, there were 42 floods in this basin in the 48 years from 1901 to 1948. In addition, the Huai River Basin is also the main water supply area and the main channel for the eastern route of South-to-North water diversion project. The hope of the project is to divert water over hundreds of kilometers from China's Yangtze River to the North China Plain (Berkoff, 2003).

Climate change of this area will affect scheduling and allocation of the entire east route project. In this paper, the area more than 120,000 km² above Bengbu station is selected as the study area. The western part and southwest part of the area are mountains and hilly area, and the rest is plain. River and weather stations in the study area are as shown in Fig. 1.

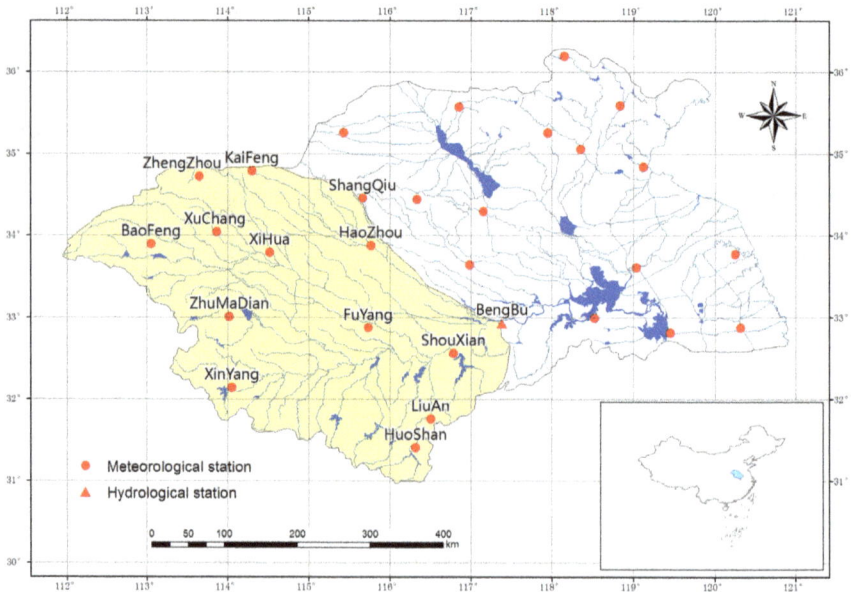

Fig. 1. The flow network and locations of the meteorological and hydrological stations in the Huai River Basin.

2.2.2 Data selection

Information used in this study includes daily precipitation, temperature, and evaporation data for the period between 1951 and 2007 at 13 meteorological stations above the Bengbu station offered by the Climate Data Office of the China National Meteorological Information Center, as well as the monthly streamflow between 1915 and 2007 at the Bengbu hydrological station. Twenty-two global climate models, provided by the Fourth Assessment Report (IPCC AR4) of the Intergovernmental Panel on Climate Change (IPCC), are used to generate monthly average temperature and precipitation data in the contemporary climate (20C3M) conditions of the reference period (1961 ~ 1990), as well as future monthly temperature and precipitation data from 2001 to 2099.

2.3 Evaluation of simulation performance of IPCC AR4 model

In order to test the simulation capability of each model on temperature and precipitation in the Huai River Basin, we analyzed differences between simulated and observed values from different spatial and temporal scales. The simulated values are achieved from the original grid monthly temperature and precipitation provided by the models. Each single model's monthly regional average value in the reference period is also calculated by the geometric average. Relative error, absolute error, correlation coefficient and determination coefficient are used as indexes for performance evaluation (Ju, 2009).

Relative error, the formula is as follows

$$\text{MRE} = \sum_{i=1}^{n} \left[(P_i - O_i) / O_i \right] \Big/ n \tag{1}$$

Where, n is total calculation periods/intervals; P_i is measured value of the ith period/interval.

Correlation coefficient, is used to evaluate the degree of fit of the measured value and the simulated values. The closer the value is to 1, the more accurate the model is. The calculation formula is,

$$R = \frac{\sum_{i=1}^{n} (O_i - \overline{O})(P_i - \overline{P})}{\left[\sum_{i=1}^{n} (O_i - \overline{O})^2 \right]^{0.5} \left[\sum_{i=1}^{n} (P_i - \overline{P})^2 \right]^{0.5}} \tag{2}$$

Where, $\overline{O}$ is the average of the observed values; $\overline{P}$ is average of the calculated values.

Determination coefficient, also called Nash efficiency coefficient (Nash and Sutcliffe, 1970), can be used as an index to evaluate the degree of fit between the calculation process and the measurement process. The closer the value is to 1, the more accurate the model is. The calculation formula is as follows:

$$E = 1.0 - \frac{\sum_{i=1}^{n} (O_i - P_i)^2}{\sum_{i=1}^{n} (O_i - \overline{O})^2} \tag{3}$$

2.3.1 Temperature

From the comparison (Table 2) between the simulated results of climate models in the reference period (1961 ~ 1990) and the observed data of regional average monthly temperature for the 13 weather stations in the Huai River Basin, it can be seen that correlation coefficients of all the models are very high, averaging above 0.97, indicating that the models are capable of simulating temperature in the Huai River Basin. The annual average temperature of Huai River Basin in the reference period is 14.2°C. The models perform differently in simulating temperature of these years. The simulated temperatures of all models are underestimated. We differentiated the models into two groups of larger deviation and smaller deviation, among which the simulated value of monthly average temperature (June to August) is consistent with the observation (Fig. 2). The IPCM4 model has the largest absolute error, which is 24.9°C lower than the observed temperature; BCM2 model has the smallest deviation of 0.2°C. As to the evaluation index of determination coefficient, the value is expected to be close to 1.0 for a good simulation of the observed temperature. There are 11 models with the determination coefficient higher than 0. Fig. 3 shows the difference between the average simulated temperature of CNCM3, CSMK3, HADGEM and NCCCSM and the observed values in the reference period.

Number	Model	Analogue value		Absolute error of temperature/°C	Relative error of precipitation/%	Correlation coefficient		Coefficient of determination	
		Temperature /°C	Precipitation /mm			Temperature	Precipitation	Temperature	Precipitation
1	BCM2	14.0	883.9	-0.2	1.25	0.977	0.459	0.952	0.092
2	CGMR	10.2	1251.0	-4.1	43.30	0.981	0.323	0.755	-0.536
3	CNCM3	13.8	850.7	-0.5	-2.55	0.978	0.486	0.943	0.134
4	CSMK3	12.2	714.5	-2.1	-18.16	0.982	0.59	0.912	0.262
5	ECHOG	-7.9	441.5	-22.1	-49.43	0.983	0.696	-4.804	0.097
6	FGOALS	-4.4	672.6	-18.7	-22.96	0.979	0.728	-3.456	0.442
7	GFCM20	9.6	963.7	-4.7	10.39	0.973	0.438	0.691	0.017
8	GFCM21	11.0	886.0	-3.2	1.49	0.974	0.439	0.831	-0.01
9	GIAOM	-7.1	726.2	-21.4	-16.81	0.976	0.756	-4.423	0.524
10	GIEH	-6.0	338.6	-20.2	-61.22	0.978	0.557	-4.231	-0.201
11	GIER	-6.3	298.4	-20.6	-65.82	0.971	0.44	-4.412	-0.354
12	HADCM3	12.9	997.4	-1.3	14.25	0.981	0.524	0.944	0.065
13	HADGEM	12.4	793.2	-1.8	-9.14	0.981	0.614	0.895	0.298
14	INCM3	-5.2	712.1	-19.5	-18.43	0.977	0.751	-3.718	0.518
15	INGSXG	-5.1	570.4	-19.3	-34.67	0.98	0.646	-3.624	0.257
16	IPCM4	-10.6	512.6	-24.9	-41.29	0.979	0.672	-6.69	0.198
17	MIHR	-6.4	531.7	-20.7	-39.09	0.982	0.712	-4.269	0.312
18	MIMR	-5.9	649.6	-20.1	-25.59	0.987	0.709	-4.044	0.421
19	MPEH5	-7.4	580.9	-21.7	-33.46	0.983	0.65	-4.698	0.284
20	MRCGCM	12.7	873.8	-1.5	0.09	0.983	0.204	0.938	-0.177
21	NCCCSM	12.8	1053.6	-1.5	20.69	0.979	0.596	0.935	0.123
22	NCPCM	11.5	1105.3	-2.7	26.61	0.972	-0.044	0.863	-1.238

Table 2. Comparison of temperatures and precipitation between simulations of climate models and observations.

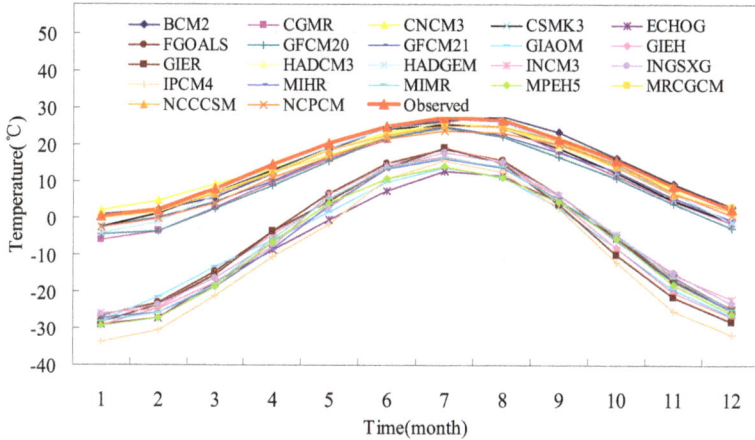

Fig. 2. Comparison of monthly mean temperatures (1961~1990) between simulation and observation in the Huai River Basin.

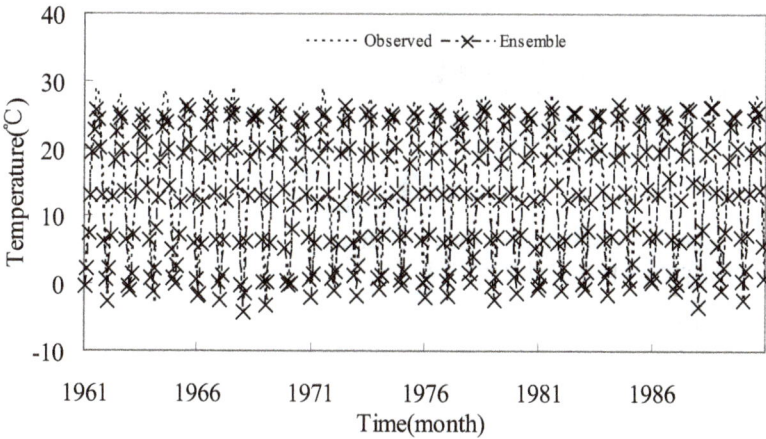

Fig. 3. Comparison of monthly temperature between simulation and observation in reference period (1961~1990) in the Huai River Basin (Ensemble denotes the average temperature of 4 models of CNCM3, CSMK3, HADGEM and NCCCSM).

2.3.2 Precipitation

The method for evaluating the models' ability in simulating precipitation is essentially similar to those for temperature. The average annual rainfall of the Huai River Basin in the reference period is 873 mm. It can be seen from Table 2 that most models slightly underestimated precipitation. Abilities of different models in simulating precipitation are remarkably different from each other. Nearly half of the models have unsatisfactory simulation results, among which GIER model has the largest relative error of 65.82%, and MRCGCM model has the smallest relative error of only 0.09% but with the negative determination coefficient value, which means that the fitness is worse than using the average. The correlation coefficient between simulation values and observed values are relatively lower than the correlation coefficient of temperature. Compared to the temperature simulation ability of climate models, precipitation results have lager significant differences (Fig. 4). Simulated results of most models are not ideal. Generally underestimating the precipitation, most models could not properly reproduce the intra-annual variation of precipitation. Maximum monthly precipitation was observed in June, but most of the simulated maximum precipitation occurs in July. In terms of the variation in four seasons, the deviation is larger in summer and autumn. In Fig. 5, the average precipitation of CNCM3, CSMK3, HADGEM and NCCCSM, which provide relatively better estimates, are taken for examples to simulate monthly precipitation in the reference period.

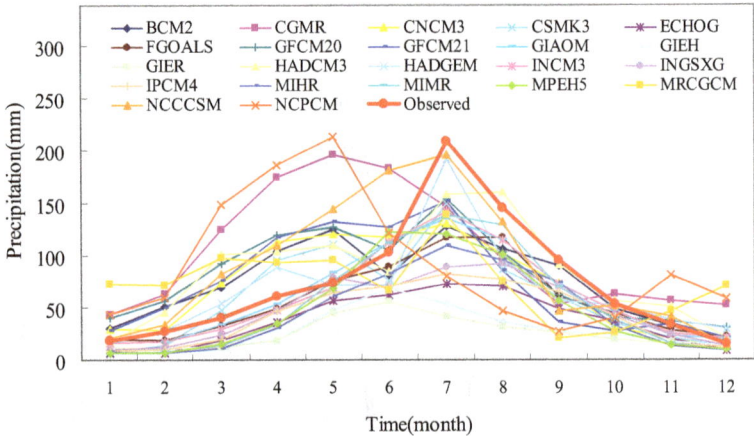

Fig. 4. Comparison of precipitation in the Huai River Basin between simulation and observation in reference period (1961~1990).

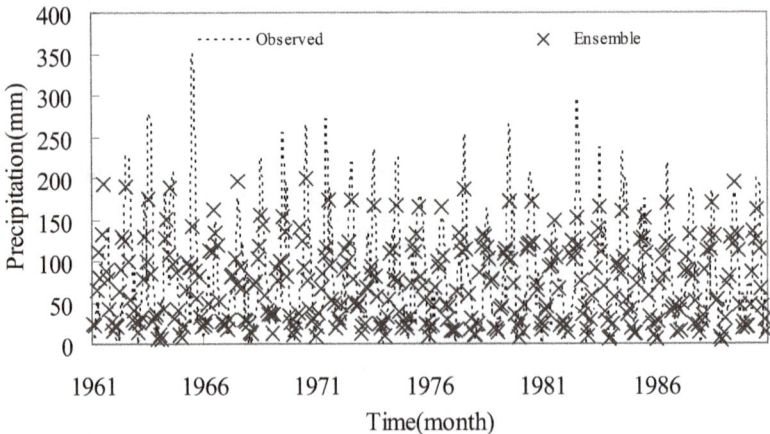

Fig. 5. Comparison of monthly precipitation in the Huai River Basin during the reference period (1961~1990) between simulation and observation (Ensemble denotes the average precipitation of 4 models of CNCM3, CSMK3, HADGEM and NCCCSM).

2.4 Future changes in temperature and precipitation

Table 3 shows the estimated linear trend of average annual precipitation and temperature for the 21st century in the Huai River Basin under different emission scenarios. Analysis shows that the values of the precipitation and temperature tend to increase in the future, except for CSMK3 and GFCM2. For the overall trend, the A2 scenario has the largest increase, followed by the A1B scenario, and the B1 scenario has minimum increase. The increase magnitude differs greatly with the models under the same scenario. The HADCM3 model has the largest increase, up to 369.08mm/100a under the A2 scenario. The models of MIMR and ECHOG have larger temperature change tendency of 7.3°C/100a and 6.9°C/100a respectively under the A2 scenario, and NCPCM has the smallest one of 2.6°C/100a, while increases predicted with other models are between 3.0 and 6.5°C/100a. Future warming predicted by B1 scenario is relatively small with a tendency rate lower than 4.0°C/100a. Future climate tends to be warm and wet.

Model	Precipitation (mm/100 years)			Temperature (°C/100 years)		
	A2	A1B	B1	A2	A1B	B1
BCM2	/	216.9	113.9	3.1	2.8	1.4
CGMR	/	189. 5	/	/	2.5	/
CNCM3	161. 3	115.2	/	4.0	3.3	1.6
CSMK3	63.1	55.2	-10.3	3.2	2.7	1.9
ECHOG	107.7	70.5	/	6.9	7.2	/
FGOALS	/	103.5	59.9	/	3.7	2.2
GFCM20	195.5	144.9	125.8	4.0	4.1	2.0
GFCM21	185.7	53.0	-29.4	3.5	3.7	2.1
GIAOM	/	65.3	58.8	/	2.8	1.5
GIEH	/	53.1	/	/	3.2	/
GIER	18.2	/	-2.1	4.3	2.6	1.2
HADCM3	369.1	273.8	177.3	5.0	4.7	3.1
HADGEM	/	/	/	/	/	/
INCM3	/	87.8	61.5	/	3.5	2.3
INGSXG	/	7.9	/	/	4.7	/
IPCM4	53.1	78.7	19.7	6.2	5.4	3.5
MIHR	/	169.6	66.5	/	5.9	3.6
MIMR	133.1	114.1	69.4	7.3	6.2	3.3
MPEH5	18.7	39.2	65.9	6.4	6.2	3.9
MRCGCM	97.6	52.9	48.9	3.8	3.8	2.2
NCCCSM	291.0	171.9	69.6	4.2	2.9	1.3
NCPCM	129.3	114.1	/	2.6	2.4	/

Table 3. Linear trend of temperature and precipitation simulated with models in 2000 to 2099 in the Huai River Basin.

2.5 Selection of climate models

By comparing the simulated and observed temperature changes (as shown in Fig. 2) of the Huai River Basin during 1961 to 1990, it can be seen that the change trends of these models during the year can be divided into two groups. One group is consistent with the measured trend and has the determination coefficient higher than 0 (Table 2); the other group has larger difference from the observed distribution process during the year (greater curvature) and negative determination coefficients. Comparison between the simulated and observed precipitation changes (Fig. 4) also shows that the greater the model's determination coefficient is, the more consistent the distribution process is with the observation. The determination coefficients larger than 0 for both the temperature and the precipitation are therefore, regarded as the first criteria for model selection in this paper. Seven models, BCM2, CNCM3, CSMK3, HADCM3, HADGEM, GFCM20, and NCCCSM meet the first requirements. The BCM2, HADCM3 and GDCM20 models are excluded while considering both the precipitation relative error and degree of fit of the intra-annual variation curve. HADGEM is also excluded because it does not provide monthly temperature and precipitation. Three climate models including CNCM3, CSMK3, and NCCCSM are therefore selected to estimate the hydrological response of future climate change under three IPCC emission scenarios of A1B, A2 and B1 in the Huai River Basin.

3. Construction of the distributed Xinanjiang model

3.1 Theory and data preparation of the Xinanjiang model

The Xianjiang model is a conceptual watershed model, which was originally developed for simulating streamflow at the daily scale or flood-event scale (Zhao, 1992; Ju et al., 2009). The key concept of the model is that runoff is only generated where the field capacities are reached. The model uses a single parabolic curve to statistically account for the non-uniform distribution of the area of runoff generation. With 14 parameters, the Xianjiang model can simulates the monthly runoff, flow routing and evapotranspiration through four main calculation processes including evapotranspiration, runoff generation, runoff dividing computation and flow routing. The runoff is composed of the surface flow, subsurface flow and groundwater flow.

The structure of Xinanjiang model is shown in Fig. 6 (Zhao, 1992; Ju, et al., 2009). All symbols outside the blocks are parameters. The meaning of all the parameters listed in Table 4. P, EM represent areal mean rainfall and measured pan evaporation, which are the input model. The outputs from the whole basin are the discharge TQ. E is the actual evapotranspiration, including three components EU, EL and ED. T is the total inflow of river network. W, S are the areal mean tension water storage, and the areal mean free water storage, respectively. The areal mean tension water W has three components. WU, WL, and WD are mean tension water for the upper, lower and deep layers, respectively. The FR is contributing area factor for runoff which is related to W. IM is the factor of percentage of impervious area in the watershed. 1-FR-IM is non-contributing area factor for runoff. The rest of the symbols inside the blocks are all internal variables. RB is the runoff directly from the small portion of impervious area. R is the runoff produced from the area and divided into three components RS, RI, and RG representing surface runoff, interflow and groundwater runoff respectively. The three components are further transferred into QS, QI, and QG, and collectively form the total inflow to the channel network of the sub-basin. The outflow of the sub-basin is Q (Ju, et al., 2009).

The principles of the monthly Xinanjiang model are similar to the daily model. This study implements the distributed monthly Xianjiang model for the Huai River Basin, developed by Hao and Su (2000). Considering the uneven spatial distribution of precipitation, the upstream area of Benbu is divided into 13 computation units through the Thiessen polygon method. Runoff generation and flow routing computations are performed for each unit. The predicted streamflow for the basin is the sum of the runoff at each unit. We calibrated the model using the data between 1961 and 1990 and verified the model for the period of 1991 and 2000. The input data are monthly observations of precipitation and temperature at each gauging station in the area, while the output is monthly streamflow.

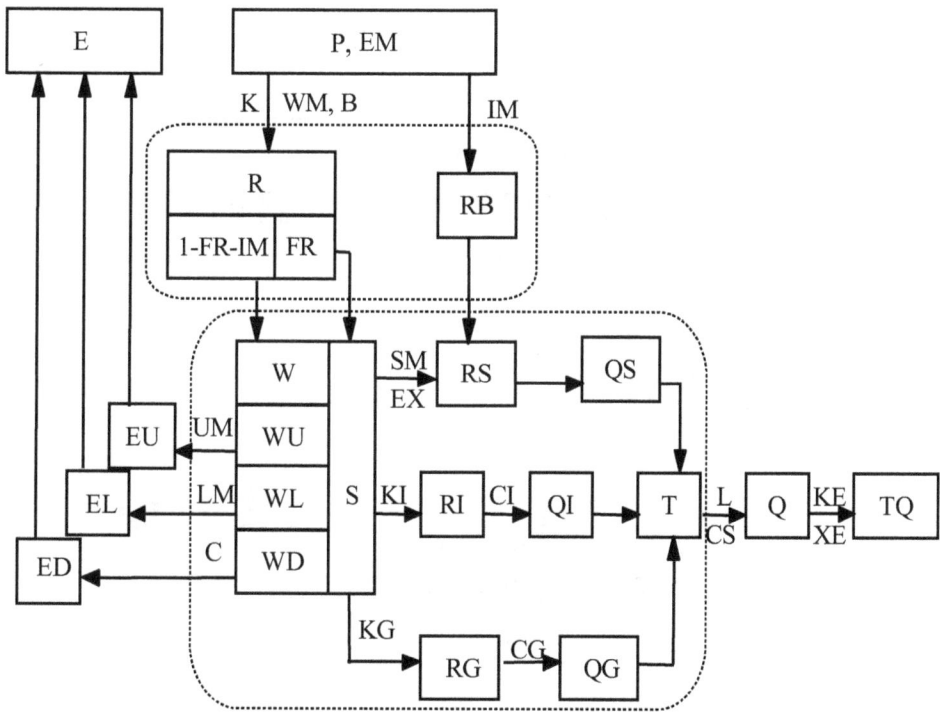

Fig. 6. Structure of the Xinanjiang model (Ju, et al., 2009).

3.2 Model calibration and validation

The 14 parameters for the model are calibrated through the trial and error method. The initial values for all the parameters are empirically determined. A set of optimum values of these parameters are determined by a comparison between the predicted and observed discharges (Table 4).

Type	Notation	Parameter Meaning	Value
Evapotrans-piration parameters	K	Ratio of potential evapotranspiration to the pan evaporation	0.56
	UM	Tension water capacity of upper layer	39
	LM	Tension water capacity of lower layer	83
	C	Evapotranspiration coefficient of deeper layer	0.12
Runoff production parameters	WM	Areal mean tension water capacity	299
	B	Exponential of the distribution of tension water capacity	0.35
	IM	Ratio of impervious area to the total area of the basin	0.01
	SM	Free water storage capacity	9.2
Runoff separation parameters	EX	Exponential of distribution water capacity	1.1
	KG	Out flow coefficient of free water storage to the groundwater flow	0.597
	KI	Out flow coefficient of free water storage to the inter flow	0.403
Runoff concentration parameters	CI	Recession constant of lower interflow storage	0.197
	CG	Recession constant of groundwater storage	0.452
	CS	Recession constant of channel network storage	0.003

Table 4. Parameters of Xinanjiang model after calibration.

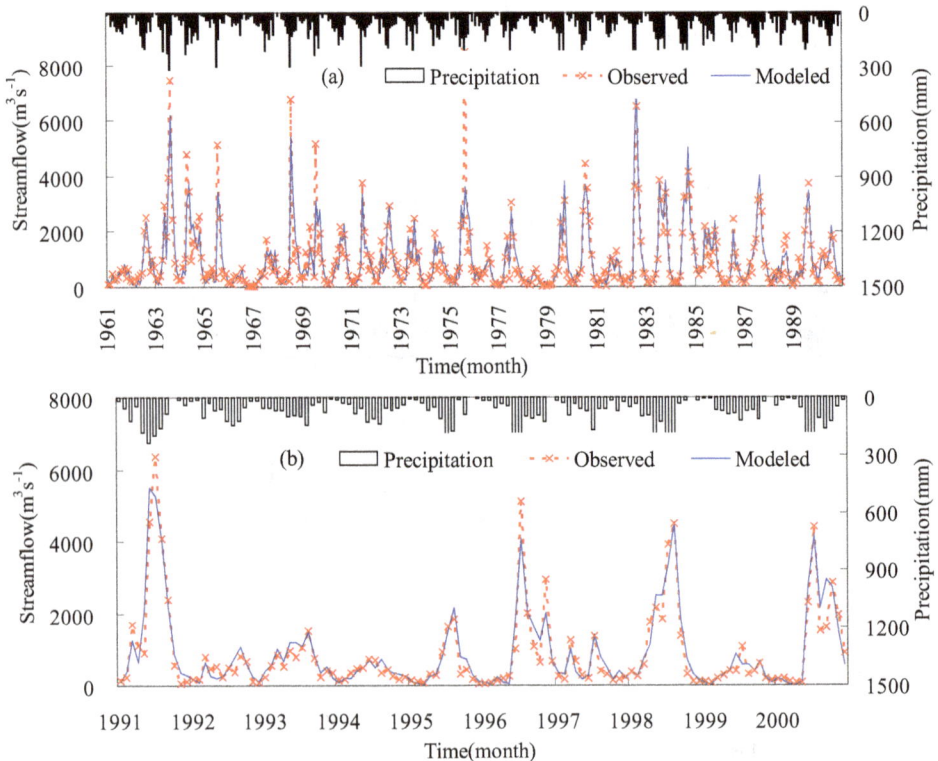

Fig. 7. Simulated and observed monthly streamflows at Bengbu gauge in Huai River Basin; (a) calibration period (b) test period.

Two error indexes, the mean relative error and the determination coefficient, are selected to evaluate the performance of the models. In calibration and validation periods, the mean relative error MRE are 2.3% and 7.8%, respectively, and the determination coefficients E are 0.828 and 0.852, respectively. A comparison of observed versus simulated monthly streamflows at Bengbu gauge for the calibration period and validation period is shown in Fig. 7. Results show that the constructed distributed monthly Xinanjiang model can reproduce the monthly average runoff process and seasonal changes. Therefore, the model can be used to carry out the assessment study of future impact of climate change on the basin discharge.

4. Impact of future climate change on water resources

4.1 Analysis for yearly runoff trend

Pearson III frequency curve is selected in this paper to analyze annual runoff frequency of the Bengbu station (1915-2007, removing a small amount of missing data year). Those years with the frequency of less than 25% on the runoff frequency curve are generally defined as wet year, those between 25% and 50% as the normal year, and those more than 75% as dry year. It can be seen from the normal annual flow frequency curve Fig. 8 that the flow limits at Bengbu station for wet years, normal years and dry years are 1184m³/s, 829m³/s, and 553m³/s, respectively. The percentage of the wet, normal and dry yea are 30.9%, 41.2%, 27.9%, respectively. Table 5 shows runoff frequency curves for the projected future climates.

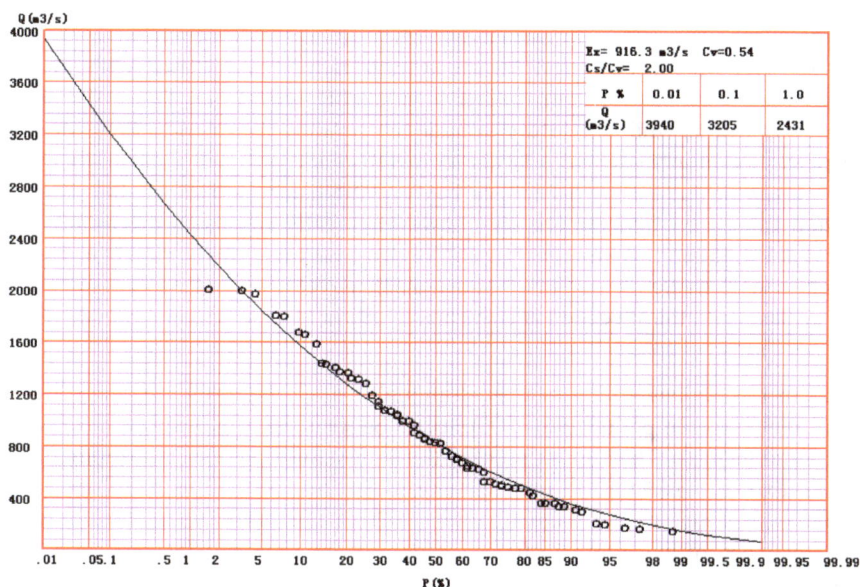

Fig. 8. The empirical frequency curves of yearly streamflows at Bengbu gauge on Huai River Basin during 1915-2007.

Model Type	A2			A1B			B1		Total
	CNCM3	CSMK3	NCCCSM	CNCM3	CSMK3	NCCCSM	CSMK3	NCCCSM	
wet year(%)	2.2	20.0	26.7	8.9	23.3	34.4	18.9	21.1	19.4
normal year(%)	47.8	54.4	63.3	55.6	57.8	60.0	54.4	68.9	57.8
dry year(%)	50.0	25.6	10.0	35.6	18.9	5.6	26.7	10.0	22.8

Table 5. The statistics for percentage of the wet, normal and dry year under the A2, A1B and B1future scenarios annual runoff with different climatic models during 2010 -2099.

Totally, there are more dry years in the A2 scenario. Normal year dominate in A1B and B1 scenarios, meanwhile wet year and dry year are roughly equal to each other (Fig 9.). In the 8 model projections involved, there are 140 wet years, accounting for 19.4% of the total, which is slightly lower when compared to the history. There has been a modest decrease in dry years, while normal years increased to 57.8%.

The annual runoff prediction process in the three emission scenarios of A1B, A2 and B1 is shown in Fig. 9. Overall, the future annual runoff are relatively stable, but the simulation results vary from each other. The CNCM3 model has evidently more dry years under the three scenarios. On the contrary, the NCCCSM model has more wet years. For the CSMK3 model, wet years and dry years are roughly equal (see statistics in Table 5). Generally, in the next 90 years, the annual average flow of the Bengbu hydrological station in the 8 situations is 866 m^3/s, decreasing by approximately 12% from the historical (1961-1990) annual average flow (987 m^3/s). Historical annual average rainfall is 873 mm, while the future average rainfall will be 884 mm, which is a slight increase of 1.1%. The historical temperature is 14.2°C, while the future average temperature increases by 2.9°C (Fig. 10). A smaller precipitation increase and a larger temperature increase will occur in the future in the Huai River Basin. The two changes have contradictory effects on the runoff. The evaporation caused by temperature rise, however, greatly surpasses the runoff increase caused by precipitation increase, leading to the decrease trend of the runoff in the future. This will be a large challenge for both the sustainable development of water resources in the Huai River Basin and the allocation and the management of water resources in the East Line of the Water Transfer Project.

As shown in the frequency distribution of annual average flow (Fig. 11) in 2020s, 2050s, 2080s and the reference period(1961-1990), it can be seen that the mean of most scenarios shift to the left in 2020s. The highest point of the annual average flow probability density is larger than that of the history in most climate projections, Similar to 2020s, most climate projections shift to the left and become thinner in 2050s and 2080s, which further indicated that the frequency of low flows in future years will increase, possibly with the trend of average annual flow reduced in the future. Most scenarios indicate that the frequency of low flows in future years will increase, possibly with a trend of reduced values. Right-shift of the probability density curve for a few climate projections indicates that the frequency of high flows in future years would increase in those cases.

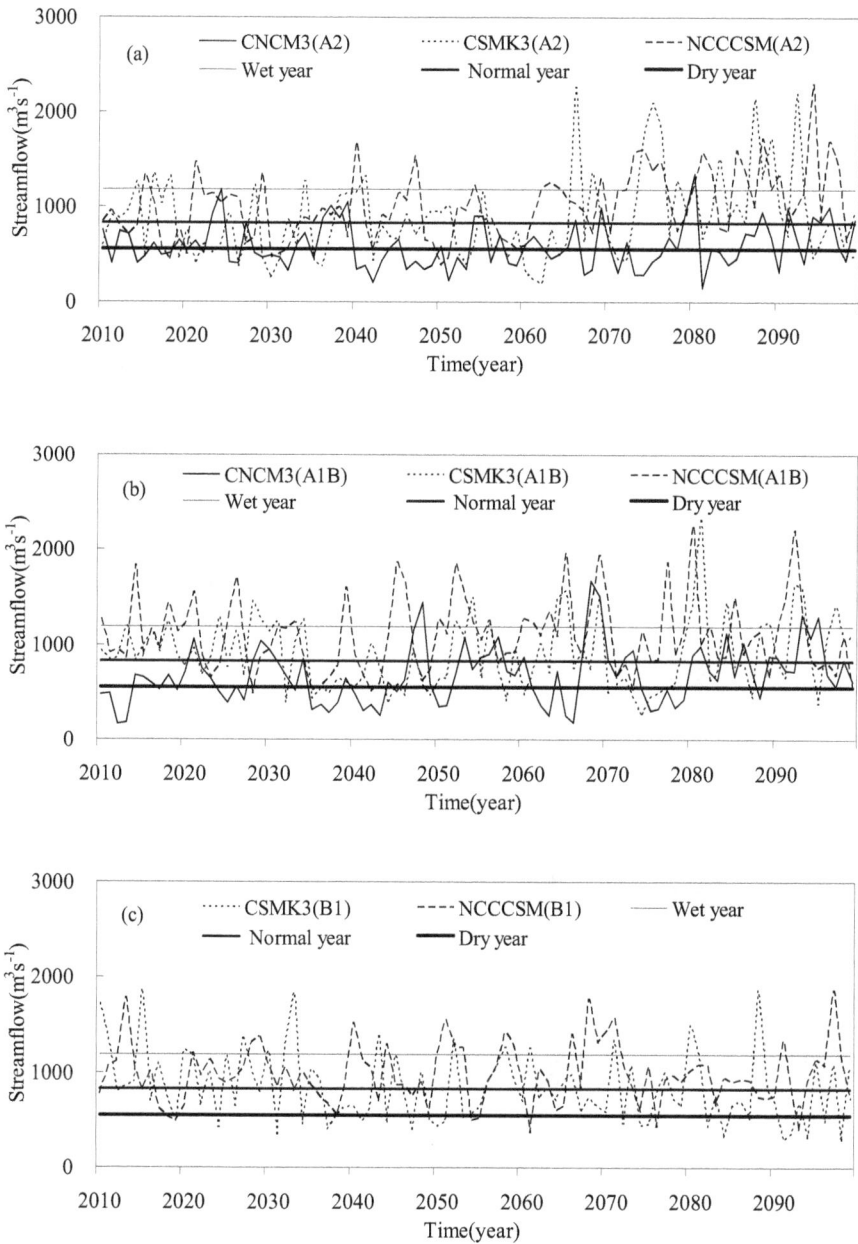

Fig. 9. Simulated annual streamflows under the different future scenarios at Bengbu gauging station in the Huai River Basin during 2010 -2099, (a)A2 scenarios; (b)A1B scenarios; (c) B1 scenarios.

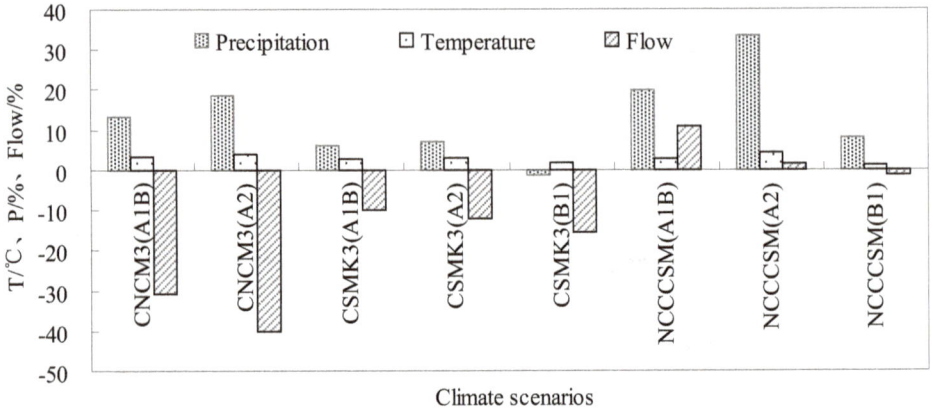

Fig. 10. Projections of annual runoff to change of precipitation and temperature under different climate scenarios during 2010 -2099 in the Huai River Basin.

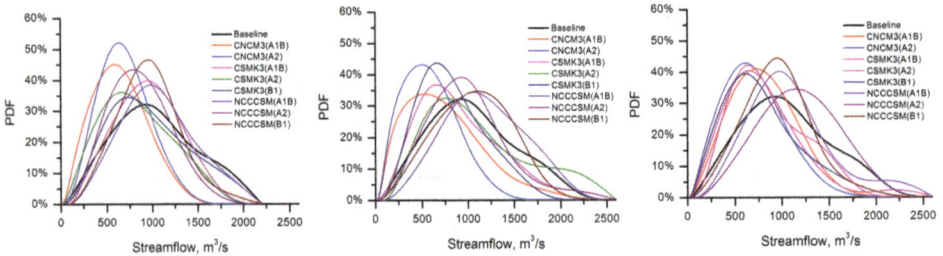

Fig. 11. Probability density function of annual mean discharge and the 90% confidence interval for 2020s(2010-2039), 2050s(2040-2069), 2080s (2070-2099) and the baseline (1961-1990).

4.2 Monthly runoff analysis

Fig. 12 is the monthly hydrograph of each climate scenario and the multi-scenario ensemble mean in different periods. The estimated maximum flow of most models appears in July, basically consistent with the historical measurement. But the estimated values are all lower than the measured ones, except that the NCCCSM model's flows in July under the A1B scenario in 2050s and under the A2 scenario in 2080 are both slightly higher than the measured flow. The average monthly flow of the NCCCSM model in B1 scenario is the most consistent with historical trends in the next three decades. For the NCCCSM model in 2050s and 2080s under scenario A1B, flows of July and August are overestimated. Among the eight scenarios, estimated flows of the CNCM3 model under the A1B and A2 scenarios are the lowest. The ensemble mean flows of July and August in the three decades tend to decrease, whereas no significant change is observed in other months. By analyzing the change ranges of all scenarios, we can see that the estimated results of June, July, August and September have much larger model-to-model differences compared with the reference period. This uncertainty increases with time. The uncertainty range of other months is relatively smaller.

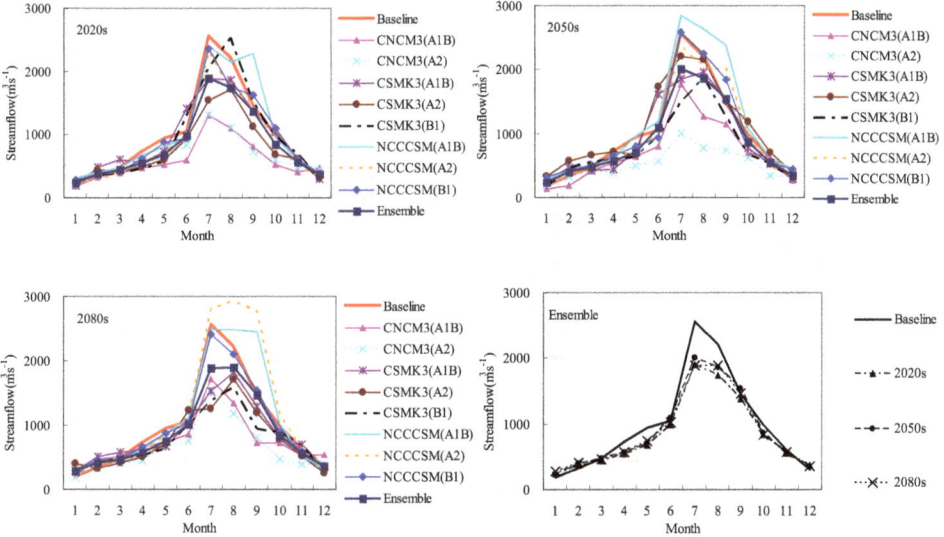

Fig. 12. Comparison of monthly flow curve for many years with different climatic models under the A2, AB and B1future scenarios at Bengbu gauge (baseline denote 1961-1990, Ensemble represent the average flow of the eight climate projections).

4.3 The runoff change analysis in flood season

The flood season of the Huai River Basin is from May to September. Other months are grouped into non-flood period. Fig. 13 shows the relative percentage change of the average flood season runoff in 2020s, 2050s and 2080s relative to that of the reference period (1961-1990). Most simulation results of the three periods were lower than the reference period. Predictions of different models vary widely from each other. Simulated results of the CNCM3 model in the three periods are all less than the reference period. Under the conditions of Special Report on Emissions Scenarios A2, the runoff has a substantial reduction of about 50%. Simulated results of CSMK3 are higher than CNCM3 but lower than NCCCSM. NCCCSM results for the A1B scenario in the 2020s, all three scenarios in the 2050s, the A1B and A2 scenario in the 2080s, produce higher streamflows than in the baseline period. Most of the simulation results in 2020s are lower than in the reference period, among which the NCCCSM model, under A1B scenario, has the highest simulated results, which is 108% of the reference period. Simulated results of both CNCM3 and CSMK3 in 2050s are lower than the reference period; part of the NCCCSM model's scenarios is above the reference period but the ratio is not significant. Simulated results of 2080s are significantly higher than those of 2020s, among which the NCCCSM model, under the A2 scenario, has the highest simulated result of 133%. The total range of streamflow rates in the flood season between the various scenarios and models, expressed as a percent of their corresponding rates in the baseline period, was 56% in the 2020s (52% to 108%), but about 80% in the 2050s and 2080s. This increase from 56% to 80% represents a growth in the uncertainty of the predictions into the more distant future.

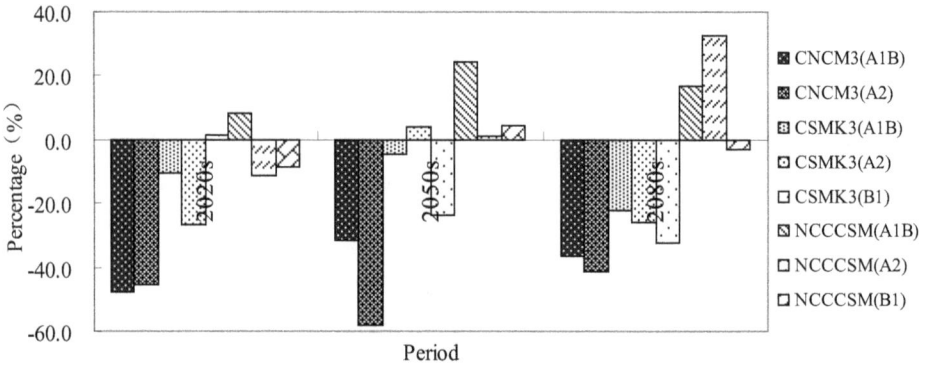

Fig. 13. The flood season Seasonal and annual runoff during 2020s, 2050s and 2080s in the Huai River Basin, as a percentage of the runoff in the baseline(1961-1990) period.

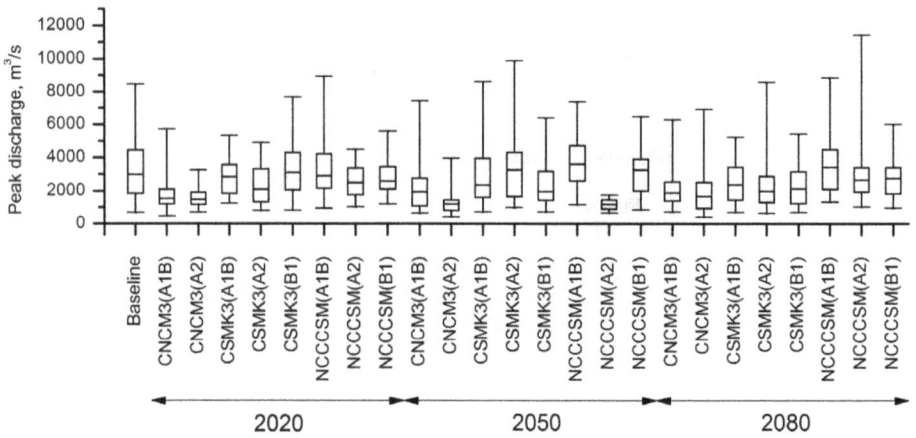

Fig. 14. Box plots of peak discharge for 2020s, 2050s, 2080s and the baseline(1961-1990).

4.4 Analysis of peak flow changes

Climate change will exacerbate the global and regional hydrological cycle process, affecting the spatial and temporal distribution of such water cycle factors as precipitation, evaporation, runoff and soil moisture, causing re-allocation of water resource in time and space, thus enhancing the possibility of various extreme hydrological events. Here, we mainly analyze the change trend of peak flow rate (annual maximum monthly flows, reference period refers to the maximum monthly flow form 1961 to 1990). Fig. 14 is the box plot diagram of the peak flow in the three estimate periods and the reference period. The median value of peak flows in the reference period is 3060 m³/s, while median values of most scenarios in the future are lower than the reference period, indicating the reduction trend of peak flow in the flowing years. Only five median values of estimated peak flows — Model CSMK3 under the B1 and A2 scenarios in 2020s, Model NCCCSM under the A1B and B1 scenarios in 2050s, and Model NCCCSM under the A1B scenario of in 2080s — are higher than the reference period. And the NCCCSM model produces highest median peak flow under the A1B scenario in 2050s. Compared with the minimum value in reference period, minimum values for the majority of future scenarios are lower, indicating the less frequent occurrence of lower peak flow in the future. Compared with maximum value of the reference period, a small part of simulations such as that under the A1B scenario predicted with NCCCSM in 2020s, the A1B and A2 scenarios with CSMK3 in 2050s, the A2 scenario with CSMK3 in 2080s, as well as the A1B and A2 scenarios with NCCCSM, have higher maximum values, indicating the less frequent occurrence of higher peak flow in the future. The uncertainty scope of the future three periods also gradually increases with time.

5. Conclusions and discussions

Based on the predicted results of 8 global climate model scenarios in the Fourth Assessment by IPCC, this study used the distributed Xin'anjiang hydrological model to simulate runoff of the Bengbu hydrological station in the Huai River Basin in the next 90 years. The monthly runoff, flood season runoff, the annual runoff and the change trends of the peak flow are analyzed comparatively and the following conclusions are made:

1. Twenty-two global circulation models released by the Fourth Assessment of IPCC were tested in the Huai River Basin. Compared with the reference period, three evaluation indexes including the mean relative error, correlation coefficient and determination coefficient are used to evaluate the performance of GCMs. Three climate models, CNCM3, CSMK3 and NCCCSM, having better simulation performances in the Huai River Basin, are selected to predict and estimate change trends of the future precipitation and temperature under three emission scenarios, A1B, A2 and B1.

2. Compared with the average annual flow of the reference period, the future average annual flow decreases and dry season runoff is lower than for the reference period. The frequency of years with low annual flow will increase in the future. The A2 scenario has the largest number of predicted dry years, and more dry years than wet years. Normal years dominate in the A1B and B1 scenarios, for which the number of wet years and dry years are roughly equal. The CNCM3 model predicts more total dry years under all three scenarios, and much more dry years than wet years. By contrast, the NCCCSM model has more wet years than dry years. For the CSMK3 model, wet years are roughly equal to dry years.

3. Compared with the monthly flow of the reference period, the maximum flow of most models is in July, which is consistent with the historical measurement, but the hydrological model estimated values are lower than the measured values. The NCCCSM model, B1 scenario, projects streamflow rates for the future three decades that are the closest from among all of the experiments to its reference rates in the baseline period. The ensemble mean flows of July and August in the future tend to

decrease, while other months will not change significantly. Predicted results in summer months of June, July, August and September have the largest differences from the reference period; the range of uncertainty increases with the length of prediction time.

4. Compared with monthly flow of the reference period, the future annual average runoff in flood season decreases. CNCM3 has the largest reduction and NCCCSM decreases the minimum. A2 scenario of NCCCSM model has the largest increase in flood season runoff, which is 133%. The estimated uncertainty range in 2050s and 2080s is larger than that in 2020s.

5. The future peak flow tends to decrease, the estimated values of only a small proportion of scenarios are higher than that of the reference period, indicating that the frequency of flood peak flow tends to decrease, but extreme flows higher than the historical maximum will also appear in 2080s. The uncertainty range of the three periods in the future also increases gradually with time.

Hydrological responses estimated from climate change scenarios can provide reference information for regional water resources management, but hydrological forecasting is a more developed approach. The scenario estimates are based on assumptions that have greater uncertainty than hydrological forecasts. Here we applied the ensemble method which constructs multi-GCMs and multi-emission scenarios to coupled meteorological - hydrological forecasts. Results show that the ensemble method can improve the accuracy of simulations to some extent. Therefore, further study is needed in model selection, the corresponding ensemble mean method, selection of uncertain quantitative assessment indexes and estimate methods when we assess the possible climate change impact.

6. Acknowledgments

This work was financially Supported by the National Natural Science Foundation of China (40830639, 50879016, 41101015), the National Basic Research Program of China (2010CB951101), the "Strategic Priority Research Program – Climate Change: Carbon Budget and Relevant Issues" of the Chinese Academy of Sciences (XDA05110102), China Postdoctoral Science Foundation funded project (20110491346) and the Special Fund of State Key Laboratory of Hydrology-Water Resources and Hydraulic Engineering (1069-50985512).

7. References

Berkoff, J., 2003. China: The South–North Water Transfer Project-is it justified? Water Policy. 5:1-28.

Beyene, T., Lettenmaier, D. P., Kabat, P, 2010. Hydrologic impacts of climate change on the Nile River Basin: implications of the 2007 IPCC scenarios. Climatic Change , 100, 433-461.

Beyene, T., Lettenmaier, D. P., Kabat, P., 2010. Hydrologic impacts of climate change on the Nile River Basin: implications of the 2007 IPCC scenarios. Climatic Change, 100, 433-461.

Christensen, N. S., Lettenmaier, D.P., 2007. A multimodel ensemble approach to assessment of climate change impacts on the hydrology and water resources of the Colorado River Basin, Hydrology and Earth System Sciences, 11(4):1417-1434.

Edwin, P. M., 2007.Uncertainty in hydrologic impacts of climate change in the Sierra Nevada, California, under two emissions scenarios. Climatic Change. 82(3-4): 309-325.

Ellis, A. W., Hawkins, T. W., Balling, R. C., et al., 2008. Estimating future runoff levels for a semi-arid fluvial system of central Arizona, USA, Climate Resear , 35: 227-239.

Hao, Z., Ju, Q., Yu Z., et al., 2010. Evaluation of the simulation performance and scenario projection of IPCC AR4 global climate models on Yangtze River Basin. Quaternary Sciences, 2010, 30(1):127-137. (in Chinese)

Hao, Z.C., Su, F.G., 2000. Modification of the Xinanjiang Grid-based Monthly Hydrological Model. Advances in Water Science 11(supplement), 80–86, (in Chinese)

Hao, Z., Wang, J., Li, L., et al., 2006. Impact of Cl imate Change on Runoff in Source Region of Yellow River. Journal of Glaciology and Geocryology. 28(1): 1-7. (in Chinese)

IPCC, 2007. Climate Change 2007: Impacts, Adaptation, and Vulnerability. Contribution of Working Group II to the Forth Assessment Report of the Intergovernmental Panel on Climate Change. Cambridge: Cambridge University Press.

Ju, Q., 2009. Research of Climate Change and Response of Water Recycling in Yangtze River Basin. Hohai University, 2009. 25~26. (in Chinese)

Ju, Q., Yu, Z., Hao, Z., et al., 2009. Division-based Rainfall-Runoff Simulations with BP Neural Networks and Xinanjiang models. Neurocomputing, 72, 2873-2883.

Nash, J.E., Sutcliffe, J.V., 1970. River flow forecasting through conceptual models, I. A discussion of principles, Journal of Hydrology, 10, 282-290.

Nash, L., Gleick, P. H, 1991. Sensitivity of Streamflow in the. Colorado Basin to climate change. Journal of Hydrology. 125, 221-241.

Nemec, J., Schaake, J., 1982. Sensitivity of water resources systems to climate variation. Science. 1982, 27(3): 327-343.

Nijssen, B., O'donnell, G. M., Hamlet, A. F., et al., 2001. Hydrologic sensitivityof global rivers to climate change. Climatic Change. 50, 143-175.

Ozkul, S., 2009. Assessment of climate change effects in Aegean river Basins: the case of Gediz and Buyuk Menderes Basins. Climatic Change, 97(1-2):253-283.

Schwarz, H. E., 1977. Climate change and water supply: how sensitive is the Northeast? Washington D C: National Academy of Science.

Young, C. A., Escobar-Arias, M. I., Fernandes, M., et al., 2009. Modeling the hydrology of climate change in California's Sierra Nevada for subwatershed scale adaptation. Journal of the American Water Resources Association. 45(6): 1409-1423.

Zhao, R.J., 1992. The Xinanjiang model applied in China. Journal of Hydrology 135, 371-381.

Zhou T, Yu R. 20th century surface air temperature over China and the globe simulated by coupled climate models. Journal of Climate. 2006, 19(22): 5843-5858.

5

Advances in Streamflow Prediction: A Multimodel Statistical Approach for Application on Water Resources Management

Sonia R. Gámiz-Fortis, María Jesús Esteban-Parra
and Yolanda Castro-Díez
University of Granada
Spain

1. Introduction

The growing demand for urban water users, industrial, environmental and agricultural, makes the task of managing water resources is becoming more complex. Obtaining accurate simulations and forecasting of water availability is a key step in efficient planning, operation and management of these resources. This is the reason because the development of reliable surface water flow forecasting methods for real-time operational water resources management becomes increasingly important. Because of this, in recent years, interest in predictability of river discharge variability has increased markedly in most of the world regions. The hydrological system acts as spatial and temporal integrator of precipitation (rain and snow), temperature, and related evapotranspiration over a specific region. Therefore, seasonal to decadal streamflow variability in many large river-basins can be controlled by corresponding changes in large scale atmospheric circulation patterns. The relationship between these climatic regimes and its hydrological response, through its streamflow, presents different grades of complexity according to the physical characteristics of the basin. Nevertheless, streamflow can be better related with important patterns of climate teleconnections than precipitation or temperature fields, since variations in precipitation are amplified in streamflow, and in general, it is easier to detect a change in discharge than directly in the basic climatic variables (Dettinger & Diaz, 2000; Trigo et al., 2004). On seasonal timescales, anomalous atmospheric conditions are often linked with seasonal variations in the rivers streamflow and reservoir storages, via variations in precipitation and temperature (Dettinger & Diaz, 2000; Cullen et al., 2002; Trigo et al., 2004). The interannual climate and hydrologic fluctuations are modulated by, or superimposed upon, lower frequency variations with decadal and longer time scales. The sources of these lower frequency climate variations are uncertain but may have roots in the tropics (Trenberth & Hoar, 1996), in the extratropical oceans (Pacific: Latif & Barnett, 1994; Atlantic: Deser & Blackmon, 1993; Tanimoto et al., 1993; Houghton, 1996), or in some interplay of the two (e.g., Graham et al., 1994; Jacobs et al., 1994). Usually, the skill of the long-range forecasts is associated with the introduction of predictors that represent the slow varying components of the climate system such as sea surface temperatures (SST) (Koster et al., 2010). Consequently, variability in SST can help provide predictive information about the hydroclimate in regions across the globe (Rimbu et al., 2005; Gámiz-Fortis, 2008b, 2010a, 2011b; Ionita et al., 2008).

Climate-based forecasts can be developed using either dynamical or statistical methods. Dynamical streamflow forecasts typically use observations of current climatic conditions, as input for large-scale general circulation model (GCM) simulations of atmospheric processes. The model output of future precipitation or runoff is then downscaled from GCM gridcell scales to river basin scales, and translated to streamflow via additional rainfall–runoff–streamflow relationhips.

In contrast, statistical streamflow forecasts typically use observations of current climate as direct predictors of future streamflow, based on historical lead–lag statistical relationships between climate predictor and streamflow values.

Regardless of the prediction method applied, the streamflow forecasts must have a strong physical foundation. For dynamical methods, this translates into reliable GCM simulations, downscaling techniques and the conversion from rainfall-runoff into streamflow in the region under study. For statistical methods, this means strong, robust, and physically justifiable statistical relationships between remote climate predictors and local streamflow predictands. In many cases, statistical models provide better results because there are still some unaccounted physical mechanisms in most coupled dynamical models.

This chapter presents a new modelling scheme that can be used for operational prediction of streamflow anomalies based on the combination of different steps: 1) a primary focus placed on the study of physical characteristics of the specific basin, in order to identify homogeneous regions of streamflow variability, 2) analysis of the temporal structure of the series for each differentiated region in the basin, 3) detection of statistically significant linkage that might exist between the streamflow series in the river basin and large scale SST variability modes, and 4) separate modelling of high and low frequency variations through the use of autoregressive-moving-average (ARMA) models and stable teleconnections between oceanic SST anomalies and river flow, respectively. Additionally, this chapter also contains a summary of the main results about the implementation of this multimodel statistical approach to different rivers of the Iberian Peninsula. Some of them has been already published (Gámiz-Fortis et al., 2008a, 2008b, 2010a, 2011b), and others constitute a new contribution (Miño river).

2. Data

This work focuses on Iberian rivers variability and predictability. For this southern European region, present and future water resources management presents a considerable number of challenges. Firstly, most Iberian rivers show relatively high coefficients of interannual streamflow variation, decreasing from rivers located in the north to those in the southern sector. These high values of variability can be explained based on the strong interannual precipitation variability observed throughout the whole western Mediterranean region (Goossens, 1985; Barry & Chorley, 1998; Esteban-Parra et al., 1998; Serrano et al., 1999). Secondly, both Portugal and Spain show an increasing demand of water supply not only for domestic use, but also for the tourist and agricultural sectors, two main economic activities accounting for more than 10% of Gross National Product (GNP) in both countries.

Monthly data flows from five different basins throughout the Iberian Peninsula have been
analyzed: the rivers Ebro, Miño, Douro, Tejo and Guadiana. Figure 1 shows the location of
the selected river basins in the Iberian Peninsula.

Fig. 1. Location of the five analyzed river basins in the Iberian Peninsula. Red dots show the
location of the selected gauging stations used in this study along with their code names.

The Ebro River (Figure 1) presents the largest catchment in Spain, with an area of 85,530 km²
in the northeast of Spain. Ebro basin is a complex region from both the topographical and
meteorological perspectives, with great physical heterogeneity, which is a deciding factor in
the annual seasonality of flow. This region forms a broadly triangular morphological unite,
which is a depression surrounded by high mountain ranges. The heterogeneous topography
contrasts the influences of the Atlantic and Mediterranean as well as of different large-scale
atmospheric patterns (Vicente-Serrano & López-Moreno, 2006). The Miño River basin is
located in the northwestern of the Iberian Peninsula. It covers a limited territory, bordered
by the Atlantic Ocean and the Cantabric Sea. And finally, the entire central Iberian plateau is
dominated by the Douro, Tejo and Guadiana River basins covering more than half of the
whole Peninsula and which are roughly oriented in a NE-SW direction, running from the
mountains in Spain down to the Atlantic Ocean across Portugal.

Monthly streamflow data of the rivers have been kindly provided by different Organisms
from both Spain and Portugal such as the River Hydrographic Administrations of Douro,
Miño and Ebro, the National Water Research Centre (CEDEX) and the Portuguese National
Electrical Supply Company (REN). These agencies control the monthly flows of rivers in
different areas by using direct gauging stations. Streamflow series distributed in the basins
have been intensively checked for inconsistencies, and only series showing less than 10% of
missing values during the period 1950-2006 have been considered in this study. Data gaps
were filled using data from neighbouring stations with a Pearson correlation coefficient of at
least 0.8. For the Ebro River, a total of 83 streamflow series, with values during the period
1950-2006, distributed in the basin, were considered. For the Miño River, the data base of
streamflows comprises monthly data from 18 stations, covering the period from 1956 to

2005. For the remainder rivers, just three stations, spanning between 1923-2004 for the Douro and Tejo, and 1947-2004 for the Guadiana, within the Portuguese section of the rivers, are considered.

Hydrological series commonly do not follow a normal distribution, being highly biased, often requiring some preliminary transformation in order to adjust the records to an appropriate distribution. Also, the magnitude of the monthly streamflow values varies greatly among the stations, due to both the different climatic regions in Ebro basin and the different drainage basin characteristics associated with each streamflow station. Some studies have shown a good adjustment of the discharge series to the log-normal distribution (Zaidman et al., 2001; Kalayci & Kahya, 2006), while Vicente-Serrano (2006) and López-Moreno et al. (2009) found that the Pearson III distribution is more appropriate over some parts of the Iberian Peninsula. Using these considerations three different theoretical distributions for modelling the monthly stream river flow were tested: the Pearson III, the log-normal and the normal distribution. Furthermore, the goodness of fit was evaluated using three different tests, the Kolmogorov-Smirnov, the Anderson-Darling and the Chi-squared tests. We find that for all the cases the log-normal theoretical distributions present good results. For this reason, and in order to facilitate the intercomparison between regions, the streamflow data were first subjected to logarithmic transformation to reduce the disparity in the magnitudes, from which monthly standardized streamflow anomalies were computed. The monthly standardized streamflow anomalies were constructed by subtracting the mean and dividing by the standard deviation for each month separately. The standardized streamflow for each station closely follows a normal distribution.

The global SST taken from the HadISSTv1.1 data set (Rayner et al., 2003) derived from the Hadley Centre for Climate Prediction and Research (UK Meteorological Office) has been used as predictor. The winter, spring, summer and autumn SST seasonal fields were generated by averaging the monthly SST anomalies (using the mean and standard deviation for the period 1961-1990) for DJF, MAM, JJA and SON, respectively.

Aditionally, some teleconnection indices such as the North Atlantic Oscillation (NAO), the East Atlantic (EA), the East Atlantic/Western Russia (EATL/WRUS), the Scandinavia (SCAND) and the Pacific Decadal Oscillation (PDO), have been obtained from NOAA, Climate Prediction Center (http://www.cpc.noaa.gov/data/teledoc/telecontents.shtml), and from the Joint Institute for the Study of the Atmosphere and Ocean (http://jisao.washington.edu/pdo/PDO.latest).

3. Methodology

This Section describes the complete modelling scheme that can be used for operational prediction of streamflow anomalies, based on the combination of different statistical techniques. As discussed in the Results Section, depending on the features of the basin and the availability of data in it, the modelling scheme can be applied in full or in part.

Firstly, for those basins that present a homogeneous spatial distribution of the gauging stations (as the case of the Ebro and Miño rivers), regionalization of streamflow series in the basins has been performed using rotated Principal Component Analysis (PCA). PCA has been applied in analyzing the spatial variability of physical fields. In climatology and hydrology, the PCA is a tool to explain the fundamental nature of streamflow, reducing a large number of interrelated variables to a few independent Principal Components (PCs) that capture much of the variance of the original dataset (Wilks, 1995). It produces a few major spatial-variability patterns (or Empirical Orthogonal Functions, EOFs), and the corresponding time series represent the time evolution of the spatial-variation patterns. Rotation (varimax) of PCA is then applied to the first few major patterns to capture better the physically meaningful and simplify spatial patterns (Barnston & Livezey, 1987). This procedure allows extracting representative stations for each of the areas obtained in the regionalization process. PCA has been used by other authors in similar contexts of the present study (Widman & Schär, 1997; Maurer et al., 2004; Kalayci & Kahya, 2006).

Second step is the study of temporal variability of the streamflow time series representative of the regions previously identify by PCA. To this end the Singular Spectral Analysis (SSA) is applied to each streamflow time series. SSA is a powerful form of the standard Principal Component Analysis based on the extensive use of the lag correlation structure of a time series (Vautard et al., 1992), which is particularly successful in isolating multiple period components with fluctuating amplitudes and trends in short and noisy series. SSA has been successfully applied to many geophysical and climatological time series to study and predict periodic activities (Ghil & Mo, 1991; Ghil & Vautard, 1991; Plaut & Vautard, 1994; Gámiz-Fortis et al., 2002, 2008a, 2010a, 2011a, 2011b; Paluš & Novotná, 2006). A comprehensive review, explaining in detail the mathematical foundations of SSA, can be found in Vautard et al. (1992) and Plaut et al. (1995).

SSA is based on the diagonalization of the lagged-autocovariance matrix of a time series. As in the case of PCA, the eigenvectors or Empirical Ortogonal Functions (T-EOFs) represent patterns of temporal behaviour, and the Principal Component series (T-PCs) are characteristic time series containing a very limited number of harmonic components. The detailed reconstruction of a set of significant components, called SSA-filtered components or reconstructed components (RCs) of the time series, is carried out by an optimal linear square fitting between the corresponding PCs and the original data. Each RC represents the contribution of its associated EOF to the variance of the time series; additionally, the RCs are additive and their sum provides the original time series. When two eigenvalues of the lagged-covariance matrix are nearly equal and their corresponding eigenvectors are orthogonal, they represent an oscillation. Therefore, we can synthesise that SSA is an eigenvalue technique particularly efficient to extract and reconstruct periodic components from noisy time series. Determining the corresponding frequencies requires, however, estimations of power spectra. The Maximum Entropy Method (MEM) is used to evaluate the spectral contents of the PC time series corresponding to the EOFs, and the Monte Carlo (MC) technique is used for the significance study (see Gámiz-Fortis et al., 2002, for further details). The analysis of the temporal structure of the series allows us to model separately their variations at different time scales: seasonal, interannual and decadal.

Next step involves identifying sectors of oceanic SST anomalies that can be used as predictors for river flow. Simple methods of statistical analysis can describe the potential relationship between the streamflow series and the SST, both contemporary and of the preceding seasons, in order to evaluate predictive ability. To do that the point linear correlation between the monthly or seasonal streamflow anomalies and the SST anomalies from previous seasons must be evaluated. Regions showing significant correlations are identified as potential predictors. A very important issue is to identify, among these regions, those that can be classified as stable predictors. This is achieved through the analysis of the variability of the correlation between flow anomalies and SST anomalies from potential predictor regions using different moving windows with lengths between 10 and 20 years. The correlation is considered to be stable for those regions where streamflow and SST anomalies are significantly correlated at 90% level for more than 80% of the 15-year windows covering the total period under study and, furthermore, where the sign of the correlation does not change with time. Regions verifying this criterion are considered as robust predictors and are used in a multiple linear regression model (hereinafter SST_model) to simulate the river flow anomalies. The method of ordinary least squares is used to estimate the regression coefficients (Draper & Smith, 1998) by minimizing the sum of squared errors (SSE), i.e. the unexplained variance part. The coefficient of multiple determination R^2, which measures the fraction of variance in the response variable that can be explained by variations in the explanatory factors also is computed. Note that a high value of R^2 does not imply that a particular model is appropriate. In fitting the model, the assumption is made that the unknown random effects are represented by "e", which is a vector of independent, normally distributed noise. The validity of this assumption should be checked.

Additionally, in order to detect some other kind of influence, the residual time series obtained by subtracting the SST_model from the raw streamflow series is studied. This residual component is firstly modelled by the significant quasi-oscillatory modes obtained from the SSA (residual_SSA_filter hereinafter) and autoregressive-moving-average (ARMA) models are fitted to the residual_SSA_filter of streamflow. The SSA filtering prior to obtaining the ARMA models considerably improves the forecasting skill of these ARMA models (Gámiz-Fortis et al., 2002, 2008a, 2010a). ARMA models can be regarded as a special case of general linear stochastic processes and provide a linear representative structure of the temporal evolution of the data. The order of the model is selected, in a preliminary approach, studying the autocorrelation function (ACF) and partial autocorrelation function (PACF). In physical terms, the best model has as few parameters as possible. The Akaike information criterion (AIC) (Akaike, 1974) has been used to select the final model among all the candidates. The AIC is based on information theory and represents a compromise between the goodness of the fit and the number of parameters of the model. A comprehensive review, explaining in detail how to fit ARMA models to datasets following the identification, estimation, and diagnostic check stages, can be found in Brockwell & Davis (1996) and Hipel & Mcleod (1994).

Finally, the combination of both SST_models and ARMA_models is evaluated in a forecasting experiment. For reliable skill assessment, a fundamental aspect of this evaluation is the separation of calibration and validation periods (Wilks, 1995). We employ data until 1989 to calibrate the different modelling, while data from 1990 onwards are used for validation purposes only. For model evaluation appropriate skill measures commonly used are employed, such as the mean absolute error (MAE) and the mean square error (MSE) as accuracy measures; the Pearson correlation coefficient, which is used to measure the relationship between the modeled/forecasted series and the original/expected series; the percentage of phase agreement, that is, the percentage of cases in which the modeled/forecasted values has the same sign as the original values; and the coefficient of multiple determination R^2.

4. Results

This section shows the main results obtained from the application of the previously described methodology to the five rivers studied in the Iberian Peninsula. Depending on the availability of data in the selected basins not always is possible to apply the full modelling scheme. For example, for Ebro and Miño rivers, that account with a relative high amount of gauging stations distributed around the basins, the regionalization process using PCA can be carried out. For the Ebro River three significant spatial EOFs, accounting for 55.4% of the total variance, were found; while for the Miño River just one significant spatial EOF with an associated variance of 80% was found. Figure 1 also shows the location of the selected streamflow time series representative of the regions identify by PCA for the Ebro and Miño basins. For the Douro, Tejo and Guadiana rivers authors do not dispose of a database homogeneously distributed over the basins, so regionalization can not be carried out. However, studies of other authors like Leite & Peixoto (1995) that use temperature, precipitation and insolation fields in the global Douro Basin reveals the existence of different climatic regions, with a western region dominated by oceanic air masses with heavy precipitation, on contrast with a central and eastern region drier and warmer. It is therefore expectable that the influence of anomalous SST on these river streamflow to be more important over the western sector of these basins. Additionally, these rivers are subjected to large regulation in the Spanish part of the basins, which can interfere the SST/river discharges relationship. Note that the expected effect of large dams is an increased time between precipitation episodes and the arrival of the corresponding flow to lower sections of the river basin. Based on these considerations it is understandable that the contribution from the SST to these river streamflows to be more important for the western part of the basins. For this reason the monthly time series of these three rivers discharge were recorded at stations situated in the Portuguese side of the border (see Figure 1), in the lower part of the catchment areas. Table 1 presents a description of all the selected gauging stations used in this study.

River	Station name	Code	Data length	Months with maximum streamflow values
	Zaragoza	EZ	1913-2006	January
Ebro	Grauss	EG	1950-2006	June
	Nonaspe	EN	1950-2006	May
Miño	Lugo	ML	1956-2005	January-February-March
Douro	Pocinho	DP	1923-2004	January-February-March
Tejo	Fratel	TF	1923-2004	January-February-March
Guadiana	Pulo do Lobo	GP	1947-2004	January-February-March

Table 1. List of individual gauging stations selected in the river basins.

As expected for these regions, different months account for the majority of runoff, depending on the climatic characteristic of these regions. It is expected that the potential relationships between the flow and the SST field will be intensified for those months showing maximum streamflow values. Taking into account the maximum streamflow values observed for the different flows (see Table 1, last column), the study was limited to those months in which discharge time series are highest. For the case of Miño, Douro, Tejo and Guadiana, where months of January, February and March appear with maximum streamflow values, the winter streamflow was computed as the average of these three monthly normalized series. For the EZ station of the Ebro River (Figure 1), January river flow accounts for the majority of runoff, being followed by a relatively long and dry summer period from July to September. The EG station is associated to the region represented by a nivo-pluvial flow type, located in the Pyreness Mountains, where snowmelt influence is very important. This area is characterized by a snowy winter, very rainy spring and rainy late summer and autumn (Bejarano et al., 2010). Finally, the EN station, located in the southeast of the Ebro basin is characterised by fairly low total annual precipitation and prolonged dry season.

4.1 Temporal variations

In this section, SSA is applied to each of the streamflow series selected in Table 1 to further identify and extract the oscillation components. Different window lengths between M = 15 and M = 32 years and the Vautard and Ghil (1989) algorithm were used in SSA. The oscillation pairs identified by SSA on streamflow series for the selected stations are summarized in Table 2 for the Ebro River and Table 3 for the reminder rivers. Because SSA cannot resolve periods longer than the window length, we identified the zero frequency as the non-linear trend, but it is important to note that this non-linear trend could be composed of quasi-oscillatory modes with periods longer than window length years.

Streamflow series	Ebro_EZ	Ebro_EG	Ebro_MN
Oscillation period (years) (explained variance)	Trend (9.1%) 23.3 y (12.7%) 2.3 y (11.1%) 9.1 y (9.5 %) 4.3 y (7.3%) 3.1 y (7.3%) 6.6 y (6.9%)	Trend (12.5%) 4.3 y (19.8%) 2.1 y (18.3%) 2.7 y (12.5%) 9.5 y (10.1%)	Trend (5.0%) 15 y (30.2%) 2.5 y (14.4%) 3.1 y (8.9%) 3.6 y (8.6%) 4.9 y (7.5%) 6.6 y (6.9%)
Total Variance	64%	73.2%	81.5%

Table 2. Significant oscillatory modes (peak period in years) identified by SSA on streamflow gauging stations of the Ebro Basin, along with the fraction of total variance (%) explained by each mode. Last row shows the total variance explained by the SSA_filter.

Note from Table 2 that for the Ebro River, along with the non-linear trends, oscillatory modes with periods longer than 9 years are explaining a big part of the variance of the series. This means that for this Mediterranean river the variance of the raw series associated to the decadal or multi-decadal variability (periods longer than 9 years) is relatively high, especially for the Ebro_EZ and Ebro_MN time series that reach 31.3% and 35.2% of the total variance, respectively.

Streamflow series	Miño	Douro	Tejo	Guadiana
Oscillation period (years) (explained variance)	7.7 y (32.7%) 2.3 y (19.5%) 2.7 y (16.1%)	Trend (24%) 5.3 y (15%) 4 y (14%) 3.3 y (13%) 8 y (11%) 2.7 y (9%)	Trend (20%) 3.6 y (21%) 5.3 y (15%) 7.5 y (12%)	Trend (15%) 2 y (19%) 4.5 y (16%) 3.4 y (16%) 6.5 y (13%)
Total Variance	68%	86%	68%	79%

Table 3. Significant oscillatory modes (peak period in years) identified by SSA on streamflow gauging stations of the Miño, Douro, Tejo and Guadiana rivers, along with the fraction of total variance (%) explained by each mode. Last row shows the total variance explained by the SSA_filter.

For the remainder rivers, no oscillatory mode with period longer than 9 years is found (see Table 3), which means that for these Atlantic rivers the interannual variability is dominating. Based on this feature the next step for the statistical modelling was reversed for the Atlantic rivers and the Ebro River. That is, for the Miño, Douro, Tejo and Guadiana rivers, interannual variability is firstly modelled by the use of ARMA models, while for the Ebro river, modelling of the decadal variability using the SST is carried out firstly. In summary, depending on which component is dominating, interannual or decadal, one or another modelling (ARMA_modelling or SST_modelling) has been firstly applied.

4.2 Ebro river modelling

This section shows the modelling of the Ebro River streamflow. The relationship between the climatic regimes and its hydrological response, through the streamflow, presents different grades of complexity according to the physical characteristics of the basin so it has been studied in detail.

4.2.1 Modelling on decadal time scales

Standardized streamflow anomalies were smoothed with a 7-year running mean filter to obtain the decadal component of the series (decadal_component hereinafter). This filter is designed to fall both near the low-frequency end of the spectrum associated with El Niño events and near the high-frequency end of the decadal climate variations of the Atlantic and North Pacific climate systems (Dettinger et al., 2001), and has been used following other authors (Ionita et al., 2011) in order to exclude any cross-spectrum between Ebro River streamflow and El Niño. Spatial correlation maps between the decadal_component of January EZ_FLOW, June EG_FLOW and May EN_FLOW and the preceding seasons for global SST data have been computed and are shown in Figure 2. For meaningful evaluation of the relationship between the SST and streamflow fields on decadal time scales, the linear trend components of all the data sets need to be removed. The fact of eliminating the trend when working with variables of different physical natures is necessary because otherwise trends could lead spurious correlations.

For January EZ_FLOW, maximum statistically significant correlations were found with the previous spring SST in the Atlantic region (see Figure 2a). Particularly, we can identify positive correlations with the SST anomalies in the tropical North Atlantic and southern Greenland, with the tropical part of the SST pattern (approximately 0-20°N) playing the most important role. This pattern represents what is commonly called North Atlantic horseshoe pattern, described e.g. in Czaja and Frankignoul (2002), and it is represented by the third spatial EOF when a PCA is applied to the global spring SST field (Figure 3). This spring SST pattern is shown to be related with the subsequent winter NAO (Rodwell and Folland, 2002).

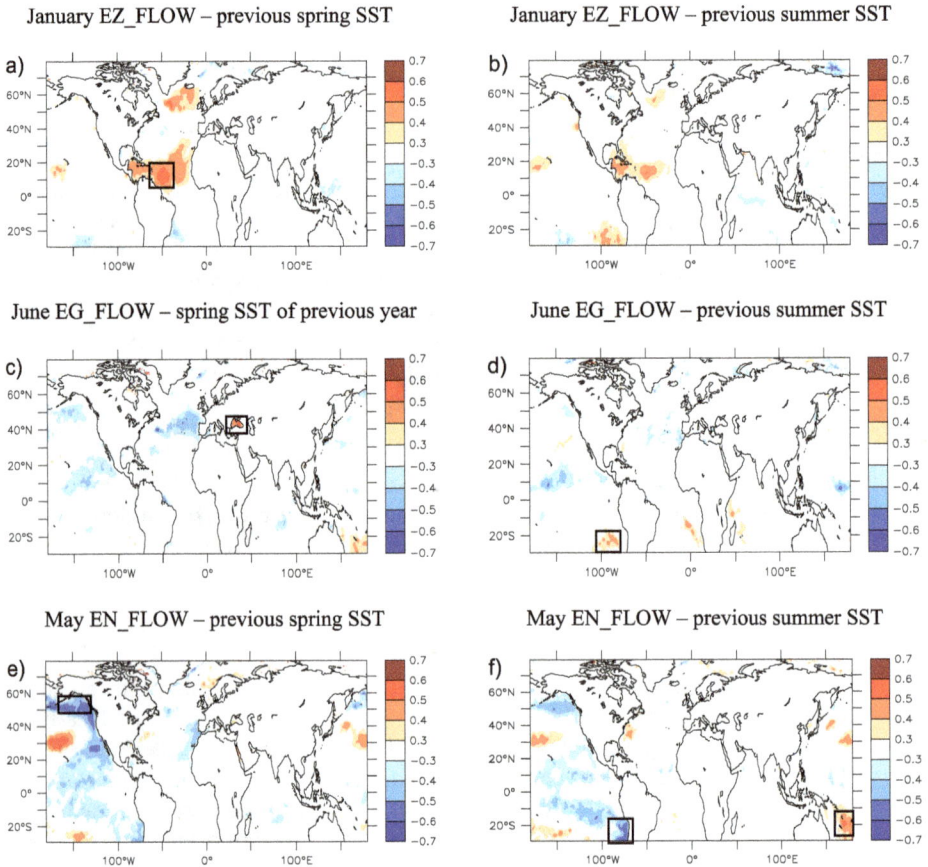

Fig. 2. Correlation maps between the decadal_component of streamflow time series of a) January EZ_FLOW and previous spring SST anomalies, b) January EZ_FLOW and previous summer SST anomalies, c) June EG_FLOW and spring SST anomalies of the previous year, d) June EG_FLOW and previous summer SST anomalies, e) May EN_FLOW and previous spring SST anomalies, and f) May EN_FLOW and previous summer SST anomalies. Only significant values at the 95% confidence level are shown. The rectangle corresponds to the stable SST regions finally considered for the decadal_component modelling.

An additional region in the south Pacific also shows significant positive correlation during previous summer (see Figure 2b). Based on this correlation maps we define three SST indices by averaging the normalized SST anomalies over the regions showing maximum correlation values: $SST1_{EZ}$ (54°W-49°W; 9°N-12°N) and $SST2_{EZ}$ (44°W-39°W; 54°N-56°N) in previous spring, and $SST3_{EZ}$ (99°W-96°W; 28°S-22°S) in previous summer.

SST indices were smoothed with a 7-year running mean filter to obtain the decadal component of the series associated to these SST regions, and correlations between the January streamflow anomalies and the $SST1_{EZ}$, $SST2_{EZ}$ and $SST3_{EZ}$ indices using a moving window of 15 years were computed. Additionally, the PC3 time series obtained from the PCA of spring SST, which represents the complete horseshoe pattern, was also considered in this analysis.

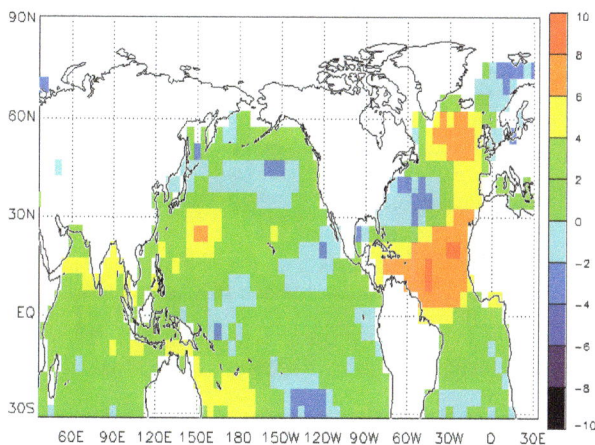

Fig. 3. Loading factors (by ten) for the third spatial mode resulting from the PCA of the global spring (March-April-May) sea surface temperature. The period of analysis is 1871-2007.

Stable positive correlations (not shown) during the period 1916-2003 are found only for the $SST1_{EZ}$ index, while the remainder indices fail the stability test and dismissed from the rest of the analysis. There are several problems when using standard climate indices for hydroclimatic prediction. This is the case of PC3 of spring SST, where the hydroclimate of the river basin is more strongly correlated with an oceanic region's SSTs, which is different from the predetermined regions that are used to calculate the standard index (Tootle and Piechota, 2006). The second problem is that the PCA, used to obtaining this index, might not preserve enough information from the original dataset (Switanek et al., 2009).

The spatial correlation maps between the decadal_component of June EG_FLOW and the preceding seasons for SST data (Figure 2c and 2d) show statistically significant correlations with global SST being maxima during the spring of the previous year and in some regions of the Pacific for the previous summer. In this case, stable correlations (not shown) during the period 1953-2003 are only found for the following defined indices: $SST1_{EG}$ = (30°E-40°E; 41°N-46°N), corresponding to the Black Sea during the spring of the previous year, and

$SST2_{EG}$ = (88°W-81°W; 25°S-20°S), associated to the south-eastern Pacific Ocean during the previous summer.

For the decadal_component of May EN_FLOW, the spatial correlation maps with the preceding seasons for SST (Figure 2e and 2f) present maximum statistically significant correlations in the Pacific region. Particularly, we can identify correlations that resemble the negative pattern of the Pacific Decadal Oscillation during the spring of the previous year. For this case, stable correlations during the period 1953-2003 are found for the following indices: $SST1_{MN}$ (158°W-154°W; 50°N-55°N), $SST2_{MN}$ (76°W-72°W; 30°S-21°S) and $SST3_{MN}$ (168°E-172°E; 21°S-18°S). Additionally, in this case, the stability of the correlation between the May streamflow and the spring PDO index of the previous year was also studied, finding negative stable correlation along the period.

Using the significant and stable indices as predictor for the streamflow time series at decadal time scales we developed a model based on linear regression, using as calibration period 1916-1989 for EZ_FLOW, and 1953-1989 for EG_FLOW and EN_FLOW. The optimal models for explaining the decadal_component of these flows can be written as:

$$SST\_model_{EZ} = 2.07 \times SST1_{EZ\_previous\_spring}$$

$$SST\_model_{EG} = 0.068 + 0.601 \times SST1_{EG\_spring\_previous\_year} + 0.53 \times SST2_{EG\_previous\_summer}$$

$$SST\_model_{EN} = -0.1 - 0.69 \times SST1_{EN\_previous\_spring} - 0.79 \times SST2_{EN\_previous\_summer} + 1.22 \times SST3_{EN\_previous\_summer} + 0.26 \times PDO_{previous\_spring}$$

These SST_models are able to explain around 56%, 71% and 81% of the variance of the streamflow decadal_component during the calibration period for the EZ_FLOW, EG_FLOW and EN_FLOW, respectively.

4.2.2 Modelling on interannual time scales

Regarding the modelling of the interannual variability (time scales less than 7 years) an additional analysis has been carried out using the quasi-oscillatory modes with periods lower than 7 years obtained from the previous SSA. Firstly, a SSA-filtered series, the interannual_SSA_filter, has been computed like the sum of the reconstructed components of these oscillatory modes, and then an ARMA_model is fitted to this SSA-filtered series. Using the sample Autocorrelation Function (ACF) and Partial Autocorrelation Function (PACF) (not shown) we find that while the interannual_component behaves as a white noise process, the interannual_SSA_filter series shows a strong autocorrelation pattern. This feature implies a higher predictability of the interannual_SSA_filter when compared to the unfiltered time series. Based on these analyses, we used the Akaike Information Criterion (AIC) to select the ARMA models for the interannual_SSA_filters, containing the following parameters for each Ebro gauging station:

ARMA(7,6)_model$_{EZ}$:

$$AR = (\Phi_1 = -1.38^*, \Phi_2 = -1.16^*, \Phi_3 = -1.15^*, \Phi_4 = -1.25^*, \Phi_5 = -1.18^*, \Phi_6 = -0.74^*, \Phi_7 = -0.09)$$

$$MA = (\theta_1 = 0.29^*, \theta_2 = 1.24^*, \theta_3 = 0.23^* , \theta_4 = -1.09^*, \theta_5 = -0.40^* , \theta_6 = 0.62^*)$$

ARMA(4,8)_model$_{EG}$:

$$AR = (\Phi_1 = -1.2^*, \Phi_2 = -1.06^*, \Phi_3 = -1.1^*, \Phi_4 = -0.5^*)$$

$$MA = (\theta_1 = 0.26^*, \theta_2 = 0.17^*, \theta_3 = 0.39^*, \theta_4 = 0.13^*, \theta_5 = 0.14^*, \theta_6 = 0.05, \theta_7 = 0.22^*, \theta_8 = -0.74^*)$$

ARMA(4,3)_model$_{EN}$:

$$AR = (\Phi_1 = -1.04^*, \Phi_2 = -0.78^*, \Phi_3 = -0.40^*, \Phi_4 = -0.37^*)$$

$$MA = (\theta_1 = -0.89^*, \theta_2 = 0.96^*, \theta_3 = 0.93^*)$$

Significance of the parameters was computed using approximate t-values, derived from the parameter standard errors. Parameters highlighted with "*" are statistically significant at 5% significance level. The selected order models indicate a certain degree of persistence.

These ARMA_models are able to explain around 67%, 76% and 64% of the variance of the streamflow interannual_components during the calibration period, for EZ_FLOW, EG_FLOW and EN_FLOW, respectively.

4.2.3 Combination of interannual and decadal modelling

This section shows the results obtained from the combination of SST_model (for decadal time scales) and ARMA_model (for interannual time scales). The observed and modelled January EZ_FLOW, June EG_FLOW and May EN_FLOW series are shown in Figure 4, for calibration and validation periods, and the statistical results are shown in Tables 4, 5 and 6.

EZ_FLOW	Calibration period 1916-1989		Validation period 1990-2003	
	ARMA (7,6)	ARMA(7,6) + SST_model	ARMA (7,6)	ARMA(7,6) + SST_model
MSE	0.49	0.35	1.02	0.73
MAE	0.56	0.47	0.80	0.72
Correlation coeft.	0.73*	0.81*	0.66*	0.79*
Phase accordance	72%	84%	43%	79%
R^2	53%	66%	44%	62%

Table 4. Statistical results for the January EZ_FLOW streamflow forecasting experiment using both the ARMA and the regression model, which includes de SST information. Results are displayed both for the calibration and validation period. For the sake of comparison, the results of the ARMA-alone forecast are also shown. Correlation coefficients with "*" are statistically significant at the 95% confidence level.

Results obtained reveal a considerable skill achieved by the combined [ARMA_model + SST_model] models (see Tables 4, 5 and 6), with good correlation coefficients between the raw series and the models for the validation period (r = 0.79, 0.90 and 0.87, respectively) and coefficients of multiple determination R^2 (62%, 81% and 76%, respectively). Moreover, the models present relatively low MSEs (0.73, 0.11 and 0.45, respectively) and MAEs (0.72, 0.26

and 0.59). From Tables 4, 5 and 6 we can also see as the overall quality of the combined model has increased significantly in both calibration and validation periods.

EG_FLOW	Calibration period 1953-1989		Validation period 1990-2003	
	ARMA (4,8)	ARMA(4,8) + SST_model	ARMA (4,8)	ARMA(4,8) + SST_model
MSE	0.34	0.22	0.18	0.11
MAE	0.46	0.37	0.35	0.26
Correlation coeft.	0.77*	0.84*	0.79*	0.90*
Phase accordance	49%	53%	86%	86%
R^2	59%	71%	62%	81%

Table 5. As Table 4 but for June EG_FLOW.

EN_FLOW	Calibration period 1953-1989		Validation period 1990-2003	
	ARMA (4,3)	ARMA(4,3) + SST_model	ARMA (4,3)	ARMA(4,3) + SST_model
MSE	0.33	0.14	0.80	0.45
MAE	0.46	0.32	0.79	0.59
Correlation coeft.	0.76*	0.90*	0.72*	0.87*
Phase accordance	82%	79%	79%	79%
R^2	58%	81%	52%	76%

Table 6. As Table 4 but for May EN_FLOW.

The combined modelling capacity can be appreciated in Figure 4 with a clear improvement in relation to the ARMA model, particularly evident during the period 1960-1993 for EZ_FLOW, 1987-1992 for EG_FLOW and during the peaks of streamflow for EN_FLOW.

4.3 Miño, Douro, Tejo and Guadiana rivers modelling

4.3.1 Modelling on interannual time scales

For these Atlantic rivers, interannual variability, which is dominating, is firstly modelled by the use of ARMA models. For the SSA_filters obtained for these rivers from the previous SSA, we have selected the following ARMA_models:

$ARMA(6,4)\_model_{Miño}$:

$$AR = (\Phi_1 = 1.31^*, \Phi_2 = 0.75^*, \Phi_3 = 0.27^*, \Phi_4 = 0.93^*, \Phi_5 = 1.34^*, \Phi_6 = 0.89^*)$$

$$MA = (\theta_1 = 0.61^*, \theta_2 = 0, \theta_3 = 0, \theta_4 = -0.35^*)$$

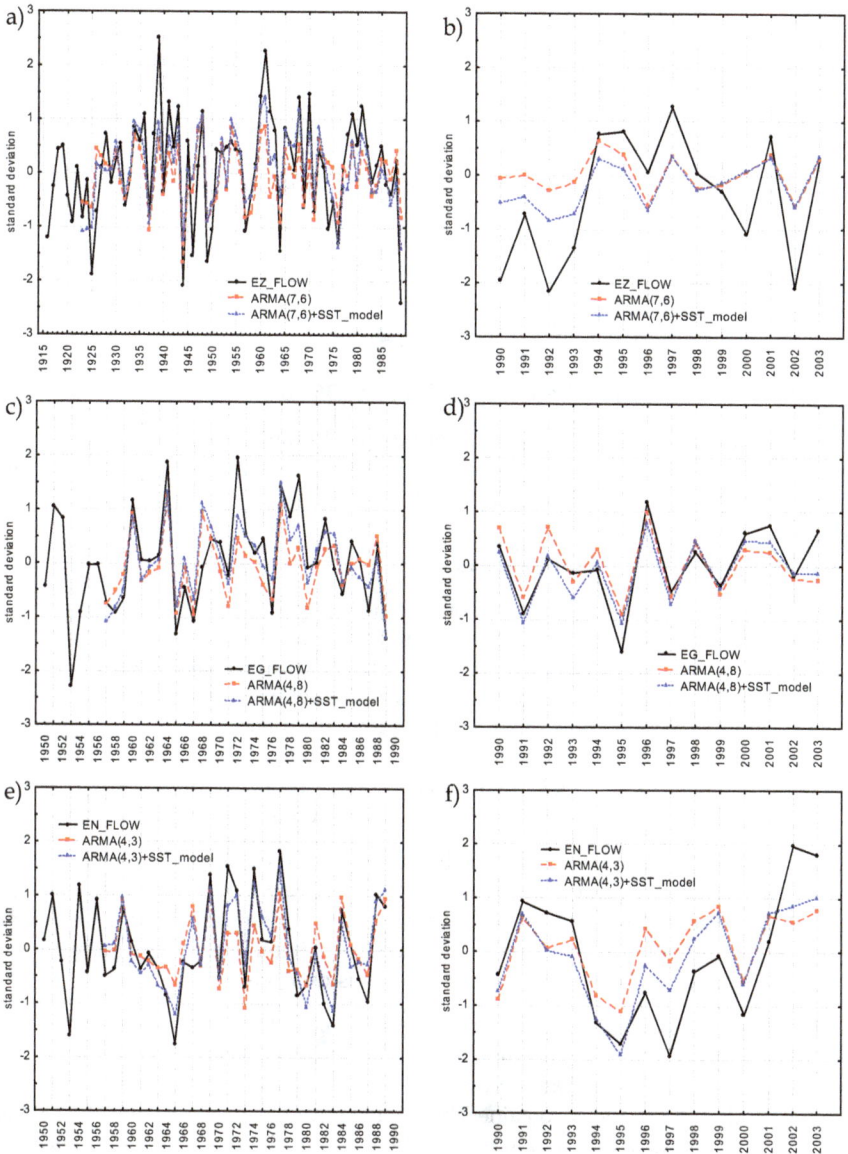

Fig. 4. a) Observed (black circles) and modelled January EZ_FLOW anomalies during the
calibration period (1916-1989) based on the interannual_component alone (red squares) and
the combined modelling [ARMA_model + SST_model] (blue triangles). b) As in a) but
during the validation period (1990-2003). c) As in a) but for the June EG_FLOW anomalies.
Calibration period in this case is 1953-1989. d) As in c) but during the validation period. e)
As in c) but for the May EN_FLOW anomalies. f) As in e) but during the validation period.

ARMA(7,3)_model$_{Douro}$:

$$AR = (\Phi_1 = 0, \Phi_2 = 0.34^*, \Phi_3 = 0.41^*, \Phi_4 = 0^*, \Phi_5 = -0.13, \Phi_6 = 0.14, \Phi_7 = -0.25^*)$$

$$MA = (\theta_1 = -0.74^*, \theta_2 = 0.50^*, \theta_3 = 0.78^*)$$

ARMA(8,4)_model$_{Tejo}$:

$$AR = (\Phi_1 = 0.42^*, \Phi_2 = 0.27^*, \Phi_3 = -0.31^*, \Phi_4 = 0.80^*, \Phi_5 = -0.44^*, \Phi_6 = 0^*, \Phi_7 = 0.31^*, \Phi_8 = -0.61^*)$$

$$MA = (\theta_1 = -0.47^*, \theta_2 = 0.76^*, \theta_3 = 0.76^*, \theta_4 = -0.18^*)$$

ARMA(6,4)_model$_{Guadiana}$

$$AR = (\Phi_1 = 0.38^*, \Phi_2 = 0, \Phi_3 = 0, \Phi_4 = 0.15^*, \Phi_5 = -0.25^*, \Phi_6 = 0.21^*)$$

$$MA = (\theta_1 = 0.18^*, \theta_2 = 0^*, \theta_3 = 0.36^*, \theta_4 = -0.67^*)$$

These ARMA_models are able to explain around the 62%, 36%, 69% and 23% of the variance of the streamflow during the calibration periods (see first column of Tables 7, 8, 9 and 10), for the Miño, Douro, Tejo and Guadiana rivers, respectively.

4.3.2 Modelling on seasonal time scales

In this section, the effort is concentrated on potential added value of the Ocean SST on the seasonal predictability of these streamflow series. The analysis is based on the results obtained with the interannual predictability experiment carried out in the previous section. Here, the main aim is to evaluate any increment in winter streamflow forecasting skill attributable to the SST. To this end, the residual time series resulting from the interannual forecasting experiment have been analyzed. This methodology allows, firstly, comparing the relative importance of the seasonal against interannual predictability, and secondly, to construct a statistical forecasting model which includes both seasonal and interannual sources of predictability.

Four new time series have been obtained by subtracting the ARMA forecasts computed in the previous section from the raw streamflow series. Note that these time series provide some kind of "residual" time series which contains the "information" that the interannual ARMA model was not able to capture. We will call hereinafter these new time series: residual_Miño, residual_Douro, residual_Tejo and residual_Guadiana.

In order to achieve this goal, the linear correlation coefficients between the previous seasonal SST and the following winter residual streamflow series for the four rivers have been computed. Additionally, time series representative of the five significant summer and autumn SST spatial modes of variability, obtained by applying a PCA to the Atlantic SST data, have been used in this analysis. Correlation results obtained for different calibration periods: 1962-1989 for the Miño River, 1930-1989 for the Douro and Tejo, and 1953-1989 for the Guadiana, show that the fourth autumn mode of Atlantic SST presents a statistically significant correlation with the Miño residual time series, meanwhile only the second autumn mode of Atlantic SST presents a statistically significant correlation with the other three residual time series. Additionally, a stability analysis shows that the correlations were stable throughout the analyzed periods. Again, the second autumn mode of Atlantic SST corresponds to the tripole pattern in the North Atlantic section described previously in Figure 3, while the fourth autumn mode of Atlantic SST (not shown) presents positive loading factors to the western North Africa and Europe and negative loading factors in the centre of the Atlantic Ocean.

Based on the previous results, we have developed linear regression models in order to study the seasonal predictability of each streamflow. The models were fitted to the residual time series: residual_Miño, residual_Douro, residual_Tejo and residual_Guadiana. To develop the models, the period 1962-1989 was used for calibration for the Miño River, 1930-1989 for the Douro and Tejo, and the shorter period 1953-1989 for the Guadiana. Again, in all the four cases, the final period 1990-2004 (or 1990-2005 for the Miño River) was used for validation purposes only. These periods correspond to the ARMA calibration and validation periods used in previous section, and were selected in order to combine both the ARMA and SST forecast (following section). The regression models fitted for the four rivers residual time-series are as follows:

$$\text{residual\_Miño} = 0.11 + 0.27 \times (\text{PC4 autumn Atlantic SST})$$

$$\text{residual\_Douro} = 0.10 + 0.64 \times (\text{PC2 autumn Atlantic SST})$$

$$\text{residual\_Tejo} = 0.020 + 0.39 \times (\text{PC2 autumn Atlantic SST})$$

$$\text{residual\_Guadiana} = -0.06 + 0.61 \times (\text{PC2 autumn Atlantic SST})$$

4.3.3 Combination of interannual and seasonal modelling

This section shows the results obtained from the combination of ARMA_models (for interannual time scales) and SST_models (for seasonal time scales). The observed and modelled Miño, Douro, Tejo and Guadiana streamflow series are shown in Figure 5 and Figure 6, for calibration and validation periods. Results obtained reveal a considerable skill achieved by the combined [ARMA_model + SST_model] models (see Tables 7, 8, 9 and 10).

As can be appreciated in Table 7, the Miño River shows a moderately improvement in the forecasting skill by using the SST information. For the validation period, the correlation coefficient is 0.81 (0.77 using only the ARMA model), which means that the combined model can explain only 65% of the variability for this river.

Miño	Calibration period 1962-1989		Validation period 1990-2005	
	ARMA (6,4)	ARMA(6,4) + SST_model	ARMA (6,4)	ARMA(6,4) + SST_model
MSE	0.58	0,42	0.56	0.52
MAE	0.45	0.27	0.30	0.60
Correlation coeft.	0.79*	0.87*	0.77*	0.81*
Phase accordance	80%	86%	83%	75%
R^2	62%	72%	61%	65%

Table 7. Statistical results for the Miño streamflow forecasting experiment using both the ARMA and the regression model, which includes de SST information. Results are displayed both for the calibration and validation period. For the sake of comparison, the results of the ARMA-alone forecast are also shown. Correlation coefficients with "*" are statistically significant at the 95% confidence level.

Figure 5 shows the results of the forecasting experiment for the Miño streamflow using the ARMA-alone and the ARMA+SST model for the calibration (1962-1989) and for the validation (1990-2005) periods. Only a very moderately improvement can be observed for individual years such as 1966, 1967, 1969, 1976, 1978 or 1981.

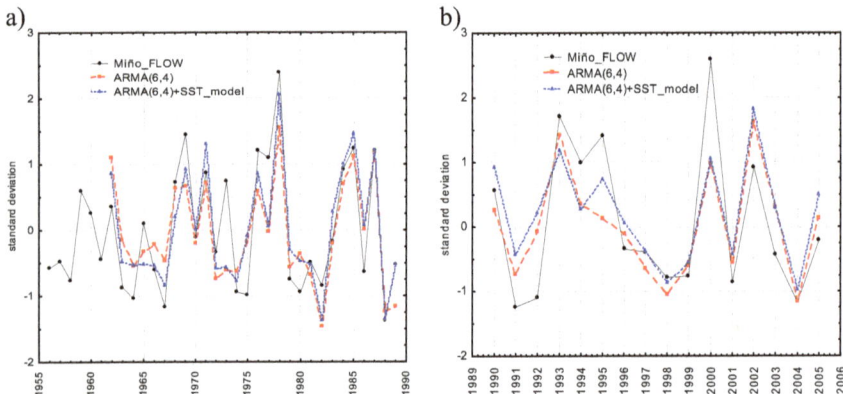

Fig. 5. a) Results of the forecasting experiment for the Miño streamflow using the ARMA-alone and the ARMA+SST model for the calibration period 1962-1989. b) As in a) but for the validation period 1990-2005.

The Douro River shows a considerable improvement in the forecasting skill by using the SST information (Table 8). Particularly, the correlation coefficient is 0.93 (0.73 using only the ARMA model), which means that the combined model can explain 86% of the variability. Finally, the phase agreement is 95% (90% using the ARMA model). Skill values are very similar during the calibration period.

Douro	Calibration period 1930-1989		Validation period 1990-2004	
	ARMA (7,3)	ARMA(7,3) + SST_model	ARMA (7,3)	ARMA(7,3) + SST_model
MSE	0.48	0.08	0.38	0.09
MAE	0.52	0.19	0.47	0.18
Correlation coeft.	0.60*	0.94*	0.73*	0.93*
Phase accordance	87%	97%	90%	95%
R^2	36%	88%	53%	86%

Table 8. As Table 7 but for the Douro River.

A close look to Figures 6a and 6b reveals the years responsible for these improvements in the skill parameters. The ARMA-alone model tends to underestimate the extreme positive values and provides relatively poor forecast for extreme negative values. The use of the SST information significantly off-set both pitfalls, providing better estimates for positive (e.g. 1996, 2001) and negative (e.g. 1989, 1992) extreme values during the validation period, or for 1975-85 during the calibration period.

The Tejo River shows lower improvement in the forecasting skill by using the SST information. As shown in Figure 6d, the improvement has to do mainly with the better capacity to forecast extreme positive values. In these cases (e.g. 1996 and 2001), the SST information provides valuable additional information in order to improve the estimates from the ARMA-alone model. This is also true for the calibration period. On the other hand, for the extreme negative streamflow values, the use of the SST information does not seem to improve model skill. Overall, the phase accordance of the model with the observation is around 90% during the validation period and the correlation coefficient is 0.89, showing that the model is able to explain around 79% of the winter Tejo streamflow variability along the period 1990-2004 (Table 9).

Tejo	Calibration period 1930-1989		Validation period 1990-2004	
	ARMA (8,4)	ARMA(8,4) + SST_model	ARMA (8,4)	ARMA(8,4) + SST_model
MSE	0.43	0.18	0.26	0.23
MAE	0.47	0.29	0.43	0.40
Correlation coeft.	0.83*	0.80*	0.85*	0.89*
Phase accordance	89%	90%	90%	90%
R^2	69%	64%	72%	79%

Table 9. As Table 7 but for the Tejo River.

Results for the Guadiana River (Table 10) are similar to those obtained for the Douro River. By using the SST information the forecast skill measures improve considerably. In particular, during the validation period, the correlation coefficient improves considerably, from 0.47 using the ARMA-alone model to 0.90 using the combined model, explaining around 80% of the streamflow variability. Again, the main reason of the improvement in the forecasting skills of the complete model is that the SSTs provide better estimates for both the extreme positive (e.g. 1996, 2001) and negative (e.g. 1989, 1992, 2000) streamflow values during the validation period, or the 1975-85 during the calibration period (Figures 6e and 6f).

Guadiana	Calibration period 1953-1989		Validation period 1990-2004	
	ARMA (6,4)	ARMA(6,4) + SST_model	ARMA (6,4)	ARMA(6,4) + SST_model
MSE	0.36	0.08	0.34	0.07
MAE	0.41	0.21	0.41	0.18
Correlation coeft.	0.48*	0.95*	0.47*	0.90*
Phase accordance	52%	93%	54%	100%
R^2	23%	90%	22%	81%

Table 10. As Table 7 but for the Guadiana River.

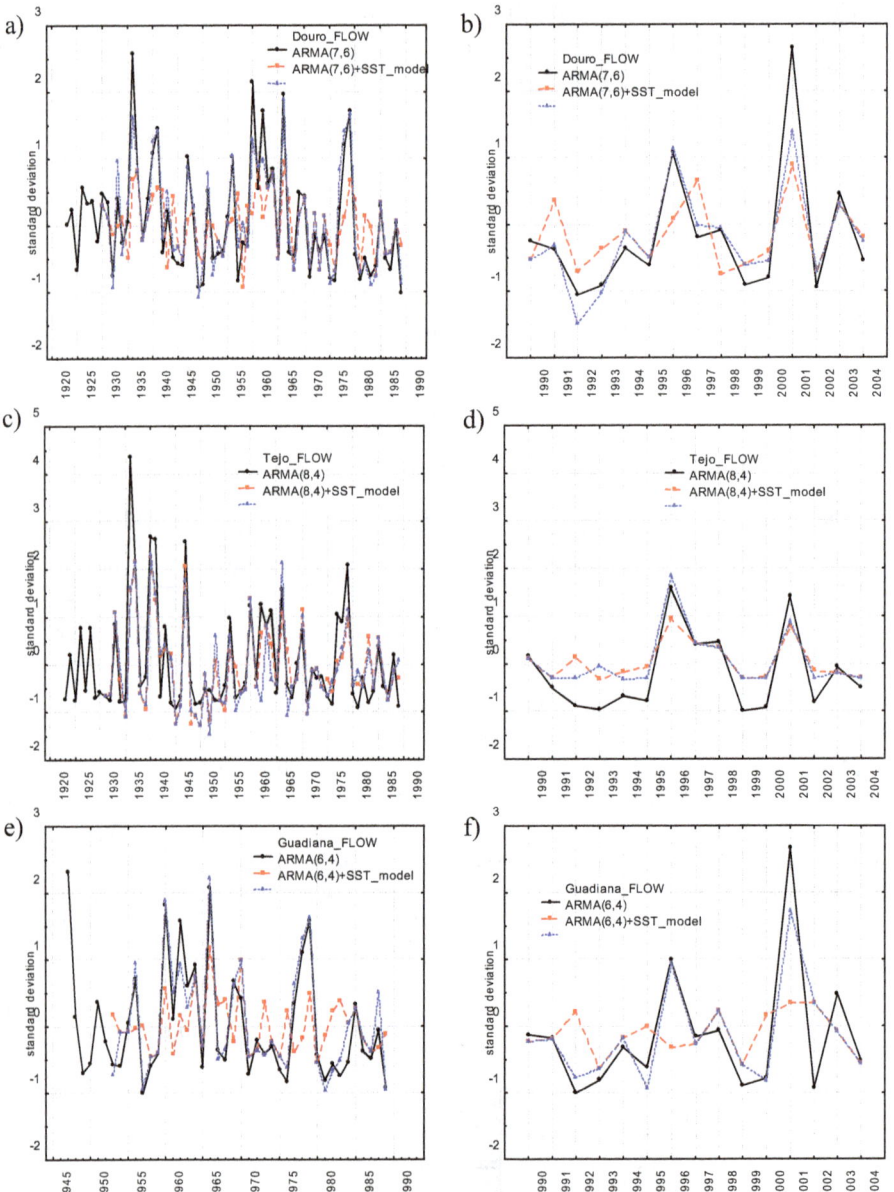

Fig. 6. a) Results of the forecasting experiment for the Douro streamflow using the ARMA-alone and the ARMA+SST model for the calibration period 1930-1989. b) As in a) but for the validation period 1990-2004. c) As in a) but for the Tejo. d) As in b) but for the Tejo. e) As in a) but for the Guadiana and for the calibration period 1953-1989. f) As in b) but for the Guadiana.

5. Conclusion

The streamflow variability and predictability of five different rivers in the Iberian Peninsula trough of a statistical multimodel approach that uses as predictor the sea surface temperatures from the preceding seasons have been investigated. Monthly data flows from the Ebro, Miño, Douro, Tejo and Guadiana Rivers have been analyzed to characterize geographic differences in terms of the streamflow seasonality and some aspects of interannual and decadal variability. The results of these analyses demonstrated that while for the Miño River, the influence of the SST on the streamflow is very moderate, for the remainder rivers, strong influences of ocean conditions are found. Particularly, for the gauging stations associated to the Basque-Cantabrian region of the Ebro basin and for the Douro, Tejo and Guadiana rivers, statistically significant linear influence was found from a tripolar spatial pattern of the anomalies in the North Atlantic SST, which is commonly called the North Atlantic horseshoe pattern (Czaja and Frankignoul, 2002). For the Basque-Cantabrian region of the Ebro, positive significant correlations between the streamflow anomalies in January and this SST pattern during the previous spring are found. This association is maximum and stable for the tropical part of the pattern (approximately 0-20°N) in previous spring, decreases in summer and autumn, and comes back strongly in December. This result is in agreement with other studies that suggest a link between spring conditions in the North Atlantic Ocean and atmospheric conditions over the same region the subsequent winter (Czaja and Frankignoul, 1999, 2002; Rodwell and Folland, 2002; Iwi et al., 2006). Rodwell and Folland (2002) showed that the North Atlantic SST in May could be used as a predictor of the subsequent winter North Atlantic Oscillation (NAO). One hypothesis is that conditions present in the North Atlantic Ocean in spring persist through the summer to influence the atmosphere in the following autumn and winter. During summer the ocean anomalies may be capped by the shallow mixed layer, subsequently to emerge as the mixed layer deepens in early winter. This mechanism is known as re-emergence (Deser et al., 2003; Cassou et al., 2004). Sutton et al. (2001) showed that a tripole pattern of SST anomalies could induce a weak NAO-like atmospheric response, and further experiments indicated the important role played by the tropical part of the pattern (Cassou et al., 2004; Peng et al., 2005).

For the Douro, Tejo and Guadiana rivers, the Atlantic horseshoe pattern in autumn has a significant influence on the variability of the following winter streamflow values. Czaja and Frankignoul (2002) also found that the negative NAO winter pattern is preceded during summer and autumn by an Atlantic SST anomaly pattern similar to that shown in Figure 3. Therefore, the positive correlation between the second autumn PC and the following winter streamflow series can be related to this phase of the NAO, which leads to positive precipitation (and streamflow) anomalies in Iberia. The opposite SST anomalies leads to the positive phase of the NAO and, then, to negative precipitation and streamflow anomalies.

The inclusion of this SST information considerable improves the skills of the forecasts compared to the ARMA-alone model forecasts. These improvements are mostly related to the ability of the SST information to provide better estimates of the extreme positive streamflow values.

These results provide information concerning the underlying physics of the teleconnection found for the streamflow and Atlantic SST, suggesting an important influence of the SLP as a main component in the SST/streamflow relationship; however, physical explanations for the connection between the SST and streamflow are not yet clear.

For the gauging stations associated to the Pyrenees region of the Ebro basin, the correlations between the June streamflow and global SST, at decadal time scales, are weaker than for those of the Basque-Cantabrian region, and only stable connections are found in the Black Sea SST during the spring of the previous year and in the south-eastern of the Pacific Ocean during the previous summer. How this connection between the SST and the flow is produced is an issue that needs further research.

For the gauging stations associated to the southeast of basin higher than average values of the Ebro river flow in May tend to be associated with cooler than average SST anomalies along the western coast of North America, and warmer than average SSTs in the central North Pacific during spring of the previous year. This Pacific SST pattern associated with the Ebro flow decadal variability bears some resemblance to the SST pattern characterizing the negative phase of the Pacific Decadal Oscillation (PDO) (Mantua et al., 1997). Currently, there are several hypotheses concerning the causes of decadal variations in different geographical regions and the influence of PDO is one of these hypotheses (Baik and Paek, 1998; Tomita et al., 2001). Other authors have also found teleconnections between the PDO and other rivers in Europe (Dettinger and Diaz, 2000; Rimbu et al., 2002; Labat, 2010). Positive phase of the PDO corresponds to the intensification of the NAO, which is accompanied by the positive anomalies of sea level pressure in the tropical and subtropical Atlantic, over central and southern Europe, and over the Mediterranean Sea. Opposite conditions correspond to the negative phase of the PDO. The basic physical proposed scheme is that the climatic anomalies of the atmosphere-ocean interaction propagate from the Pacific Ocean toward the neighbouring continents and other oceans by means of stationary Rossby waves and synoptic atmospheric formations (Ambrizzi and Hoskins, 1997).

Overall, the analysis reveals the existence of a valuable predictability of the streamflows at seasonal, interannual and decadal time scales, a result that may be useful to water resources management. The combination of different modelling gives as result a methodology that has the capacity to provide basin-specific hydroclimatic predictions for the Iberian Peninsula river streamflows, which may improve the management of the increasing limited water resources in this region. Although the research is site-specific, its conceptual basis and lessons learned should be transferable to other areas of the world facing similar problems.

From this work several issues that require further analysis can be raised. First, our results are encouraging since that only information contained in the SST has been used to explain the streamflows variability. Statistical forecasting models that use multiple variables such as land surface temperature, precipitation or SLP as potential predictor variables can improve forecast skill by increasing the robustness of the methodology. Second, at interannual time scales the direct influence of SST variability on streamflows has not been considered. However, the periodicities associated to El Niño, which oscillate between 2 and 7 years, could be related in some way with the interannual quasi-oscillatory modes found in the streamflows. Finally, further research should be conducted in order to clarify physical mechanisms that produce the connections between the SST and the streamflow in each specific basin.

6. Acknowledgment

The Spanish Ministry of Science and Innovation, with additional support from the European Community Funds (FEDER) and CGL2010-21188/CLI, financed this study. S.R. Gámiz-Fortis is supported by the University of Granada under a postdoctoral contract.

7. References

Akaike, H. (1974). A new look at the statistical model identification. *IEEE Transactions on Automatic Control Systems*, Vol.19, pp. 716– 7233, ISSN 0018-9286.

Ambrizzi, T. & Hoskins, B. (1997). Stationary Rossby-wave propagation in a baroclinic atmosphere. *Quarterly Journal of the Royal Meteorological Society*, Vol.123, No.540, pp. 919–928, ISSN 0035-9009.

Baik J.J. & Paek, J.S. (1998). A climatology of sea surface temperature and the maximum intensity of western North Pacific tropical cyclones. *Journal of Meteorologycal Society of Japan*, Vol.76, No.1, pp. 129–137, ISSN 0026-1165.

Barnston, A.G. & Livezey, R.E. (1987). Classification, seasonality, and persistence of low-frequency atmospheric circulation patterns. *Monthly Weather Review*, Vol.115, pp. 1083-1126, ISSN 0027-0644.

Barry, R. G. & Chorley, R.J. (1998). *Atmosphere, Weather and Climate*. Routledge, ISBN 0-203-87102-2, New York, EE.UU.

Bejarano, M.D.; Marchamalo, M.; García de Jalón, D. & González del Tánago, M. (2010). Flow regime patterns and their controlling factors in the Ebro basin (Spain). *Journal of Hydrology*, Vol.385, pp. 323-335, ISSN 0022- 1694.

Brockwell, P.J. & Davis, R.A. (1996). *Introduction to Time Series and Forecasting*. Springer-Verlag, ISBN 978-0-387-95351-9, EE.UU.

Cassou, C.; Deser, C.; Terray, L.; Hurrell, J.W. & Drevillon, M. (2004). Summer sea surface temperature conditions in the North Atlantic and their impact upon the atmospheric circulation in early winter. *Journal of Climate*, Vol.17, pp. 3349-3363, ISSN 0894-8755.

Cullen, H.M.; Kaplan, A.; Arkin, P. & DeMenocal, P.B. (2002). Impact of the North Atlantic Oscillation on Middle Eastern climate and streamflow. *Climate Change*, Vol. 55, pp. 315-338, ISSN 0165-0009.

Czaja, A. & Frankignoul, C. (2002). Observed impact of Atlantic SST anomalies on the North Atlantic Oscillation. *Journal of Climate*, Vol.15, No.6, pp. 606-623, ISSN 0894-8755.

Deser, C. & Blackmon, M. (1993). Surface climate variations over the North Atlantic Ocean during winter: 1900-1989. *Journal of Climate*, Vol. 6, pp. 1743-1753, ISSN 0894-8755.

Deser, C.; Alexander, M.A. & Timlin, M.S. (2003). Understanding the persistence of sea surface temperature anomalies in midlatitudes. *Journal of Climate*, Vol.16, pp. 57-72, ISSN 0894-8755.

Dettinger, M.D. & Diaz, H.F. (2000). Global characteristics of streamflow seasonality. *Journal of Hydrometeorology*, Vol.1, pp. 289-310, ISSN 1525-755X.

Draper, N.R. & Smith, H. (1998). *Applied Regression Analysis*. J. Wiley and Sons, New York.

Esteban-Parra, M. J.; Rodrigo, F.S. & Castro-Díez, Y. (1998) Spatial and temporal patterns of precipitation in Spain for the period 1880-1992. *International Journal of Climatology*, Vol.18, pp. 1557-1574, ISSN 1097-0088.

Gámiz-Fortis, S.R.; Pozo-Vázquez, D.; Esteban-Parra, M.J. & Castro-Díez, Y. (2002). Spectral characteristics and predictability of the NAO assessed through Singular Spectral Analysis. *Journal of Geophysical Research*, Vol.107, No.D23, pp. 4685, ISSN 0148-0227.

Gámiz-Fortis, S.R.; Pozo-Vázquez, D.; Trigo, R.M. & Castro-Díez, Y. (2008a). Quantifying the predictability of winter river flow in Iberia. Part I: interannual predictability. *Journal of Climate*, Vol.21, pp. 2484-2502, ISSN 0894-8755.

Gámiz-Fortis, S.R.; Pozo-Vázquez, D.; Trigo, R.M. & Castro-Díez, Y. (2008b). Quantifying the predictability of winter river flow in Iberia. Part II: seasonal predictability. *Journal of Climate*, Vol.21, pp. 2503-2518, ISSN 0894-8755.

Gámiz-Fortis, S.R.; Esteban-Parra, M.J.; Trigo, R.M. & Castro-Díez, Y. (2010a). Potential predictability of an Iberian river flow based on its relationship with previous winter global SST. *Journal of Hydrology*, Vol.385, pp. 143-149, ISSN 0022- 1694.

Gámiz-Fortis, S.R.; Esteban-Parra, M.J.; Pozo-Vázquez, D. & Castro-Díez, Y. (2010a). Variability of the monthly European temperature and its association with the Atlantic sea-surface temperature from interannual to multidecadal scales. *International Journal of Climatology*, Vol.31, pp. 2115-2140, ISSN 1097-0088.

Gámiz-Fortis, S.R.; Hidalgo-Muñoz, J.M.; Argüeso, D.; Esteban-Parra, M.J. & Castro-Díez, Y. (2011b). Spatio-temporal variability in Ebro river basin (NE Spain): global SST as potential source of predictability on decadal time scales. *Journal of Hydrology*, Vol.409, pp. 759-775, ISSN 0022- 1694.

Ghil, M. & Mo, K. (1991). Intraseasonal oscillations in the global atmosphere-Part I: Northern Hemisphere and tropics. *Journal of the Atmospheric Sciences*, Vol.48, pp. 752-779, ISSN 0022-4928.

Ghil, M. & Vautard, R. (1991). Interdecadal oscillations and the warming trend in global temperature series. *Nature*, Vol.310, pp. 324–327, ISSN 0028-0836.

Goossens, C. (1985). Principal component Analysis of the Mediterranean rainfall. *International Journal of Climatology*, Vol.5, pp. 379-388, ISSN 1097-0088.

Graham, N.E.; Barnett, T.P.; Wilde, R.; Ponater, M. & Schubert, S. (1994). On the roles of tropical and midlatitude SSTs in forcing interannual to interdecadal variability in the winter Northern Hemisphere circulation. *Journal of Climate*, Vol.7, pp. 1416-1441, ISSN 0894-8755.

Hipel, K.W. & Mcleod, A.I. (1994). *Time Series Modelling of Water Resources and Environmental Systems*. Elsevier, ISBN 978-0444892706.

Houghton, R.W. (1996). Subsurface quasi-decadal fluctuations in the North Atlantic. *Journal of Climate*, Vol. 9, pp. 1363-1373, ISSN 0894-8755.

Ionita, M.; Lohmann, G. & Rimbu, N. (2008). Prediction of Spring Elbe Discharge Based on Stable Teleconnections with Winter Global Temperature and Precipitation. *Journal of Climate*, Vol.21, No.23, pp. 6215-6226, ISSN 0894-8755.

Ionita, M.; Rimbu, N. & Lohmann, G. (2011). Decadal variability of the Elbe River streamflow. *International Journal of Climatology*, Vol.31, No.1, pp. 22–30, ISSN 1097-0088.

Iwi, A.M.; Sutton, R.T. & Norton, W.A,(2006). Influence of May Atlantic Ocean initial conditions on the subsequent North Atlantic winter climate. *Quarterly Journal of the Royal Meteorological Society*, Vol.132, pp. 2977-2999, ISSN 0035-9009.

Jacobs, G.A.; Hurlbert, J.C.; Kindle, J.C.; Metzger, E.J.; Mitchell, J.L.; Teague, W.J. & Wallcraft, A.J. (1994). Decade-scale trans-Pacific propagation and warming effects of an El Niño anomaly. *Nature*, Vol.370, pp. 360-363, ISSN 0028-0836.

Kalayci, S. & Kahya, E. (2006). Assessment of streamflow variability modes in Turkey: 1964-1994. *Journal of Hydrology*, Vol.324, pp. 163-177, ISSN 0022- 1694.

Koster, R.D.; Mahanama, S.P.P.; Livneh, B.; Lettenmaier,D.P. & Reichle, R.H. (2010). Skill in streamflow forecasts derived from large-scale estimates of soil moisture and snow. *Nature Geoscience*, Vol.3, pp. 613-616, ISSN 1752-0894.

Labat, D. (2010). Cross wavelet analyses of annual continental freshwater discharge and selected climate indices. *Journal of Hydrology*, Vol.385, pp. 269–278, ISSN 0022- 1694.

Latif, M. & Barnett, T.P. (1994). Causes of decadal climate variability over the North Pacific and North America. *Science* , Vol.266, pp. 634-637, ISSN 0036-8075.

Leite, S.M. & Peixoto, J.P. (1995). Spectral analysis of climatological series in Duero Basin. *Theorical and Applied Climatology*, Vol.50, pp. 157–167, ISSN 0177-798X.

López-Moreno, J.I.; Vicente-Serrano, S.M.; Beguería, S.; García-Ruiz, J.M.; Portela, M.M. & Almeida, A.B. (2009). Dam effects on droughts magnitude and duration in a transboundary basin: The Lower River Tagus, Spain and Portugal. *Water Resources Research*, Vol.45, pp. W02405., ISSN 0043–1397.

Mantua, N.J.; Hare, S.R.; Zhang, Y.; Wallace, J.M. & Francis, R.C. (1997). A Pacific decadal climate oscillation with impacts on salmon. *Bulletin of the American Meteorological Society*, Vol.78, pp. 1069-1079, ISSN 1520-0477.

Maurer, E.P.; Lettenmaier, D.P. & Mantua, N.J. (2004). Variability and potential sources of predictability of North American runoff. *Water Resources Research*, Vol.40, pp. W09306, ISSN 0043–1397.

Peng, S.; Robinson, W.A.; Li, S.L. & Hoerling, M.P. (2005). Tropical Atlantic SST forcing of coupled North Atlantic seasonal responses. *Journal of Climate*, Vol.18, pp. 480-496, ISSN 0894-8755.

Plaut, G.; Ghil, M. & Vautard, R. (1995). Interannual and Interdecadal Variability in 335 years of Central England Temperatures. *Science*, Vol.268, pp. 710-713, ISSN 0036-8075.

Plaut, G. & Vautard, R. (1994). Spells of low-frequency oscillations and weather regimes in the northern hemisphere. *Journal of the Atmospheric Sciences*, Vol.51, No.2, pp. 210-236, ISSN 0022-4928.

Rayner, N.A.; Parker, D.E.; Horton, E.B.; Folland, C.K.; Alexander, L.V.; Rowell, D.P.; Kent, E.C. & Kaplan, A. (2003). Globally complete analyses of sea surface temperature, sea ice and night marine air temperature, 1871-2000. *Journal of Geophysical Research*, Vol.108, pp. 4407, ISSN 0148–0227.

Rimbu, N.; Boroneant, C.; Buta, C. & Dima, M., 2002. Decadal variability of the Danube river flow in the lower basin and its relation with the North Atlantic Oscillation. *International Journal of Climatology*, Vol.22, pp. 1169–1179, ISSN 1097-0088.

Rimbu, N.; Dima, M.; Lohmann, G. & Musat, I. (2005). Seasonal prediction of Danube flow variability based on stable teleconnection with sea surface temperature. *Geophysical Research Letters*, Vol.32, pp. L21704, ISSN 0094-8276.

Rodwell, M.J. & Folland, C.K. (2002). Atlantic air-sea interaction and seasonal predictability. *Quarterly Journal of the Royal Meteorological Society*, Vol.128, pp. 1413-1443, ISSN 0035-9009.

Serrano, A.; Garcia, A.J.; Mateos, V.L.; Cancillo, M.L. &. Garrido, J. (1999) Monthly modes of variation of precipitation over the Iberian Peninsula. *Journal of Climate*, Vol.12, pp. 2894-919, ISSN 0894-8755.

Sutton, R.T.; Norton, W.A. & Jewson, S.P. (2001). The North Atlantic Oscillation – What Role for the Ocean? *Atmospheric Science Letters*, Vol.1, pp. 89-100, ISSN 1530-261X.

Switanek, M.B.; Troch, A. & Castro, C.L. (2009). Improving Seasonal Predictions of Climate Variability and Water Availability at the Catchment Scale. *Journal of Hydrometeorology*, Vol.10, pp. 1521-1533, ISSN 1525-755X.

Tanimoto, Y.; Iwasaka, N.; Hanawa, K. & Toba, Y. (1993). Characteristic variations of SST with multiple time scales in the North Pacific. *Journal of Climate*, Vol. 6, pp. 1153-1160, ISSN 0894-8755.

Tomita, T.; Wang, B.; Yasunari, T. & Nakamura, H. (2001). Global patterns of decadal-scale variability observed in sea-surface temperature and lower-tropospheric circulation fields. *Journal of Geophysical Research*, Vol.106, No.C11, pp. 26805–26815, ISSN 0148-0227.

Tootle, G.A. & Piechota, T.C. (2006). Relationships between Pacific and Atlantic ocean sea surface temperatures and U.S. streamflow variability. *Water Resources Research*, Vol.42, pp. W07411, ISSN 0043-1397.

Trenberth, K.E. & Hoar, T.J. (1996). The 1990-1995 El Niño-Southern Oscillation event: Longest on record. *Geophysical Research Letters*, Vol. 23, pp. 57-60, ISSN 0094–8276.

Trigo, R.M.; Pozo-Vázquez, D.; Osborn, T.J.; Castro-Díez, Y.; Gámiz-Fortis, S. & Esteban-Parra, M.J. (2004). North Atlantic Oscillation influence on precipitation, river flow and water resources in the Iberian Peninsula. *International Journal of Climatology*, Vol.24, pp. 925-944, ISSN 1097-0088.

Vautard, R. & Ghil, M. (1989). Singular spectrum analysis in non-linear dynamics with applications to paleoclimatic time series. *Physica D*, Vol.35, pp. 395– 424, ISSN 0167-2789.

Vautard, R.; Yiou, P. & Ghil, M. (1992). Singular spectrum analysis: A toolkit for short, noisy chaotic signal. *Physica D*, Vol. 58, pp. 95–126, ISSN 0167- 2789.

Vicente-Serrano, S.M. (2006). Differences in spatial patterns of drought on different time scales: an analysis of the Iberian Peninsula. *Water Resources Management*, Vol.20, pp. 37-60, ISSN 0920-4741.

Vicente-Serrano, S. & López-Moreno, J.I. (2006). Influence of atmospheric circulation at different spatial scales on winter drought variability through a semiarid climatic gradient in Nort-east Spain. *International Journal of Climatology*, Vol.26, No.11, pp. 1427-1453, ISSN 1097-0088.

Widmann, M.L. & Schär, C. (1997). A principal component and long-term trend analysis of daily precipitation in Switzerland. *International Journal of Climatology*, Vol.17, pp. 1333-1356, ISSN 1097-0088.

Wilks, D.S. (1995). *Statistical Methods in the Atmospheric Sciences*. Academic Press, San Diego, CA 467.

Zaidman, M.D; Rees, H.G. & Young, A.R. 2001. Spatio-temporal development of streamflow 25 droughts in north-west Europe. *Hydrology and Earth System Sciences*, Vol.5, No.4, pp. 733–751, ISSN 1027-5606.

6

Nonparametric Statistical Downscaling of Precipitation from Global Climate Models

Paul Nyeko-Ogiramoi[1,2,3], Patrick Willems[2],
Gaddi Ngirane-Katashaya[3] and Victor Ntegeka[2]
*[1]Ministry of Water and Environment, Directorate of Water Development,
Rural Water and Sanitation Department, Kampala,
[2]Katholieke Universiteit Leuven, Civil Engineering Department,
Hydraulics Laboratory, Leuven,
[3]Makerere University, Civil Engineering Department, Kampala,
[1,3]Uganda
[2]Belgium*

1. Introduction

Assessment of climate change impacts on hydrometeorological variables such as rainfall and temperature at regional or local (catchment) scale requires projected future time series. One of the common sources of such future time series are Global Climate Model experiments (GCM runs). However, direct use of GCM runs may not be appropriate for climate change impacts assessment at catchment scale because the scales in GCMs are not at par with the scale at catchment level. For example, if the magnitude of the biases in rainfall and temperature is very high, there is a tendency for the impact signals in the GCM runs to be amplified under very wet and dry conditions (Christensen et al., 2008). Thus, the need for circumventing the biases in or downscaling the GCM runs. Once projected future time series are derived through downscaling, they can either be assessed for impacts by comparing them with the observed or used as inputs into a rainfall-runoff model in order to obtain future streamflow time series. The latter can be compared with the present day control streamflows; hence impacts on streamflows can be assessed. Therefore, methods are needed to downscale output from GCM to represent local climate variables.

Downscaling can be dynamical through the use of an RCM (Regional Climate Model) with output from GCM as the boundary condition (e.g. Christensen et al., 2007) or through statistical (empirical) methods conditioned on large-scale predictors (e.g. Fowler et al., 2007). In most cases, outputs from RCM are also biased and bias-removal (e.g. Piani et al., 2010) is often employed. Often, this method involves some form of transfer function derived from the observed and simulated cumulative distribution functions (e.g. Ines & Hansen, 2006). This method is given a wide range of names in literature such as statistical downscaling, quantile mapping, histogram equalizing and rank matching among others. In applying a hindcast-derived correction to simulations of projected climate, it is assumed that the correction still holds for the projected climate. This is a non trivial assumption (Trenberth et al., 2003). However, the assumption is more plausible provided the transfer function between the raw and the corrected RCM output is robust. In many regions of the world (e.g. Africa) data limitation will continue to constrain the calibration of RCMs (Anyah & Semazzib, 2007; Christensen et al., 2007) and use of other methods is sought. Bias correction method is sometimes applied to GCM data (e.g. Ines & Hansen, 2006).

In a statistical regression method, a purely statistical relation is sought between a model field that is well-represented on the large-scale (predictor), e.g., sea-level pressure or the height of the 500 hPa level, and the local quantity of interest (predictant), e.g., precipitation or temperature (Benestad et al., 2008). Assuming that the relation still holds in the changed climate, the changes of the large-scale circulation are translated into the local changes that are of interest. Note that the predictant needs not be a variable of the global model, but can be anything related to climate. The great advantage of the statistical method is that it is easy and computationally simple. Nevertheless, (long) time series of both the predictor and the predictant are needed to calibrate the regression model. While the first requirement is usually not a problem, the second limits the applicability of the method to a limited number of places and to surface variables only.

The "delta" downscaling technique uses the concept of change factors (multiplicative or additive) extracted from the climate models and applied to observed time series. The former has been assessed by several researchers (e.g. Diaz-Nieto and Wilby, 2005; Lenderin et al., 2007). The traditional delta technique applies the changes to a time series without considering the variability of the time series. The technique assumes that relative changes obtained from the climate models are more representative than the absolute ones. Furthermore, it is assumed that the biases in the control (present) simulations are similar to the biases in the future simulations. Moreover, the temporal structure of the derived time series is maintained. With significant changesin the variability of time series under climate change the delta method may not be suitable. The earlier attempts to improve on the approach included examining various scenarios (Prudhomme et al., 2002), and applying quantile scaling techniques (Harrold and Jones, 2003) to account for the variability in the time series. Olsson et al. (2009) and Willems and Vrac (2011) demonstrated that deriving future time series that considers changes in extremes is possible through the use of exceedance probabilities. This approach ensures that there is increased variability in heavy rainfall amounts compared to meek rainfall events. This approach is simple, robust and can be applied to any set of data without worrying about the length of a previous record. However, the omission of other changes such as the wet spells, which has very strong association with the extremes in precipitation downscaling, makes the approach further faulty. Accounting for change in wet spells can improve precipitation downscaling and impacts results focusing on extremes.

Previous studies have used and compared different downscaling techniques and have concluded that the choice of the method is dependent on the nature of the study and that more research is needed to improve on available downscaling techniques (Fowler et al., 2007). The tenet of this chapter is centered on the quantile perturbation technique because it is often used to assess the impact of climate change on extremes of the hydrometeorological variable such as precipitation. Improvement of the technique is here suggested and applied to data from a catchment in a tropical climate.

2. Case study

The downscaling technique described in this chapter is illustrated based on a case study of the upper River Nile basin, Lake Victoria basin (Figure 6.1). The River Nile basin (Figure 6.1(a)) is situated between 8°S to 33°N and 20°E to 42°E covering an area of approximately 3,762,000 km² (Figure 6.1(b)). Lake Victoria (Figure 6.1(c)) is the largest fresh water body in Africa and a constituent of the most upper part of the River Nile basin. Lake Victoria is geopolitical in nature and characterized by Kenya, Uganda and Tanzania with Rwanda and Burundi as key sources of the famous River Kagera, the major contributor to the lake Figure 6.1(a). Lake Victoria is, on average, 68,800 km², and is politically shared as follows: Kenya (6%), Uganda (43%) and Tanzania (51%).

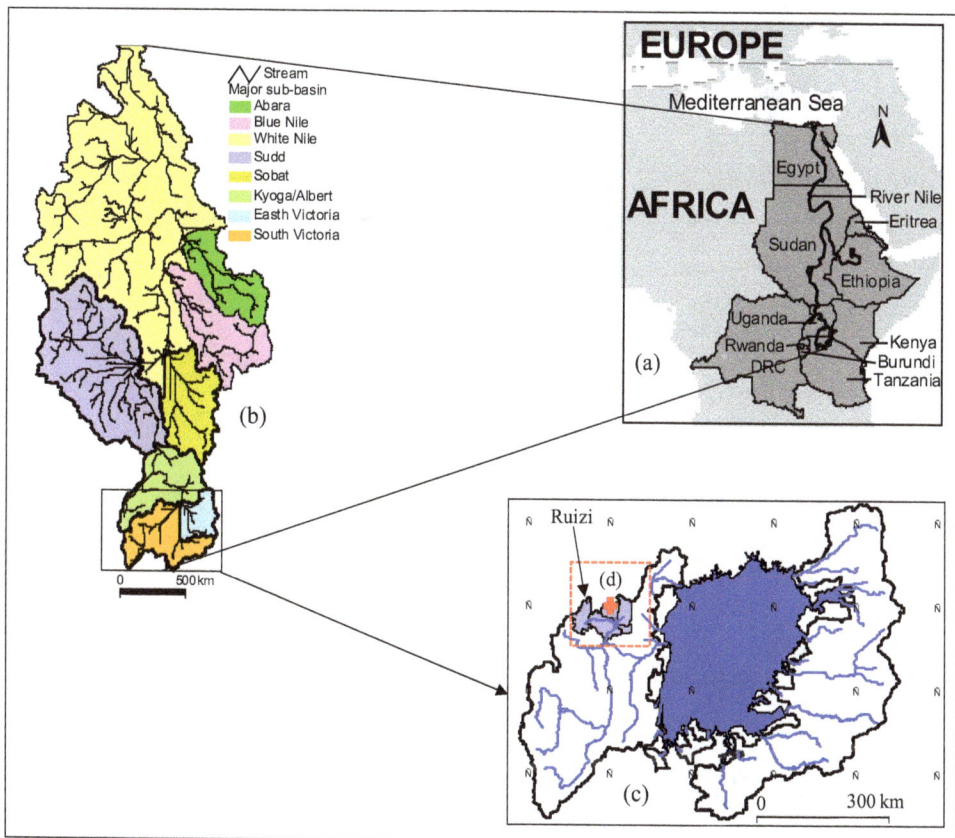

Fig. 6.1. Location of River Nile in Africa, the countries in which the Nile basin takes part (a), the major streams and major catchments of the River Nile (b), the Lake Victoria basin (c) and River Ruizi catchment. The plus (+) signs, superimposed over Lake Victoria basin and River Ruizi catchment indicate the mean grid size of the GCMs used. The dotted square grid with a center at • indicates the typical GCM grid used in the case study.

The basin area in Uganda is 59,858 km² out of about 196,000 km². Lake Victoria stretches approximately 415 km from north to south; between latitudes 0°30′ N and 3°12′ S and from west to east between longitudes 31°37′ and 34°53′ E. It is situated at an altitude of about 1,130 m above sea level, and has an estimated volume of about 2,750 km², and an average and maximum depth of 40 m and 80 m, respectively. The area of the River Ruizi catchment indicated in Figure 6.1(d) is about 6,000 km² and the observed daily rainfall data of the measuring stations falling within the catchment was used to demonstrate how the GCM data can be downscaled to catchment scale.

3. GCM data

The information of the GCM data (Table 6.1) used here were obtained from the Programme for Climate Model Data Intercomparison (PCMDI) database[1]. The data were used for the Fourth Assessment Report of the Intergovernment Pannel on Climate Change (AR4 IPCC), 2007. There are, however, several public domain databases from which one can obtain either observed climate data and/or climate model data.

Modeling Institution, Country	Center acronym	Model	Model Resolution Lon	Lat	20C3M	A2	A1B	B1
Bjerknes Centre for Climate Research, Norway Norway	BCCR	BCM2.0	2.813	2.791	✔	✔	✔	✔
Canadian Centre for Climate Modelling & Analysis, Canada	CCCma	CGCM3.1(T47)	3.750	3.711	✔	✔	✔	✔
		CGCM3.1(T63)	3.750	3.711	✔	✘	✔	✔
Centre National de Recherches Meteorologiques, France	CNRM	CM3	2.813	2.791	✔	✔	✔	✔
Australia's Commonwealth Scientific and Industrial Research Organisation, Australia	CSIRO	Mk3.0	1.875	1.865	✔	✔	✔	✔
		Mk3.5	1.875	1.865	✔	✔	✔	✔
Max-Planck-Institute for Meteorology, Germany	MPI-M	ECHAM4-OM	3.750	3.711	✔	✔	✘	✔
		ECHAM5-OM	1.875	1.865	✔	✔	✔	✔
Research Institute of KMA, Germany/Korea	MIUB	ECHO-G	3.750	3.711	✔	✔	✔	✔
Institute of Atmospheric Physics, China	LASG	FGOALS-g1.0*	2.500	2.022	✔	✘	✔	✔
Geophysical Fluid Dynamics Laboratory, USA	GFDL	CM2.0	2.500	2.022	✔	✔	✔	✔
		M2.1	2.500	2.022	✔	✔	✔	✔
Goddard Institute for Space Studies, USA	GISS	AOM	3.750	3.711	✔	✘	✔	✔
		E-H	5.000	4.000	✔	✘	✔	✘
		E-R	5.000	4.000	✔	✔	✘	✘
Institute for Numerical Mathematics, Russia	INM	CM3.0	5.000	4.000	✔	✔	✔	✔
Institut Pierre Simon Laplace, France	IPSL	CM4	3.750	2.535	✔	✔	✔	✔
National Institute for Environmental Studies, Japan	MRI	CGCM2.3.2a*	2.813	2.791	✔	✔	✔	✔
Meteorological Research Institute, Japan	NIES	MIROC3.2(h)	2.813	2.791	✔	✘	✔	✔
		MIROC3.2(m)	2.813	2.791	✔	✔	✔	✔
National Center for Atmostpheric Research, USA	NCAR	CCSM3.0*	1.406	1.401	✔	✔	✔	✔
		PCM1*	2.813	2.791	✔	✔	✔	✔
Hadley Centre for Climate Prediction and Research Met Office, United Kingdom	UKMO	HadCM3	3.750	2.750	✔	✘	✘	✘
		HadGEM1	1.875	1.250	✔	✘	✘	✘

Table 6.1. Information on the GCMs and scenarios for the data available at PCMDI. 20C3M is the simulation of the 20th century climate using historical GHG (Green House Gases) concentrations for the period 1871-2000.

[1] The website for PCMDI database is "http://www2-pcmdi.llnl.gov/esg_data_portal", last accessed 13 June 2011.

For its Third Assessment Report, the IPCC commissioned a Special Report on Emissions Scenarios (SRES) which developed about forty different emissions scenarios (Nakicenovic et al., 2000). Of these forty emissions scenarios, six have been chosen as illustrative or marker scenarios: A1FI, A1B, A1T, A2, B1 and B2. Of these six marker scenarios most global climate modelling groups completed climate change simulations using A2, A1B and B1 emissions scenarios in the AR4 IPCC, 2007 (Table 6.2). The choice of the domain is dependent on the purpose for which the climate data is required, the spatial coverage of the data, the temporal resolution, user interest, data completeness and restrictions. The GCM data are coded (mainly in netcdf format, ".nc") and some have quality problems (e.g. error in naming of files, some precipitation values being negatives, etc.). The data were processed and checked for their quality improvements using a cross-pollination of techniques, experiences and tools. For example, CDO (Climate Data Operator) (Schulzweida & Kornblueh, 2011) is a powerful tool for manipulating grid data of different formats. In addition, many other climate data processing tools, such as the NCDF, can be annexed to MATLAB (MathWorks Inc., 2008) and are very instrumental in climate model data processing.

Scenario	Data set	Description	Simulation period	Data available for
20CM3	20th Century simulation	Model input forcings or initial conditions (e.g., solar irradiance, ozone, sulfates, greenhouse gases) are temporally and spatially varied	1870-2000	1961-2000
SRES B1	550 ppm CO_2 maximum (SRES B1)	Atmospheric CO_2 concentrations reached 550 ppm in the year 2100 in a world characterized by low population growth, high GDP growth, low energy use, high land-use changes, low resource availability and medium introduction of new and efficient technologies.	2001-2100	2046-2065 2081-2100
SRES A1B	720 ppm CO_2 maximum (SRES A1B)	Atmospheric CO_2 concentrations reach 720 ppm in the year 2100 in a world characterized by low population growth, very high GDP growth, very high energy use, low land-use changes, medium resource availability and rapid introduction of new and efficient technologies.	2001-2100	2046-2065 2081-2100
SRES A2	850 ppm CO_2 maximum (SRES A2)	Atmospheric CO_2 concentrations reach 850 ppm in the year2100 in a world characterized by high population growth, medium GDP growth, high energy use, medium/high land-use changes, low resource availability and slow introduction of new and efficient technologies.	2001-2100	2046-2065 2081-2100

Table 6.2. Definitions of the SRES scenarios (Nakicenovic et al., 2000).

Although the IPCC (2001a) recommended a 30-year period (1961-1990) as sufficient to measure climate and detect climate change, challenges in climate modelling and archiving of global climate data have forced the global community to consider a 20-year period as plausible for impacts assessment. This is due to the fact that deviation from using a 20-year period from that of using a 30-year period is not very significant. A number of fixed time horizons in the future are considered in literature, especially in the recent AR4 IPCC, e.g., the 2020s (2010-2039), the 2050s (2040-2069), and the 2080s (2070-2099). These are future periods for which any assessment of climate change impacts can be made. Similar to the adjustment of future periods, reference periods have followed suit and quite often studies have adopted 1961-1980, 1971-1990 and 1981-2000. However, in the case where there are sufficient data, it is strongly recommended that a 30-year period for the current (control) and a 30-year period for the future (scenario) climate be considered.

The GCMs provide area averaged data and this means that for the evaluation purpose, the rainfall measured at a point, for example, requires scaling if areal rainfall measurement is not available. The scaled point rainfall intensity accounts for the expected systematic difference between the point rainfall and the grid averaged GCM rainfall. Point rainfall can be scaled to area averaged rainfall by either applying areal reduction factor (Svensson & Jones, 2010) or by spatial interpolation of point rainfalls using the technique such as Thiessen polygon to obtain areal rainfall. The points (measurement locations) under consideration should all fall within the GCM grid boundary. Thiessen polygon is a preferred method for estimating areal rainfall and was used in this study (Figure 6.1).

Statistics of errors, biases, correlations and trends have been used to quantify statistical inconsistencies between the model simulation and the historical time series. (e.g. Nyeko-Ogiramoi et al., 2010). The bias in the GCMs can not be ignored; especially regarding their selection and application in impact assessment. First, models that are extremely biased can be sieved out of impact assessment. Secondly, the bias in the GCMs means that it is inappropriate to use outputs from GCMs directly for impact assessment at local scale. Furthermore, Christensen et al. (2008) noted that significantly biased models for the current and past climate have potential for transferring significant bias to the future. However, because the pattern of the current and past climate seems to be well represented in the GCM, its pattern prediction (or signal) may be reliable. Conventionally, GCMs are most often assessed for the historical performance alone. However, for a more robust impact assessment, the inter-comparison of the future projections is also vital (Nyeko-Ogiramoi, et al., 2010). This is because models that have good ability in estimating the observed rainfall may not necessarily produce robust predictions (close to or in same direction as other models). Models which perform well for historical periods but are projecting disparate future changes should be further examined for performance. In intermittent variables such as rainfall, the complex climate system may introduce inconsistencies for the future climate. Disqualifying a model, however, from further analysis because it is inconsistent with other models is only vindicated if further examination shows a previously overlooked bias against the observed data (Nyeko-Ogiramoi et al., 2010). The inter-comparison of the projections aims to increase the confidence in the GCM projections while eliminating spurious projections. In addition, the differences in the future projections (controls and projections) if combined with the differences in the current simulation (controls and observation) can enhance the understanding of the effects of the model bias on the future projections. Evaluation studies are valuable as they identify the weaknesses and spot the models whose performance is questionable. The inconsistencies of the AR4 GCMs with the observation over the Lake Victoria basin suggest further tasks for the climate model scientist. That is, further improvements for the GCMs are necessary to increase on the confidence for the assimilation of their outputs. However, the performance of a climate model is regional or catchment based and should be treated as such; the conclusions of performance are mainly valid for the studied area. The models marked with "*" (Table 6.1) are the most biased with respect to rainfall over the Lake Victoria basin (Nyeko-Ogiramoi et al., 2010).

4. Synchrony between wet extremes and wet spells

Analysis of projected changes in climate by Meehl et al. (2007) showed that the type, frequency and intensity of extreme events are expected to change even with relatively small mean climate changes. Meehl et al. (2007) further noted that in a warmer world, precipitation tends to be concentrated into more intense events, with longer periods of light precipitation in between. Thus, intense and heavy downpours would be interspersed with longer relatively dry periods. Furthermore, Meehl et al. (2007) noted that wet extremes are projected to be more severe where mean precipitation is expected to increase and dry extremes are projected to become more severe in areas where mean precipitation is projected to decrease. In concert with the results of projected increased extremes of intense precipitation, even if the wind strength of storms in a future climate did not change, there would be an increase in rainfall intensity (Meehl et al., 2007). Kharin & Zwiers (2005) and Barnett et al. (2006) noted also that the increase in extreme events may be perceived most through the impacts of extremes. The implications are that changes in wet extremes are, quite often, associated with longer wet spells. The need to consider the characteristics of the wet spells is paramount in improving the precipitation downscaling techniques that employ the quantile perturbation approach. Thus, this chapter mainly explores the technique of employing the characteristics of wet spells in and for the improvement of the quantile perturbation approach for downscaling precipiation.

5. Perturbations

In this chapter, perturbation refers to any change that can be obtained from the GCM scenario run in comparison with the control or from the observed time series. That is, the properties (or statistics) of the GCM control run (series) are compared with those of the GCM future series to obtain changes (perturbations) that are projected under the different climate change scenarios. Note that changes that occur in quantitative hydrometeorological variables such as rainfall and evapotranspiration can be obtained in many forms (e.g. ratios, percentage, difference, etc.). Meanwhile, changes that occur in qualitative meteorological variables such as temperature can mainly be derived as differences. Different perturbations for rainfall and temperature time series can be extracted and analysed for different months or seasons and for different GCMs. In the context of this study, emphasis is put on analysing perturbations for daily rainfall because it is an essential variable of climate which is needed for hydrological impacts of climate change. The main principle behind perturbation is that several perturbations can be isolated or extracted from GCM paired (control and scenarios) data at different aggregation levels and can be analysed for their respective properties. In the subsequent sections perturbations for rainfall are analysed for daily rainfall for each month. The results are presented only for the months of January and April for typical illustrations. This is because of the fact that January and April are considered representative of dry season (dry months) and wet season (wet months), respectively, of the climate of the case study.

5.1 Rainfall perturbations

5.1.1 Rainfall wet-days quantile perturbations

Derivation and analysis of rainfall perturbation can be performed for rainfall time series at different time scales. However, in this case, the focus is on a daily time scale and the perturbations were derived by considering daily rainfall time series for each month, separately. In other words, the time series for each month for all the considered years are pooled together before perturbations are derived. Perturbations are derived only for the wet-day frequency of the rainfall series. The definition of a wet day is given in the next section. If $q_{s1} \geq q_{s2} \geq q_{s3} \ldots q_{sn}$ represent the scenario quantiles and $q_{c1} \geq q_{c2} \geq q_{c3} \ldots q_{qn}$ represents the control quantiles then quantile perturbations are derived as q_{si} / q_{ci}, for $i = 1$, 2, 3, ...n. Alternatively, perturbations can be derived as $P_p = Q_{s,p} / q_{c,p}$, where P_p is the perturbation corresponding to probability p, $Q_{s,p}$ is the future scenario value (corresponding to probability p) and $q_{c,p}$ the control value (for the same probability p). The plot of quantile perturbations versus the return period (or exceedance probability) can reveal the effect of projected possible warming for the future.

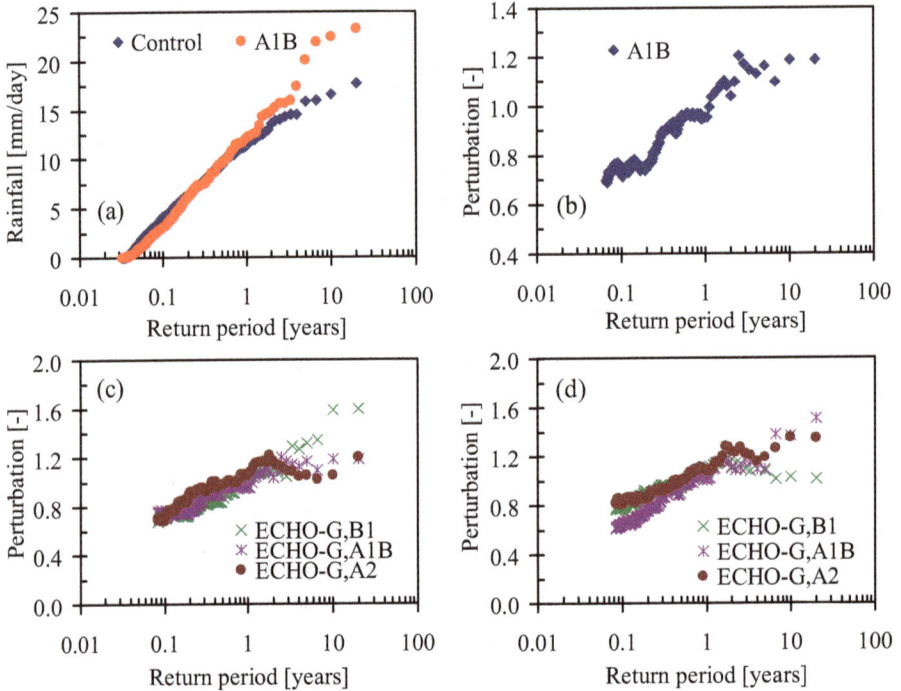

Fig. 6.2. Distributions for January rainfall for control and scenario (A1B) (a), and the corresponding perturbations under A1B scenario (b), and 3 scenarios for periods: 2050s (c), and 2090s (d). The control and the scenarios are for one GCM (ECHO-G) to illustrate the typical perturbations for the two periods. The results are for the GCM data extracted from a grid over the River Ruizi catchment.

Fig. 6.2 shows an illustration of the January daily rainfall distributions for a 20-year period (Fig. 6.2(a)) and the corresponding plot of perturbation versus return period (Fig. 6.2(b)). Perturbations for different scenarios and the corresponding plots for different future scenarios, projected for the periods 2050s (Fig. 6.2(c)), and 2090s (Fig. 6.2(d)), can similarly be obtained. Note that the daily rainfall distribution and the perturbation plots, provided in

Fig. 6.2, represent evolutional stages of rainfall perturbations analysis of an example of one GCM (ECHO)-G). Perturbations analysis for other GCMs can also be obtained and analysed in a similar way.

5.1.2 Dependency of rainfall wet-days quantile perturbations

The effects of climate change on the mild and heavy rainfall events can be exposed by examining the perturbation plot. Such effects can be analysed for different temporal scales such as daily, weekly, monthly and seasonal. For a tropical climate, where the interest of this study lies, the climate is characterized mainly by wet and dry seasons. However, it is important to note that for a given season, daily rainfall variations among the months can be very high. Thus, analysis of perturbations for daily rainfall series for each month is particularly important.

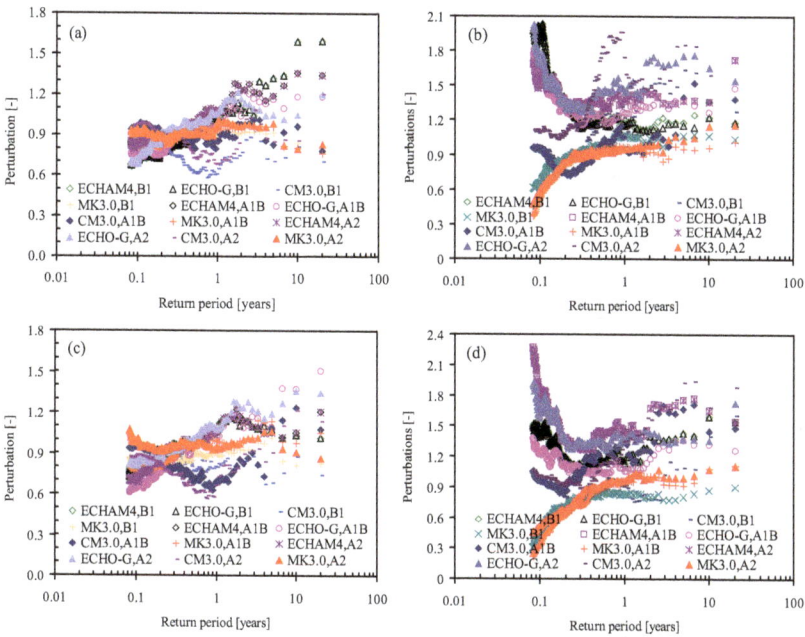

Fig. 6.3. Perturbations for the daily rainfall: (a) January, 2050s, (b) April, 2050s, (c) January, 2090s and (d) April, 2090s. The perturbations are for 4 GCMs to illustrate the typical perturbations for a relatively dry month (January) and a relatively wet month (April). The results are for data extracted from a grid over the Ruizi catchment

Fig. 6.3(a) and (b) shows plots of perturbation versus return period for the months of January and April, respectively, for the period 2050s. For January, the perturbations are generally greater than 1 for the heavy rainfall events (> 1 year). However, the perturbations are less than 1 for lighter rainfall events (< 0.2 years). The perturbations for the mean rainfall events are fairly constant (0.2-0.4 years). It is important to note that for some GCMs (models) the perturbations are less than 1 between 1.1-1.4 years return periods. This variation of the perturbations for the heavy, mean and light rainfall events has implications for the changes in the rainfall intensity and frequency. First, it is expected that given a rainfall intensity of a return period less than 1 year, the intensity is projected to decrease from the current to the future. Secondly, given a rainfall above one year return period, the intensity is projected to increase from the current value to the future for the most heavy rainfall events. In the latter case, the rainfall intensity for the medium heavy rainfall events is expected to decrease as projected by some GCMs. Thus, for the dry months, the dry days will become moderately dryer and the very wet days will become much wetter. In some cases the frequency of such wetter events will increase.

For the month of April (Fig. 6.3(b)) the perturbations are dramatically higher than 1 for return periods < 0.2 years. This is the case for 9 out of 12 selected GCM runs irrespective of the scenarios. For return periods > 0.2 years the perturbations are generally > 1 but increase with return period moderately. For three models, perturbations are less than 1 for return periods < 1 years. The variation of the perturbations with return period for the month of April has implication for the projected changes in the rainfall intensity and frequency. The rainfall events are expected to strongly increase compared to the current but with moderate increase in their frequencies. Thus, more wet days are projected for the wetter months and the rainfall intensity is projected to increase compared to the current. However, the projections by some models indicate that as the intensity and frequency of the heavy events strongly increase and the magnitude of the mean rainfall events remains constant, the rainfall intensity and frequency of the light rainfall events is projected to decrease compared to the current. In the latter case, it means that lesser amounts of rainfall is projected (dry days becoming dryer) coupled with a decrease in frequency. It is noteworthy, however, that perturbation become very sensitive (very high factors) as the return period decreases and the lighter events for the control series tend to zero. This is particularly important to note because further discussions are given in section 8.1.1.

The differences in the variation of the perturbations for the low (B1), middle (A1B) and high (A2) scenarios, and for the two different future periods 2050s and 2090s, can only be discerned by examining Fig. 6.2(c) and (d) or isolating a case for a model from Fig. 6.3. For return period < 1.1 years, the differences in the magnitude of the perturbations among the scenarios are not very eminent and are consistent for both the 2050s and 2090s. However, for a return period > 1.1 years, the perturbations are eminently different for low, middle and high scenarios and inconsistent for both the 2050s and 2090s. If many models are compared (Fig. 6.3), the differences in the variation of the perturbations for the low, middle and high scenarios become trivial and the intermodal variation becomes fundamental. An important point to note is that perturbations are very sensitive when light rainfall events of very low return periods are considered. That is, the values of the perturbations can be dramatically low or large for meek events with intensity close to naught. In the latter case, it is recommended to derive perturbations while considering only the days above a certain-carefully-selected threshold intensity value or particularly wet days which buffers the sensitivity of the perturbations.

5.1.3 Wet days and wet-days frequency perturbation

Schmidli & Frei (2005) defined wet days as the annual count of days with daily precipitation greater than or equal to a certain threshold precipitation value (e.g. ≥ 1 mm). In contrast, dry days are defined as the annual count of days with daily precipitation less than a certain threshold precipitation value (e.g. < 1 mm). Thus, wet days of a given month can be defined as the monthly count of days with daily precipitation greater than or equal to a certain threshold precipitation value (e.g. ≥ 1 mm or ≥ 0.1 mm). Given the fact that the frequency and intensity of precipitation are projected to change, as learned from the perturbations of rainfall intensity, wet days are also projected to change and it is possible to obtain wet-day frequency perturbations and analyze its variation for the different scenarios and models. Thus, the monthly wet-day frequency perturbation is, hereafter, simply referred to as wet-day perturbation. The wet-day perturbation is calculated as the ratio of the projected (scenario) total number of wet days to the corresponding total number of wet days in the control period and can be calculated as follows:

Let $(X_p)_{i,m}$ be the model daily time series of a given month, m, corresponding to the control of the pth GCM, $p \in \{1,...,M\}$, M is the number of control simulations or the GCMs, and let $(Z_{p,s})_{i,m}$ be the model time series for the projected scenario, s, $s \in \{1, ... ,W\}$, where W is the total number of projected scenarios, for i = 1, 2,..., n, and for month m = 1, 2, ..., 12. Note that n is the total number of years of the model simulations. If, $t_{c,m}$ is the threshold rainfall intensity for a given month, m, the wet-day perturbation is given by

$$\Delta d_{p(m)} = 100 * \left(\frac{d_{p(s,m)} - d_{p(c,m)}}{d_{p(c,m)}} \right) \tag{6.1}$$

where, $\Delta d_{p(m)}$ is the projected percentage change in wet-day for a given month, m, $d_{p(s,m)}$ and $d_{p(c,m)}$ are the respective total number of wet days with intensity > $t_{c,m}$ in scenarios and control. Note that the control and future periods considered are 1971-1990 (Control) and 2045-2065 (2050s) and 2081-2100 (2090s), respectively. Thus, n = 20 years.

Fig. 6.4 shows the wet-day perturbations for 16 GCM runs for high, middle, and low scenarios and for two different projected future periods (e.g. A2, 2050s for Fig. 6.4 (a)). For the months of January-May the wet-day perturbations are generally > 1 but < 2. In contrast, the wet-day perturbations are also generally > 1 for the months of October-December but > 2 for some models. Meanwhile, for the months of June-September the wet-days perturbations are generally < 1. It can further be seen that the perturbations, for the months of April-May and October-November; are relatively higher for the months where perturbation values > 1. Also for the months where perturbations < 1 its values for the months of June-August are relatively lower. Note that the wetter months are April, May, October and November; and the drier months are mainly June-August. Thus, the implications of the differences in the perturbations for the different months are that the wet days in the wet and dry seasons are projected to increase and decrease respectively in the future. The increase in the wet days will vary with the high, middle, and low scenarios. The high scenario (first row of charts in Fig. 6.4) reveals more increase in wet days than the middle scenario (second row of charts in Fig. 6.4); and middle scenario (second row of charts in the Fig. 6.4) reveals more increase in wet day than for the low scenario. If the charts in left column (for 2050s) and the ones in the right column (2090s) of Fig. 6.4 are compared, it can be seen that the increase and decrease in the wet days for the wet and dry seasons, respectively, are projected to be relatively higher for the 2090s than for the 2050s.

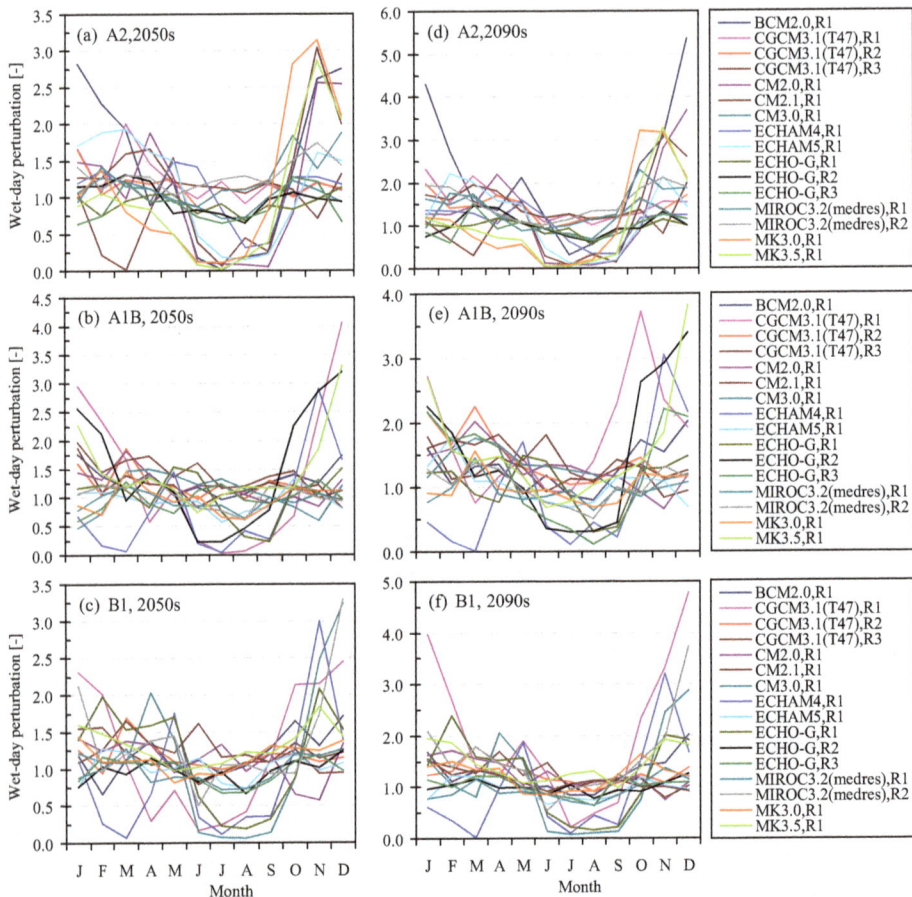

Fig. 6.4. Typical monthly wet-day perturbations for GCM runs calculated for a grid over the Ruizi catchment.

5.1.4 Wet spells and mean wet-spell perturbation

A wet spell is defined as the number of consecutive wet days in a time series in which precipitation intensity exceeds a certain threshold precipitation value (Lall et al., 1996). The length of a wet spell is measured in days.

Wet spells are considered to be one of the most important indicators of extreme precipitation indices. The basic indices of wet spells include, but are not limited to: (1) the maximum number of consecutive wet days in which the total precipitation is greater than or equal to a certain amount, and (2) the mean wet-spell length (mean wet spell), which is the average length of the wet spells in a month, season or year (Schmidli & Frei, 2005). These indices represent characteristics of the duration of consecutive wet-day sequences (Schmidli & Frei, 2005). The latter is particularly of interest to this study. Analysis of the projected changes in wet spells provides insight into how the future rainy days, as projected by the climate models, will be like. As established in the analysis of the wet-day perturbations that the wet days are projected to change, it is important to assess how the increase in the wet days, for example, affects the mean length of wet spells. Study by Yue-Cong and Barry (2010) on how climate change may influence the demand for water in the future under different climate change scenarios, showed that changes in wet spells will have significant implications for water supply.

Given a GCM run (control and scenarios) and if 1 mm/day is the threshold precipitation that define a wet day in the control run ($t_{c,m}$), then the mean wet spell for a given month can be obtained. Similarly, the corresponding mean wet spell for the scenario is also calculated based on $t_{c,m}$. This procedure can be repeated for all the months, scenarios, and for all the considered GCMs as well as for the periods under consideration. Fig. 6.5 shows an example from one GCM run under the high, middle and low scenarios for periods 2050s and 2090s to illustrate the typical mean wet spell in control and scenario.

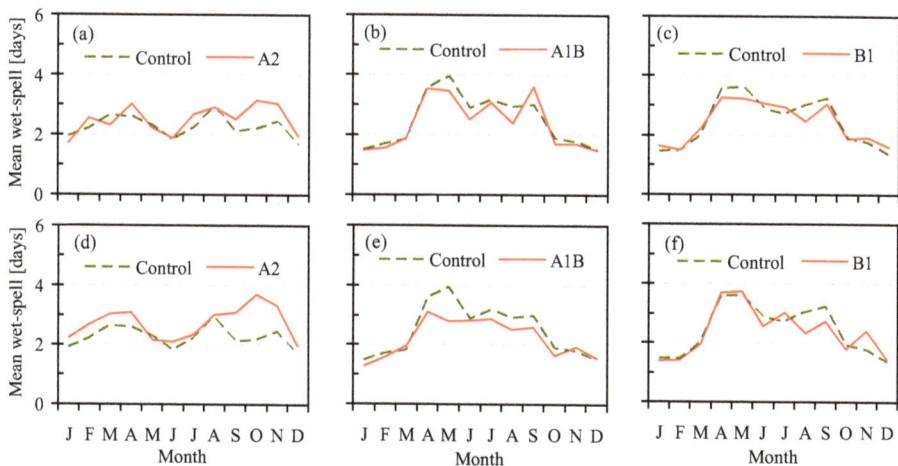

Fig. 6.5. Typical monthly mean wet-spell length for rainfall data of a GCM (MIROC3.2(medres),R1) run calculated for a grid over the River Ruizi catchment for the periods: (a)-(c) 2050s, and (d)-(f) 2090s.

It can be seen that, under the high scenario (2050s), the mean wet spell for the months of January and March is projected to decrease with respect to the present whereas for the months of February, April, July and September-December it is projected to increase also with respect to the present (control) (Fig. 6.5(a)). Fig. 6.5(d) shows that the increase in the mean wet-spell for the months of March-April and September-November will relatively be higher.

The implications are that the increase in mean wet spell for the wet months is probably a manifestation of longer wet spells under the high scenario. Also the increase in the mean wet spell in the 2090s will be higher than that in the 2050s under the high scenario. Under the middle scenario (Fig. 6.5(b) and (e)), the plots for the scenario lie below that of the control run for both the 2050s and 2090s mainly for the very dry (June-August) and very wet months (April-May). This implies that the GCM run generally projects a decrease in mean wet spells for both the very dry and very wet months with a relatively no change for the other months. The model, under low scenarios, however, projected little change (Fig. 6.5(c) and (f)).

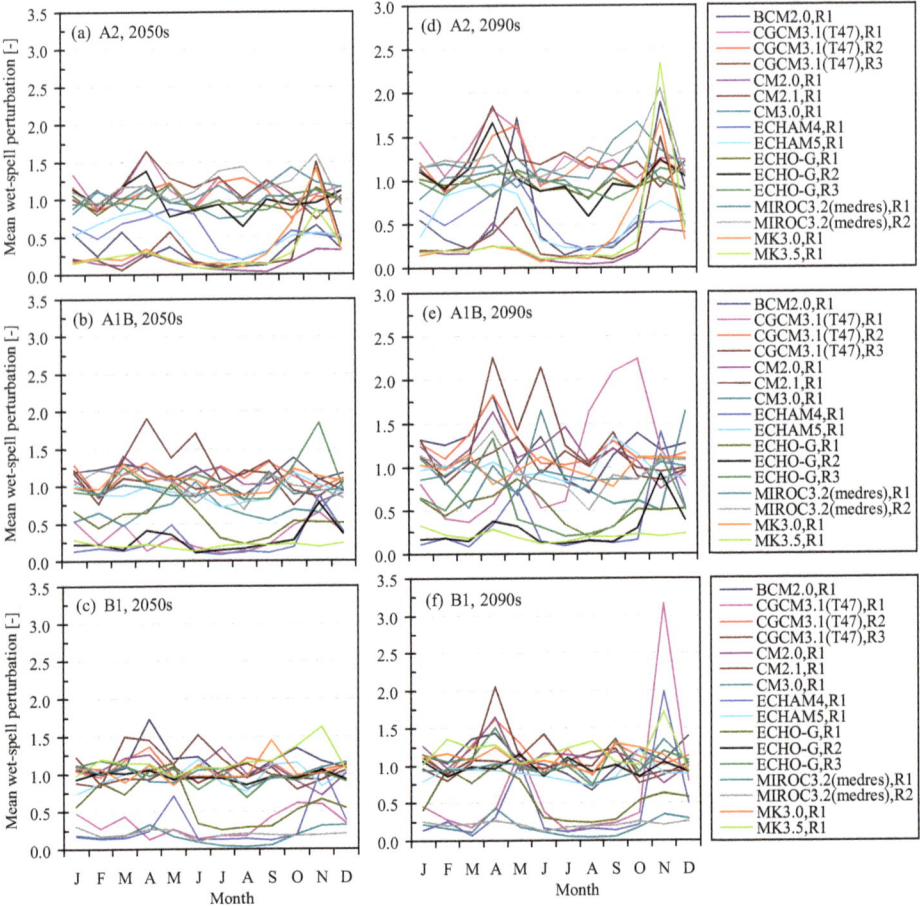

Fig. 6.6. Typical monthly mean wet-spell perturbations obtained from daily rainfall of 16 qualified GCM runs extracted from a grid over the River Ruizi catchment.

Perturbation of mean wet spell is calculated as the ratio of the mean wet spell for the scenarios series to that of the control and can easily be represented in terms of percentage. If $w_{p(s,m)}$ and $w_{p(c,m)}$ is the mean wet spell for scenario and control series, respectively, the mean wet-spell perturbation is given

$$\rho_{p(m)} = \left(\frac{w_{p(s,m)}}{w_{p(c,m)}} \right)$$

(6.2)

where $\rho_{p(m)}$ is the mean wet-spell perturbation and the equivalent percentage change is given by

$$\Delta\rho_{p(m)} = 100 * \left(\frac{w_{p(s,m)} - w_{p(c,m)}}{w_{p(c,m)}} \right) \tag{6.3}$$

Fig. 6.6 shows monthly mean wet-spell perturbations of 16 model runs under high, middle and low scenarios for the 2050s and 2090s. It can be seen that there is a "band of plots" with mean wet-spell perturbations that resonates around 1, with some completely below 1 and some, for the month of November, willowing out (Fig. 6.6(a)). A similar pattern of the former can be seen in Fig. 6.6(d) but with a shift in the upper "band of plots". Fig. 6.6(d)-(f) further shows that for models with mean wet-spell perturbations < 1, the mean wet-spell perturbations are generally < 0.5 for most models, except for the months of May and November. Similarly, Fig. 6.6(d)-(f) also shows that for the models with mean wet-spell perturbations > 1 the mean wet-spell perturbations generally lie in the range 1-1.25 except for the months of April and November. Fig. 6.6(a)-(f) also reveals that the mean wet-spell perturbations for the months of February and August < 1 for most models. It can be seen also that the "depression" in the plots for models with mean wet-spell perturbations < 1 is eminent for the months of June-September. The preceding discussions have implications for the changes in the wet- and dry spells. First, the mean wet spell for the wet seasons is projected to increase and the wet spells for the wetter months will increase more than that for the less wet months. Secondly, for dry season where the mean wet spell is projected to decrease, the wet spells in the drier months will decrease more than that in the less dry ones. Furthermore, the change in the mean wet spell implies that the distributions of both the wet- and dry spells will alter.

Fig. 6.7 shows the distributions of the wet/dry spell lengths for the months of January and April. The ordinates and the abscissa of the plots (Fig. 6.7) represent the relative frequency (pmf) and the days, respectively. It can be seen that the proportion of wet spells with length of one day is projected to reduce and the proportion of wet spells of length 2 days will dominate (Fig. 6.7(a)). Meanwhile the proportions of wet spells of length 2-3 days will reduce in the future. Wet spells of length greater than 4 days are projected to increase. The implication of the latter is that increase in the frequency of longer rainy days may be linked to river flooding. Fig. 6.7(c) shows that, generally, the proportion of the dry spells will reduce for the dry months. Fig. 6.7(c) shows that the proportions of wet spells of length between 3-10 days will increase and this means that the current longer wet spells are projected to be longer. For both the dry and wet months the wet spells of length 4-10 days are projected to increase. Thus, the frequency of wet days and dry days are projected to increase and reduce, respectively. Furthermore, the increase in the mean wet spell may be a manifestation of the increase in the longer wet spells.

Note that perturbations of the mean and coefficient of variation of the intensity for the wet days can also be obtained by comparing the one of the scenarios to that of the control. This enables analysis of the possible change in the mean and also the variability of the wet day intensity as a result of the influence of climate.

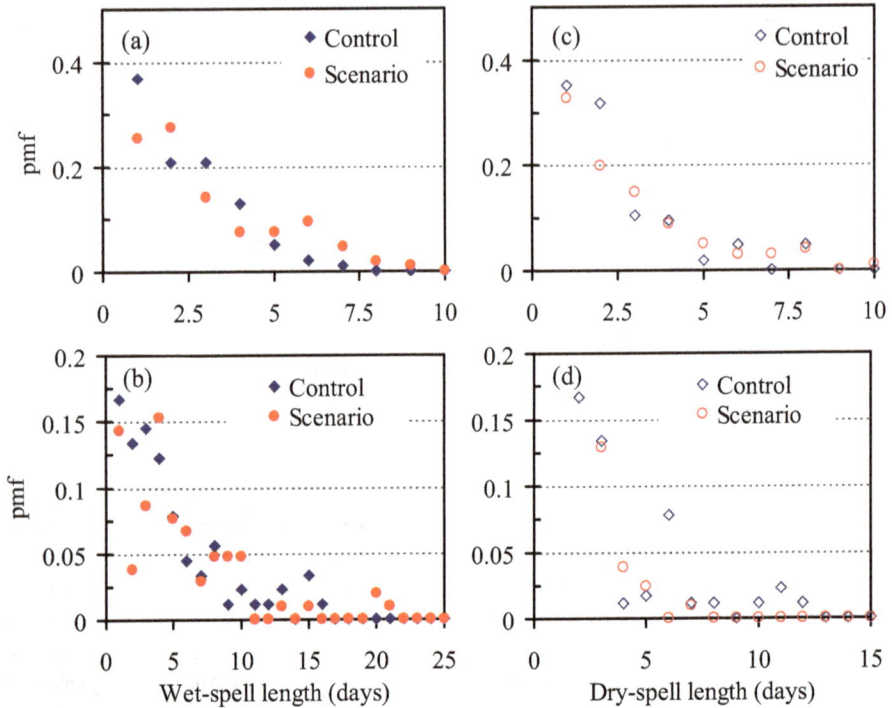

Fig. 6.7. Typical distributions for: (a)-(b) wet spells, (c)-(d) dry spells for: (a) and (c) January, and (b) and (d) April for an example data set for a GCM run (CGCM3.1(T47),R2) under A2 scenarios, 2090s, extracted for a grid over the River Ruizi catchment.

6. Climate change signals

We refer to perturbations extracted from GCM control and scenario runs, such as for rainfall intensity, wet-day, mean wet-spell, mean and C_v of the daily rainfall for each month as climate change signals (CCS). If we assume that the bias in the control runs is similar to the bias in the scenario runs then CCS are bias-free. Transfer of the CCS to observed time series (OS) at local scale would thus produce perturbed observed series (POS) which have similar statistical properties as the projected future time series. In the context of this study, we refer to the process of transferring the CCS to OS as a nonparametric statistical downscaling of GCM runs based on adaptive perturbation approach. The procedure for transferring the climate change signals is discussed in section 8.0

7. Conclusions on perturbations and climate change signals

The change (perturbation) between GCM scenario and control runs provides useful information on climate change signals in the hydrometeorological time series. Such climate change signals if carefully extracted can provide substantial preliminary information on climate change impacts assessment at local scale. A methodology that can transfer CCS to OS at local scale with minimum error margins would constitute an important nonparametric statistical downscaling of GCM runs. The resulting downscaled time series can be used in hydrological climate change impacts at local scale.

8. Perturbation approach for statistical downscaling

The CCS, considered important for capturing change, can be transferred to OS using a perturbation approach without explicit assumption of the underlying distribution to obtain POS. A number of nonparametric approaches are used in stochastic hydrology to generate weather variables (Lall, 1995). An approach is considered nonparametric if (1) it is capable of approximating a large number of target functions, (2) it is "local" in that estimates of the target function at a point use only observations located within some small neighborhood of the point, and (3) no prior assumptions are made as to the overall functional form of the target function. In the perturbation approach, some CCS is used to perturb OS to obtain POS such that the other CCS are used to validate POS. Once POS is obtained and validated it can be employed in climate change impacts assessment at local scale.

8.1 Application to rainfall

In the downscaling of the daily rainfall time series, two CCS are considered: (1) wet-day rainfall intensity perturbations, and (2) mean wet-spell perturbations, as the most important signals to transfer to OS. The mean and the coefficient of variation of the wet-spell rainfall intensity, the mean wet-spell and the distributions of the wet-spells are the statistics used in the validation of POS. In the following sections we discuss how each of the selected CCS are transferred and validated.

8.1.1 Wet-day climate change signals

The wet-day intensity perturbations (perturbation factors) are calculated based on the procedure described in section 5. However, a methodology for choosing a threshold intensity value that defines a wet day in the GCM control series (CS) is revised by involving OS. Let t_o be the threshold wet-day rainfall intensity selected to define a wet day for OS. The corresponding wet-day rainfall intensity value for CS, t_c, is the value that makes the number of wet days in the OS equals the number of wet days in the CS, provided CS is positively biased. If CS is negatively biased, the value of t_c is taken to be the same as t_o. This is to ensure that all the wet days' properties for OS, POS, CS and scenarios series are correctly estimated and to eschew "chaos" from wet-day intensity perturbations with low return periods. For example, if the GCM is positively biased and t_o = 1 mm, $t_c > t_o$ (e.g. t_c = 8.2 mm). Thus, all the properties of the wet days in both the CS and the GCM scenario series (SS) are calculated by considering the value of t_c. Fig. 6.8 shows an example of the threshold wet-day intensity values for 16 different CS for all the months compared against t_o = 0.1 mm of an observed time series. It can be seen that the CS can be highly and positively biased with, for example, 0.1 mm in observed time series being simulated to be 16 mm in the CS.

The values of the wet-day intensity for the OS are ranked and the exceedance probability of each data point is calculated based on its rank number. Similarly, the values of the wet-day intensity for SS are ranked and the exceedance probability of each data point is calculated based on its rank number. The perturbation is therefore the ratio of the wet-days intensity for SS to that of CS with same probability. Note that if the SS has more wet days than the control, the exceedance probability points are also more than that of the CS and OS. Thus, the perturbations of the extra wet-days intensity values in the SS are obtained by interpolating over the wet-day intensity values for CS. The perturbations are then applied to the wet-day intensity for the OS to obtain POS. In the case where more wet days are projected by the SS the extra wet day perturbations are also applied to the wet-day intensity values interpolated over wet-days intensity values for OS. Note that this methodology ensures that the extra wet days (projected increase in wet days) are added during the application of perturbation factors to OS. If the GCM projects decrease in the number of wet-day then the wet-day intensity perturbations are only applied to the OS based on exceedance probability "equation" and the extra wet days in the OS are converted to dry days. Figure 6.9. (a) and (b) shows the wet-day intensity before and after application of perturbations, respectively.

The procedure described above is carried out for each month but for all the years of the time series record. The POS is then resorted to enable transfer of CCS for the mean wet spell. The properties or the statistics of the resulting POS are compared with the statistics of the OS to obtain climate change signal at local scale (CSL).

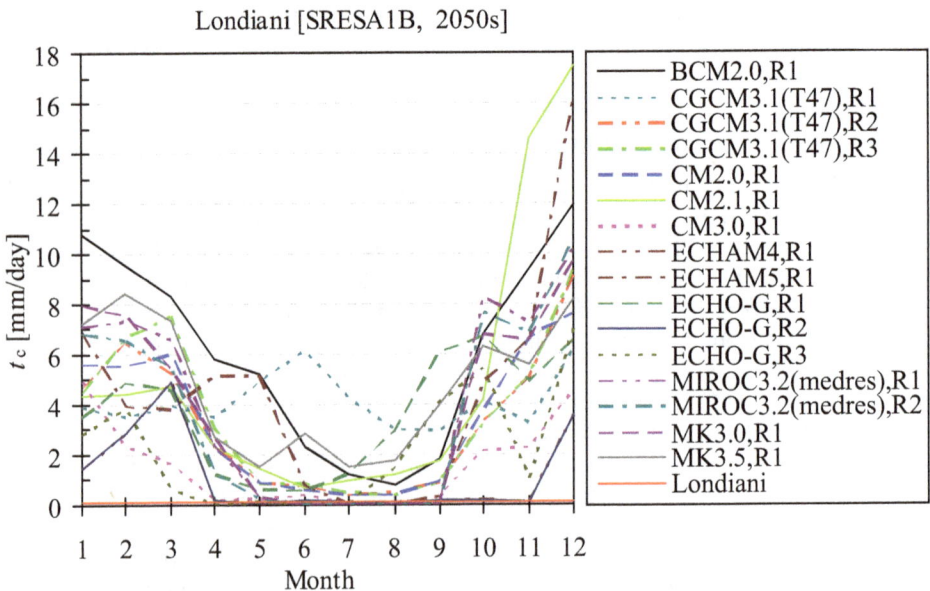

Fig. 6.8. Typical threshold for monthly wet-days daily intensity for different CS compared to $t_0 = 0.1$ mm for an observed time series (Londiani) for a GCM grid over the River Ruizi catchment.

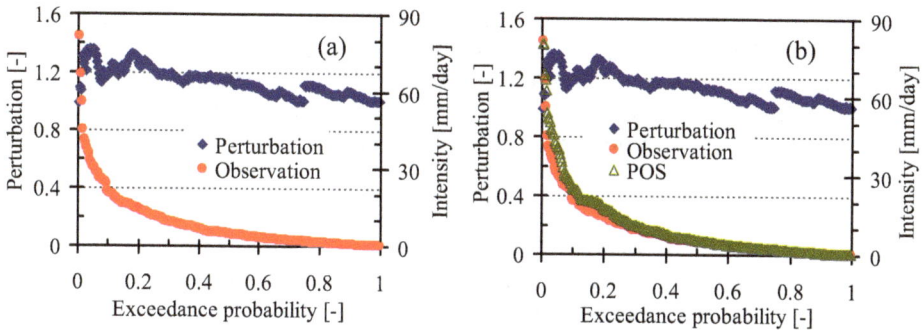

Fig. 6.9. Typical wet-days daily rainfall intensity for the OS before (a), and after (b), application of perturbations. The OS is the areal rainfall data over Ruizi catchment and the GCM data is extracted from a grid superimposed over the River Ruizi catchment.

8.1.2 Wet-spell climate change signal

The transfer of the mean wet-spell perturbation is only considered after the perturbations of wet-days intensity. The wet spells to be adjusted are those in POS. There are only two cases of change to be considered in the mean wet spell: an increase or a decrease in the mean wet spell. In the case of an increase in the mean wet-spell, each of the wet spells in POS greater than their mean value is adjusted by the mean wet-spell perturbation. That is, each of the wet spells in POS greater than their mean value is extended by giving it additional wet days. The value(s) of the wet-day intensity to be added, in order to extend a wet spell, is/are obtained through a non-parametric resampling technique using kernel density estimates (Lall et al., 1996, also see next section). For the case of a decrease in the mean wet spell, each of the wet spells in POS greater than their mean value is reduced by the mean wet-spell perturbation. The reduction of each of the wet spells with length greater than their mean value is carried out by extending the dry spells bounding the target wet spell. That is, extra wet days in the target wet spell are converted into dry days by removing the required number of wet days located at the ends of the wet spells and replacing them with dry days. The statistics of POS are calculated and compared with the statistics of OS to obtain adjusted climate change signal at local scale (CSLa). CSLa is validated against CCS and if the error margins are small enough, the modified POS (POSa) is the downscaled GCM run which represents the projected time series at the local scale.

Note that the wet days to be added are considered days with missing value(s) and is/are added at the ends of the target spells in a proportional way. That is, the wet and dry spells are assumed to be independent where no transition to the same spell is possible. The application of wet-spell perturbation to OS, in downscaling GCM runs, is similar to the nonparametric approach used for generating weather variables in which the wet/dry spell approach is used. There are two major advantages of considering perturbation of wet spells in the perturbation approach for downscaling GCM runs. First, compared to the quantile perturbation approach by Ntegeka (2011), the rainfall spell structure, in the OS and the change in the mean of the OS wet spells lengths, as projected by GCM, are both considered concurrently. This makes it an improvement and advancement of the quantile perturbation approach by Ntegeka (2011) where only the wet-day frequency and wet-day rainfall intensity are considered. Secondly, the distribution of the spells can easily be validated against that of the GCM. That is, the distribution of the spell change signal can be checked graphically. However, as noted by Lall et al. (1996) the justification of the independence between the wet and dry spell lengths at short time scales is difficult. Nevertheless, the data are allowed to inform the wet spells perturbation process to ensure that very long wet spells, separated by a very short dry spell, are not merged.

8.1.3 Kernel density estimation method

Kernel density estimation is a robust nonparametric way of estimating the Probability Density Function (PDF) of a random variable. The technique does not assume any functional form of the PDF and allows its shape to be entirely determined from the data. The paper by Lall et al. (1995) presents a comprehensive nonparametric approach to a stochastic model for generating daily precipitation based on Kernel density estimation. The salient features of the model were the consideration of the alternating wet/dry spells of a daily rainfall structure within the wet spells. Kernel density estimates (k.d.e) were espoused as effective methods for recovering univariate, multivariate (conditional), discrete and/or continuous probability densities that were directly required from the histogram record. Furthermore, in Lall et al (1996), kernel density estimators of continuous and discrete variables were critically reviewed and tested with various data sets. The k.d.e methods have garnered favours for generating weather time series for various applications with Rajagopalan et al. (1997), and Rajagopalan & Lall (1999) expanding the methods to "k-nearest neighbours resampling". Lall et al. (1996) stated that sampling from k.d.e., compared to sampling from the empirical distribution of the data itself, can lead to a reduced variance of the popular Monte Carlo design. The aim here is to take advantage of the flexibility and robustness of the kernel density estimator for generating daily precipitation from which it can be resampled to extend the wet spells in POS. The normal kernel (Lall et al., 1996) is a robust estimator and is often recommended for use when dealing with real-valued random variables that tend to cluster around a single mean value and was used in this study.

8.1.4 The overall flow chart for the downscaling

The downscaling process applied in this study involves the following seven steps. (1) Choose the threshold intensity that defines a wet day from OS and obtain the corresponding value that defines a wet day in CS and SS. Select and calculate the CCS from CS and SS needed for perturbation. (2) Calculate the wet-day intensity perturbations from CS and SS by considering rainfall quantiles with same exceedance probability. Note that the exceedance probability is calculated based only on the wet days. If there are more wet days in SS than in CS, the additional quantiles for CS are obtained by interpolation over its ranked series. (3) Modify each of the OS quantiles by wet-day intensity perturbation to obtain POS. Note that additional wet days are added through interpolating over OS for additional quantiles and are modified by the extra perturbation factors obtained in step (2). (4) Calculate CSL and validate them against CCS obtained in step (1). (5) Transfer the mean wet-spell perturbation through extending or reducing the length of each of the wet spells greater than their mean value. Use k.d.e to generate intensity values from which you can sample to extend the required wet spells. (6) Calculate the new CSL after application of the wet-spell perturbation (CSLa) and validate it against CCS. If the respective errors between

CSLa and CCS are small enough then the final POS is POSa (POS modified by wet-spell perturbation), if not, repeat step (5)–(6). (7). Examine the distribution of the wet spells graphically for visual satisfaction.

8.1.5 Validation of results

The key aspect in the wet-spell technique of the perturbation approach for downscaling GCM runs to local scale is the validation of the climate change signals. Four important characteristics of a rainfall time series for deriving climate change signals needed for the validation of the results are: (i) mean wet-day daily intensity, (ii) mean monthly volume of the wet-day intensity, (iii) mean wet spell and (iv) coefficient of variation of the wet-day intensity. This section presents an example of one GCM to illustrate the typical validation results.

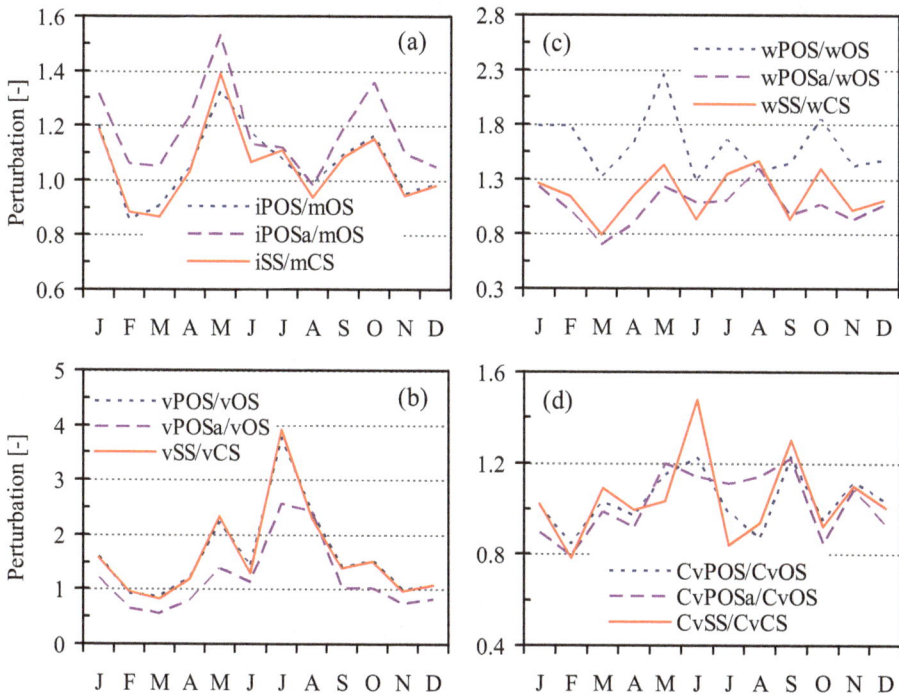

Fig. 6.10. Typical validation results for the climate change signals: (a) mean wet-day daily rainfall intensity, (b) monthly mean volume for the wet-day rainfall intensity, (c) mean wet-spell length, (d) coefficient of variation for the wet-days daily rainfall intensity for an example GCM run (CGCM3.1(T47), R2). The results are for the data extracted from a GCM grid over the River Ruizi catchment.

Fig. 6.10 shows the validation results for the selected climate change signals. The perturbation represents the change or ratio of the time series feature between the control and the target series for the different months. The plots represented by continuous dot and dash lines are for CCS(xSS/CS), CSL(xPOS/OS) and CSLa(xPOSa/xOS), respectively, where x represents the time series feature or statistic under consideration. The time series features, $x = i$, v, w, C_v, represent intensity, volume, mean wet spell and coefficient of variation, respectively. It can be seen that application of wet-day intensity and wet-day frequency perturbations to OS can perfectly transfer the CCS for intensity and volume (Fig. 6.10(a)-(b)). In addition the change in variability (C_v) is also well transferred simply by applying wet-day intensity and wet-days frequency perturbations (Fig. 6.10(d)). However, the transfer of wet-day intensity and wet-day frequency perturbations alone do not "honor" in any way the CCS for the wet spells except or perhaps by chance (Fig. 6.10(c)). Fig. 6.10(a) and (c) reveals that application of wet-day intensity and wet-day frequency perturbations to OS alone results in a perturbed observed series which has "inflated" mean wet spell above the one projected by the climate model. Fig. 6.10 (a)-(b) indicates that adjustment of POS series to reflect change in the spells, as projected by the model, results in the reduction of the volume of the time series, especially in the case where the model projected decrease in the mean wet spell. However, the reduction in the mean wet-day is an indication that the mean wet-days intensity will increase (Fig. 6.10 (a)). In general, Fig. 6.10 shows that the mean wet spell is projected to reduce but the intensity of the rainfall will increase. The latter is projected to influence variability in the intensity and volume of rainfall for the very dry months or season (June-August).

(a) Distribution of wet spells

Figure 6.11 (a)-(b) shows the distribution of the wet spells for the months of January and April. The ordinate represents the relative frequency or the mean proportion of the wet spell of a given length (1-10 days) in that month in a period of 20 years. Figure 6.11. (c)-(d) shows the change in the distribution of the wet spells. From Figure 6.11.(c)-(d) it can be seen that application of the perturbation for the mean wet spell substantially improves the distribution of the wet spells in the POS. Furthermore, it can be seen from Figure 6.11. that there are more wet-spell events with length < 3 days for the month of January compared to that of April which has less wet-spell events of length less than 3 days. The wet spells with length 3-5 days dominate the wet-spell events for the month of April. Figure 6.11. shows that the proportions of wet spells of length 5-7 days for January, and 4-7 days for April, are projected to increase for both the months of January and April. Taking into account that intensity is projected to increase; the increase in longer wet spells has implications for the land areas with very high runoff coefficient and poor drainage. The longer wet spells events may have some influence on flooding than shorter ones. Note that, albeit the changes in the distribution of the wet spells of several GCM runs were analyzed, for both for the period 2050s and 2090s and also for different observed data, the results presented here are not in any way exhaustive.

(b) Distribution of rainfall series

Fig. 6.12 shows the distribution versus retrun period for the daily rainfall for eight different months. Generally, it can be seen that the rainfall intensity with lower exceedance probability are projected to increase more than those with higher exceedance probability. Also, it can also be seen from Fig. 6.12 that generally the intensity of POSa, with exceedance probability between 0.08 and 0.1, will not change so much from that of OS. However, from Fig. 6.12(b), (f) and (h), it can be seen that the distributions for the POS for the mean events are above that of OS. This implies that application of intensity and wet-day frequency perturbations alone without considering perturbation for the mean wet spell results in, probably, overestimation of the mean rainfall events (also see Fig. 6.2). Consider Fig. 6.12(b)-(c) and given an exceedance probability less than 0.08; projected distributions for POS and POSa lie above the distribution of OS. Similarly, the distributions for POS and POSa lie below the distribution of OS for exceedance probability > 0.1. These changes imply that the values of the intensity for the mild events are projected to decrease and that for heavy events are projected to increase for that month. Fig. 6.12(f)-(h) also shows similar changes. Thus, the mean of the wet-day intensity will increase for the wet months, which is consistent with the results of the increase in the mean of the wet-day intensity shown in Fig. 6.10(a).

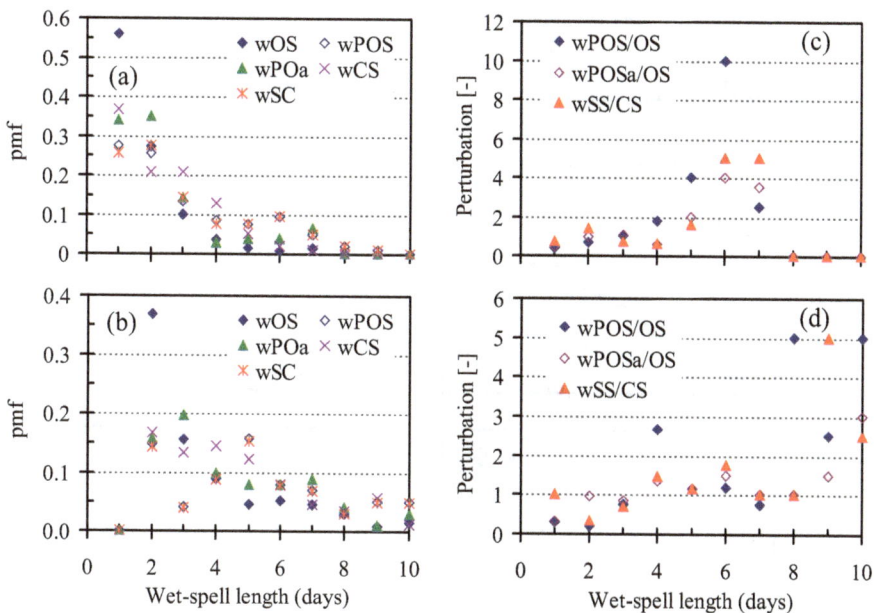

Fig. 6.11. Typical validation results for the wet spell distribution: (a)-(b) January (b)-(c) April for an example GCM run (CGCM3.1(T47), R2). The results are for the data extracted from a GCM grid the River Ruizi catchment.

However, Fig. 6.12(a) shows that the intensity for both the mild and heavy events will decrease. In contrast, Fig. 6.12(e) shows that the intensity of both the mild and heavy events will increase. Further more, Fig. 6.12(d) shows that the intensity of the medium events will decrease and the intensity for the heavy events will increase. Thus, for the dry months consistent change pattern is not eminent. Fig. 6.13(a) shows the distributions for OS, POS and POSa. Meanwhile, Fig. 6.13(b) shows the original perturbations derived from control and scenarios and that which is derived from observed and the downscaled series for the month of January. Fig. 6.13(a) can be compared with that of similar plot for CS and SS (e.g. Fig. 6.2(a)). Note that the result is for the same model and same scenarios (ECHO-G). It can be seen that the distributions for OS and POSa follows similar change pattern as the distributions for CS and SS. This may suggest that Fig. 6.10(a) gives a false impression of a perfect match when iPOS/mOS iSS/mCS are compared. Fig. 6.13(b) shows that the actual perturbations needed to be transferred to the OS is not the same as that derived from CS and SS if the change in wet spells is to be considered. In the latter case, the perturbations for intensity with heavy events are actually higher and those for mild events are lower if change in wet spells is taken into account.

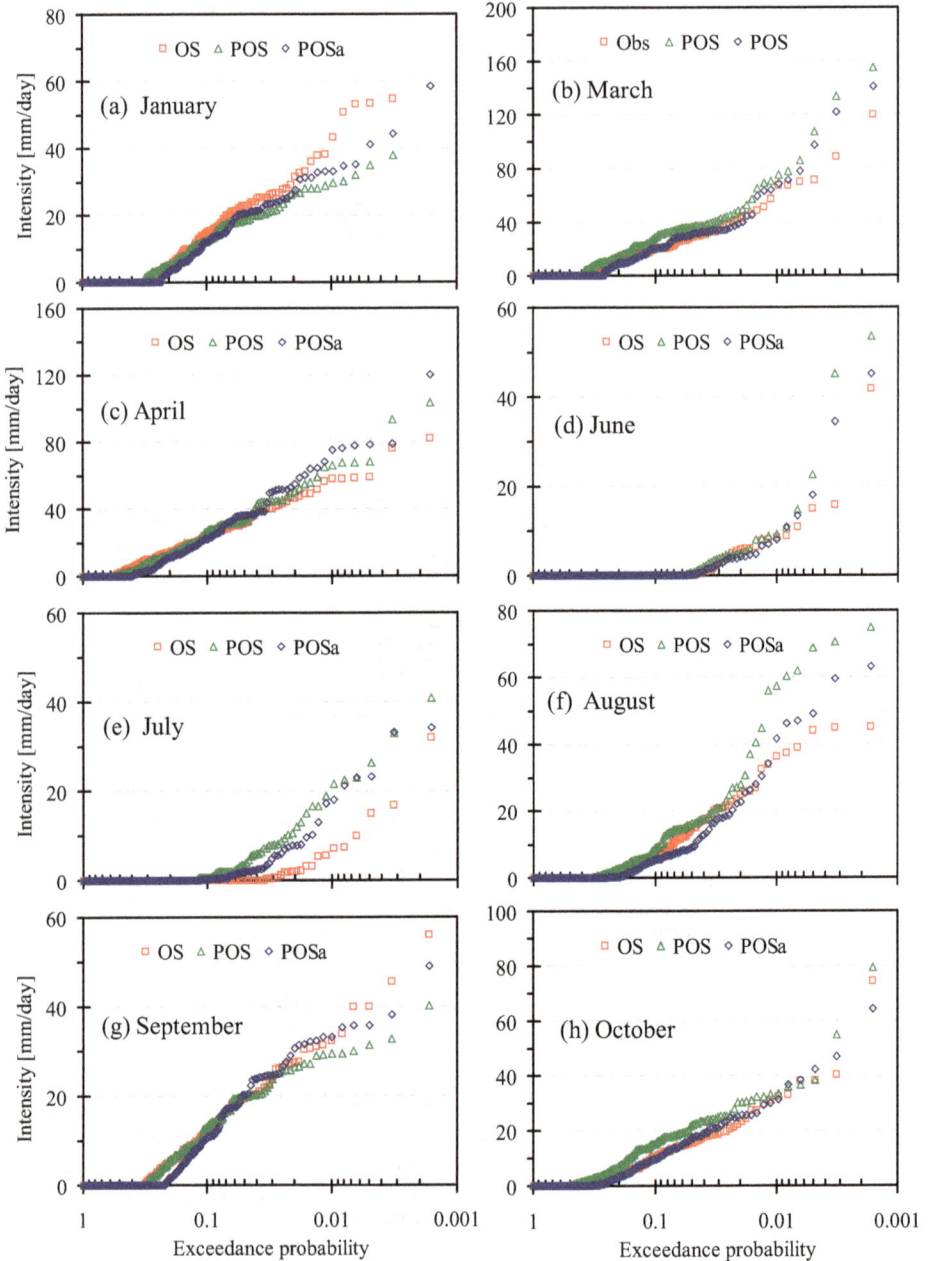

Fig. 6.12. Typical distribution of daily rainfall for an example dataset for different months for an example GCM run (CGCM3.1(T47), R2). The results are for the data extracted from a GCM grid over the Ruizi catchment. The OS is the areal rainfall over the River Ruizi catchment.

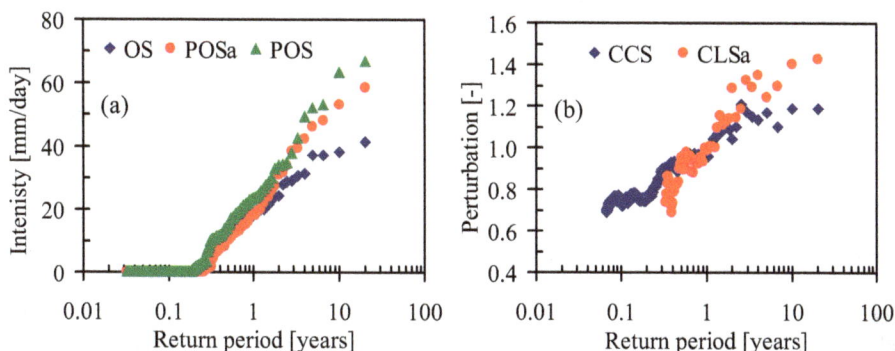

Fig. 6.13. Classical distribution for January daily rainfall (a), and the perturbations (b) derived from SS and CS (CCS) and from OS and POSa (CLSa) for a grid over the River Ruizi catchment.

9. Conclusion on downscaling

A nonparametric statistical downscaling of daily rainfall time series from GCM runs that uses a perturbation approach is formulated and demonstrated. The core of the principle lies in the verity that climate change signals can be extracted from the GCM runs (control and scenarios) in an empirical way without explicit assumption of the underlying probability distribution and applied to the observed series. The modified observed time series are the downscaled GCM results, which are plausible for climate change impacts assessment at local scale. Among the important features of the rainfall time series are wet spells, wet-day intensities, wet-day frequencies and coefficients of variation and are considered in the perturbation approach. If only wet-day intensity and wet-day frequency perturbations are considered, the resulting time series can still have similar signals of coefficient of variation. However, it leads to an overestimate of the change in wet spells and intensity. Thus, the changes in the structure of both the dry/wet spells, which are very important temporal features of rainfall, are not captured in the perturbed series. Since hydrological models are very sensitive to rainfall, overestimates of the change in the wet spells and rainfall intensity events may significantly influence hydrological extremes. In order to eschew and buffer these problems, consideration of the changes in the wet spells is crucial. One other advantage of the wet-spell approach for statistical downscaling using quantile perturbation approach is that the changes in the distributions of both the wet and dry spells can be validated through a graphical approach. In addition, the wet-spell approach preserves the changes in extremes of rainfall as projected by the climate models. Thus, hydrological impacts of climate change on extremes can appropriately be estimated given the fact that rainfall time series is an important input into the hydrological model.

Example of studies that have focused on the impacts of climate change on extremes include that of Taye et al. (2011), Nyeko-Ogiramoi et al. (2011), Willems and Vrac (2011), and Willems et al. (2011 in press). A conclusive statement in Nyeko-Ogiramoi (2011) states that for the Lake Victoria basin, the impact of climate change on the intensity and frequency of precipitation extremes and daily maximum temperature are projected to be significant in the 2050s and 2090s. It further states that water professionals should take into account the expected impact of climate change on the precipitation extremes as it will significantly affect the design statistics, which is very important for many engineering applications. The importance of the assessment of the possible impacts of climate change on precipitations extremes and the implications for hydraulic engineering practices can indeed not be underlooked.

10. Acknowledgements

This research was supported by the Flemish Interuniversity Council (VLIR) and is linked to the FRIEND/Nile project of UNESCO and the Flanders in Trust Fund of the Flemish Government of Belgium. We also acknowledge this support which has initiated collaboration between Departments of Civil Engineering, Katholieke Universiteit Leuven, Belgium and Makerere University Kampala, Uganda. Thanks to Ministry of Water and Environment, Uganda for availing some of the data which facilitated the works.

11. References

Anyah, R.O. & Semazzib, F.H.M. (2007), Variability of East African rainfall based on multiyear RegCM3 simulations, *Int. Jol. Climatol.*, Vol.27, (September 2006), pp. 357–371, DOI: 10.1002/joc.1401.

Arai, T. & Kragic, D. (1999). Variability of Wind and Wind Power, In: *Wind Power*, S.M. Muyeen, (Ed.), 289-321, Scyio, ISBN 978-953-7619-81-7, Vukovar, Croatia.

Barnett, D. N., Simon, J., Murphy, B.J.M., Sexton, D.M.H., & Webb, M.J. (2006). Quantifying uncertainty in changes in extreme event frequency in response to doubled CO2 using a large ensemble of GCM simulations, *Clim. Dyn.*, Vol.26, (January 2006), pp. 489–511.

Benestad, R.E., Hanssen-Bauer, I. & Chen, D. (2008). Empirical-statistical downscaling, *World Scientific, Singapore*, ISBN: 978-981-281-912-3.

Christensen, J.H, Carter T.R. & Rummukainen, M. (2007). Evaluating the performance and utility of regional climate models: the PRUDENCE Project, *Clim Change 2007*, Vol.81, Issue S1, pp. 1-6, ISSN: 01650009, DOI: 10.1007/s10584-006-9211-6.

Christensen, J.H., Boberg, F., Christensen, O.B. & Lucas-Picher P. (2008). On the need for bias correction of regional climate change projections of temperature and precipitation, *Geophys. Res. Lett* Vol.35, L20709, (October 2008), 6pp. doi:10.1029/2008GL035694.

Diaz-Nieto, J. & Wilby, R.L. (2005). A comparison of statistical downscaling and climate change factor methods: impacts on lowflows in the river Thames, united kingdom, *Climatic Change*, Vol.69, No.2-3, pp. 245–268, DOI: 10.1007/s10584-005-1157-6.

Fowler, H.J., Blenkinsopa, S. & Tebaldib, C. (2007). Review: Linking climate change modelling to impacts studies: recent advances in downscaling techniques for hydrological modelling. *Int. J. Climatol.* Vol.27, (September 2007), 1547–1578, DOI: 10.1002/joc.1556.

Harrold, T.I. & Jones, R.N. (2003). Downscaling GCM rainfall: a refinement of the perturbation method, In *International Congress on Modelling and Simulation*, pp. 14–17, MODSIM2003, Townsville, Australia, 4-17 July, 2003.

Ines, A.V.M. & Hansen, J.W. (2006). Bias correction of daily GCM rainfall for crop simulation studies, *Agric For Meteorol,* Vol.138, (March 2006), pp. 138:44–53, doi:10.1016/j.agrformet.2006.03.009.

IPCC (2001a). Climate change: the scientific basis, pp. 996, Cambridge University Press, Cambridge and New York, NY, USA, available online.

Kharin, V.V. & Zwiers, F.W. (2005). Estimating extremes in transient climate change simulations, *J. Clim.,* Vol.18, (September 2004), pp. 1156–1173.

Lall, U. (1995). Nonparametric function estimation: Recent hydrologic applications. *Reviews of Geophysics,* Vol.33, Iss.S1, pp. 1093–1102, doi: 10.1029/95RG00343.

Lall, U., Rajagopalan, B. & Tarboton, D.G. (1996). A nonparametric wet/dry spell model for resampling daily Precipitation. *Water Resour. Res.* Vol.32, No.9, (September 1996), pp. 2803–2823.

Lenderink, G., Buishand, A. & van Deursen, W. (2007). Estimates of future discharges of the river Rhine using two scenario methodologies: direct versus delta approach, *Hydrol. Earth Syst. Sci.,* 11, 1145–1159.

Li, B.; Xu, Y. & Choi, J. (1996). Applying Machine Learning Techniques, *Proceedings of ASME 2010 4th International Conference on Energy Sustainability,* pp. 14-17, ISBN 842-6508-23-3, Phoenix, Arizona, USA, May 17-22, 2010

Lima, P.; Bonarini, A. & Mataric, M. (2004). *Application of Machine Learning,* InTech, ISBN 978-953-7619-34-3, Vienna, Austria

MathWorks Inc, MATLAB (2008). Application programme interface reference version 7.0. *The MathWorks Inc.,* Natick, MA, USA, pp. 01760-2098.

Meehl, G.A., Stocker, T.F., Collins, W.D., Friedlingstein, P., Gaye, A.T., Gregory, J.M., Kitoh, A., Knutti, R., Murphy, J. M., Noda, A., Raper, S. C. B., Watterson, I. G., Weaver A.J., & Zhao, Z.-C. (2007). Global Climate Projections. In: Climate Change 2007: The Physical Science Basis. Contribution of Working Group I to the Fourth Assessment Report of the Intergovernmental Panel on Climate Change [Solomon, S., D. Qin, M. Manning, Z. Chen, M. Marquis, K.B. Averyt, M. Tignor and H.L. Miller (eds.)]. *Cambridge University Press, Cambridge,* United Kingdom and New York, NY, USA, pp. 747-846.

Nakicenovic, N., Alcamo, J., Davis, G., de Vries, B., Fenhann, J., Gaffin, S., Gregory, K., Grübler, A., Jung, T. Y., Kram, T., La Rovere, E. L., Michaelis, L., Mori, S., Morita, T., Pepper, W., Pitcher, H., Price, L., Riahi, K., Roehrl, A., Rogner, H.-H., Sankovski, A., Schlesinger, M., Shukla, P., Smith, S., Swart, R., van Rooijen, S., Victor, N. & Dadi Z. (2000). IPCC Special Report on Emissions Scenarios. Cambridge University Press, Cambridge, United Kingdom and New York, NY, USA. 599pp.

Ntegeka, V., (2011). Assessment of the observed and future climate variability and change in hydroclimatic and hydrological extremes. *PhD dissertation,* Arenberg Doctoral school of Science, Engineering and Technology, Katholieke Universiteit Leuven, Belgium.

Nyeko-Ogiramoi, P, Willems, P. & Ngirane-Katashaya, G. (2011). Assessment of the impact of climate change on extreme precipitation and temperature events over the upper River Nile basin, *Proceeding of the Advances in Engineering & Technology international conference,* 1410AET2011-(109), January 30 – February 1, 2011, Entebbe, Uganda.

Nyeko-Ogiramoi, P., Ngirane-Katashaya, G., Willems, P. & Ntegeka, V. (2010). Evaluation and inter-comparison of Global Climate Models' performance over Katonga and Ruizi catchments in Lake Victoria basin, *Phy. Chem. Earth*, Vol.35, Iss.13-14, (August 2010), pp. 618–633, doi: 10.1016/j.pce.2010.07.037.

Olsson, J., Berggren, K., Olofsson, M., & Viklander, M. (2009). Applying climate model precipitation scenarios for urban hydrological assessment: A case study in Kalmar City, Sweden. *Atmos. Res.*, Vol.92, (May 2009), Iss. 3, pp. 364–375, doi: doi:10.1016/j.atmosres.2009.01.015

Piani, C., Haerter, J. O. & Coppola, E. (2010). Statistical bias correction for daily precipitation in regional climate models over Europe, *Theor Appl Climatol.*, Vol.99, No.1-2, (April 2009), pp. 187–192, doi: 10.1007/s00704-009-0134-9.

Prudhomme, C., Reynard, N. & Crooks, S (2002). Downscaling of global climate models for flood frequency analysis: Where are we now?, *Hydrol. Process.*, Vol.16, No. (March 2002), pp.1137–1150.

Rajagopalan, B. & Lall, U., 1999. A *k*-nearest-neighbor simulator for daily precipitation and other variables, *Water Resour. Resear.* Vol.35, No.10, (October 1999), 3089–3101.

Rajagopalan, B., Lall, U., Tarboton, D. G. & Bowles, D. S. (1997). Multivariate nonparametric resampling scheme for generation of daily weather variables, *Stochastic Hydrol. Hydraul.* Vol.11, Iss.1, (1997) pp. 65-93.

Schmidli J. & Frei, C. (2005). Trends of heavy precipitation and wet and dry spells in Switzerland during the 20th century. *Int. J. Climatol.* Vol.25, Iss.6, (May 205), pp. 753–771, DOI: 10.1002/joc.1179.

Schulzweida, U., & Kornblueh, L. (2011). Climate Data Operators, Ver.1.5., pp. 1-164, *MPI for Meteorology*, Germany.

Siegwart, R. (2001). Indirect Manipulation of a Sphere on a Flat Disk Using Force Information. *International Journal of Advanced Robotic Systems*, Vol.6, No.4, (December 2009), pp. 12-16, ISSN 1729-8806.

Svensson, C. & Jones, D.A. (2010). Review of methods for deriving areal reduction factors, *J. Flood Risk Management* Vol.3, Iss.3,(June, 2010), pp. 232–245.

Taye, M. T., Ntegeka, V., Nyeko-Ogiramoi, N. P. & Willems, P. (2011). Assessment of climate change impact on hydrological extremes in two source regions of the Nile River Basin, *Hydrol. Earth Syst. Sci.*, Vol. 15, No.1 (Januanry 2011), pp. 209-222.

Trenberth, K. E, Dai, A., Rasmussen, R. M. & Parsons, D. B. (2003), The changing character of precipitation. *Bull Am Meteorol Soc* Vol.84, (March 2003), pp. 1205–1217, doi: 10.1175/BAMS-84-9-1205.

Van der Linden, S. (June 2010). Integrating Wind Turbine Generators (WTG's) with Energy Storage, In: *Wind Power*, 17.06.2010, Available from

Willems, P. & Vrac, M. (2011). Statistical precipitation downscaling for small-scale hydrological impact investigations of climate change. *J. Hydrol.* Vol.402, (April 2011), pp.193–205.

Willems, P., Arnbjerg-Nielsen, K., Olsson, J., Nguyen, V.T.V. (2011). Climate change impact assessment on urban rainfall extremes and urban drainage: methods and shortcomings. *Atmospheric Research,*Vol.103(January 2012), pp.106-118.

Yue-Cong W. & Barry A. (2010). Climate change and its impacts on water supply and demand in Sydney, *Summary report*: NSW office for water, Sydney, NSW, pp. 12.

7

Modeling Extreme Climate Events:
Two Case Studies in Mexico

O. Rafael García-Cueto and Néstor Santillán-Soto
Universidad Autónoma de Baja California
Instituto de Ingeniería
México

1. Introduction

The most severe impacts of climate on human society and infrastructure as well as on ecosystems and wildlife arise from the occurrence of extreme weather events such as heat waves, cold spells, floods, droughts and storms. Recent years have seen a number of weather events cause large losses of life as well as a tremendous increase in economic losses. According to the IPCC (2007), an extreme weather event is an event that is rare at a particular place and time of year. Definitions of *rare* vary, but an extreme weather event would normally be as rare as or rarer than the 10th or 90th percentile of the observed probability density function. Changes in frequency or/and intensity of extreme events can affect not only human health, directly through heat and cold waves and indirectly by floods or pollution episodes, but also for example, on crops or even insurance calculations. Climate extremes associated with temperature (heatwaves) and precipitation (heavy rain, snow events, droughts) can also affect energy consumption, human comfort and tourism and are responsible for a disproportionately large part of climate-related damages (Easterling et al., 2000; Meehl et al., 2000). Extreme weather events recorded in recent years, and associated losses of both, lives and economics goods, has captured the interest of the general public, governments, stakeholders and media. The scientific community has responded to this inquiry and has raised interest in studying with more attention to detail. Our understanding of the mean behavior of climate and its normal variability has been improving significantly during the last decades. In comparison, climatic extreme events have been hard to study and even harder to predict because they are, by definition, rare and obey different statistical laws than averages. In particular, extreme value analysis usually requires estimation of the probability of events that are more extreme than any that have already been observed, and they are linked to small probabilities. Climate extremes can be placed into two broad groups: (i) those based on simple climate statistics, which include extremes such as a very low or very high daily temperature, or heavy daily or monthly rainfall amounts, that occur every year; and (ii) more complex event-driven extremes, examples of which include drought, floods, or hurricanes, which do not necessarily occur every year at a given location. Katz & Brown (1992) first suggested that the sensitivity of extremes to changes in mean climate may be greater than one would assume from simply shifting the location of the climatological distributions. Since then, observations of historical changes as well as future projections confirm that changes in the distributional tails of climate variables may not occur in proportion to changes in the mean, particularly for precipitation, and may not be symmetric in nature, as demonstrated by differential changes in maximum vs. minimum temperatures (e.g., Kharin & Zwiers, 2005; Deguenon & Barbulescu, 2011).

With respect to changes in climatic extremes, the Fourth Assessment Report of IPCC (2007) noted that since 1950 the number of heatwaves has increased and widespread increases have occurred in the numbers of warm nights. The extent of regions affected by droughts has also increased as precipitation over land has marginally decreased while evaporation has increased due to warmer conditions. Generally, numbers of heavy daily precipitation events that lead to flooding have increased, but not everywhere. Tropical storm and hurricane frequencies vary considerably from year to year, but evidence suggests substantial increases in intensity and duration since the 1970s. In the mid-latitudes, variations in tracks and intensity of storms reflect variations in major features of the atmospheric circulation, such as the North Atlantic Oscillation.

As extreme events are, by definition, rare and unusual, the statistical quantification of potential change in their trends and intensity becomes a very difficult task (Palmer & Räisänen, 2002). So this chapter begins with a background section covering the origins of the statistics of extremes and applications of this statistics toward weather extremes. An overview of the statistical theory of extreme values with emphasis on block maxima approach and peaks over threshold is then provided, followed by two case studies in two cities of Mexico: a climatic application of modeling summer maximum temperatures in an arid city, and modeling of daily rainfall over a threshold in a humid city.

For contrasting the results obtained with extreme value theory, scenarios of summer maximum temperature and daily rainfall, with climate forcing caused by anthropogenic effect, were projected. With this aim, we used a statistical-dynamic model with two emission scenarios, A2 and B2.

2. Historical origins of statistics of extremes

The astronomers were the first to be interested in establishing a criterion for the acceptance or rejection of an outlying value. One of the first researchers that studied statistics of extremes was Nicolaus Bernoulli, in 1709; he answered the question: if n men of equal age die within t years, what is the mean duration of life of the last survivor? In 1852, Benjamin Peirce published the first significance test for eliminating outliers from data sets. He determined the expected value of the largest of any given number of independent life-times, uniformly distributed on an interval (Gumbel, 1958). In 1922, Ladislaus von Bortkiewicz was the first to study extreme values that dealt with the distribution of range in random samples from a normal distribution. The importance of the work of Bortkiewicz is because he introduced the concept of distribution of largest value for the first time.

The first significant contribution to the field of Extreme Value Theory (EVT) was made by Fisher & Tippett (1928) who attempted to find the distribution of the maximum, or minimum of the data. The Fisher-Tippett Theorem states that, if the distribution of the normalized maximum of a sequence of random variables converges, it always converges to the Generalized Extreme Value (GEV) distribution, regardless of the underlying distribution; this result is very similar to the central limit theorem. Maurice Fréchet in the year 1927, was the first in obtain an asymptotic distribution of the largest value. He introduced the stability postulate according to which the distribution of the largest value should be equal to the initial one, except for a linear transformation. The problem of finding the limiting distribution of the maximum of a series of random variables was also later solved by B. Gnedenko (1943), who continued Fisher's research, and gave necessary and sufficient conditions under which three asymptotic distribution are valid. Emil J. Gumbel developed new distributions in the 1950s; his book *Statistics of Extremes* (Gumbel, 1958) was important contribution to EVT above all in the field of engineering. In 1970, L. de Hann provided a rigorous mathematical framework about theory of *regular variation*, which has played a crucial role in EVT. Pickands (1975) generalized them classic limit laws proposing model exceedances above a large threshold, and data above that threshold were fit to the Generalized Pareto Distribution (GPD). The Generalized Extreme Value distribution and the Generalized Pareto Distribution are just the tip of the iceberg of an entire new and quickly growing branch of statistics. The first application was to answer environmental questions, quickly followed by the finance industry. In 1990's multivariate and other techniques explored as a means to improve inference, and 2000's interest in spatial and spatio-temporal applications, and in finance. Textbooks on EVT include to Leadbetter et al., (1983) treat the general theory of extreme values of mainly one-dimensional stochastic sequences and processes; Embrechts *et al.* (1997) combine theoretical treatments of maxima and sums with statistical issues, focusing on applications from insurance and finance; Kotz & Nadarajah (2000) give a comprehensive and down-to-earth survey of the theory and practice of extreme value distributions; Coles (2001) has more emphasis on applications; Beirlant *et al.* (2004) covers a wide range of models and application areas, including risk and insurance, and contains also material on multivariate and Bayesian modeling of extremes; Finkenstädt & Rootzén (2004) explore the application and theory of extreme value in finance, insurance, environment and telecommunications; De Haan & Ferreira (2006) focuses on theoretical results along with many applications; and Reiss & Thomas (2007) is a book with an introduction to parametric model, exploratory analysis and statistical inference for extreme values.

3. Extreme Value Theory

Most statisticians aim to characterize typical behavior and focus on the center of data. EVT aims to characterize rare events by describing the tails of the underlying distribution. Probabilistic EVT deals with the asymptotic stochastic behavior of extreme order statistics of a random sample, such as the maximum and the minimum of independent identically distributed (iid) random variables. EVT has been one of the most quickly developing areas in the last decades (Hadživuković & Emilija, 2005). It has found many applications (Berning, 2010) in different areas such as: *Environmental* (Floods, wave heights, wind speeds, heat waves, cold spells, pollutant concentrations), *Engineering and Reliability* (material strength, metal fatigue, corrosion), *Finance* (portfolio risk, value-at-risk, insurance risk, financial econometrics), *Sociology* ((human longevity, flood risk, management strategies, sport records), *Statistical Methodology* (multiple testing, simultaneous inference), *telecommunications* and *biostatistics*.

The historical cornerstone of EVT is the Generalized Extreme Value (GEV) distribution which classically models block maxima data (or minima) over certain slices of time (such as annual maximum precipitation, monthly maximum/minimum temperature). According to EVT, identically distributed block maxima can be modeled with a GEV distribution defined by equation (1), in which G is the Cumulative Distribution Function (CDF) of block maximum z.

$$G(z; \mu, \sigma, \xi) = \exp\left[-\left\{1 + \xi(z-\mu)/\sigma\right\}_+^{-1/\xi}\right] \tag{1}$$

where $x_+ = \max(x, 0)$. The parameter μ represents the location parameter ($-\infty < \mu < \infty$), determining the location of the peak of f; σ the scale parameter ($\sigma > 0$), determining the "wideness" of the distribution, and ξ the all-important shape parameter ($-\infty < \xi < \infty$) which determines the nature of tail behavior of the maximum distribution. The justification for the GEV distribution arises from an asymptotic argument. As the sample size increases, the distribution of the sample maximum, say X, will asymptotically follow either Fréchet ($\xi > 0$), Weibull ($\xi < 0$), or Gumbel ($\xi = 0$) distribution (Naveau et al., 2005). Each of the three types of distributions has distinct forms of behavior in the tails. The Weibull is *bounded* above, meaning that there is a finite value which the maximum cannot exceed. The Gumbel distribution yields *light tail*, meaning that although the maximum can take on infinitely high values, the probability of obtaining such levels become small exponentially. The Fréchet distribution with a heavy tail, decays polinomially, so that higher values of the maximum are obtained with greater probability, that would be the case with a lighter tail (Gilleland & Katz, 2006). The flexibility of the GEV to describe all three types of tail behavior makes it a universal tool for modeling block maxima. As Naveau et al. (2005) say…"it is important to stress that the GEV is the proper fit to maxima not only from a parent distribution like the Gaussian one, but also from any continuous distribution (e.g., exponential, Cauchy, etc.). Hence the methodology is general and independent of specific numerical values"…

3.1 The three types of extreme value distributions

There are three extreme value distributions — one for ordinary parent distributions (the Gumbel type), another for many parent distributions that are truncated on the right (the Weibull type) and a last for parent distributions that lack all or higher moments (the Fréchet type). The Gumbel type includes most ordinary distributions — for example, the normal, lognormal, gamma, exponential, Weibull and logistic — and was the focus of classical EVT. The Gumbel extreme value distribution is often referred to as the extreme value distribution. Collectively, these three classes of distribution are termed the extreme value distribution, with types I, II and III widely known as the Gumbel, Fréchet and Weibull families respectively. Each family has a location and scale parameter, μ and σ, respectively; additionally, the Fréchet and Weibull families have a shape parameter ξ.

Type I: Gumbel-type distribution

$$\Pr[X \leq x] = \exp[-e^{(x-\mu)/\sigma}] \tag{2}$$

Type II: Fréchet-type distribution

$$\Pr[X \le x] = \begin{cases} 0, & x < \mu \\ \exp\left\{-\left(\frac{x-\mu}{\sigma}\right)^{-\varepsilon}\right\}, & x \ge \mu \end{cases} \quad (3)$$

Type III: Weibull-type distribution

$$\Pr[X \le x] = \begin{cases} \exp\left\{-\left(\frac{\mu-x}{\sigma}\right)^{\varepsilon}\right\}, & x \le \mu \\ 0, & x > \mu \end{cases} \quad (4)$$

4. Threshold models

One statistical method for analyzing extreme values is to fit data to an extreme-value distribution. This method is carried out by two alternative approaches: block maxima and peaks over threshold (POT). The approach leading to distribution (1) assumes data are maxima from finite-sized blocks, such as annual maximum temperature. However, if daily observations are available, models which use only each year's annual maximum discards other extreme data that could provide additional information. The POT approach allows for more data to inform the analysis. The POT models are generally considered to be the most useful applications due to their more efficient use of the data on extreme values. For the POT approach, a threshold is first determined, and data above that threshold are fit to the Generalized Pareto Distribution (GPD), which is based on the excesses above a threshold, and it also has an asymptotic justification, such as EVT. The amounts by which observations exceed a threshold u (called exceedances) should approximately follow a GPD as u gets large and sample size increases. In this case, the tail of the distribution is characterized by the equation 5.

$$G(x; \tilde{\sigma}, \xi, u) = 1 - \left[1 + \xi(x-u) / \tilde{\sigma}\right]^{-1/\xi} \quad (5)$$

where $x-u > 0$, $1+ \xi (x - u)/ \tilde{\sigma} > 0$ and $\tilde{\sigma} = \sigma + \xi (u-\mu)$. The parameter μ represents the location parameter, and $\tilde{\sigma}$ the scale parameter. This function gives the cumulative probability for X exceeding the value of x, given that it already exceeds the threshold u. The duality between the GEV and generalized Pareto families means that the shape parameter ξ is dominant in determining the qualitative behavior of the GPD, just as it is for the GEV distribution. In particular, the values of ξ are common across the two models. Furthermore, the value of $\tilde{\sigma}$ is found to be threshold-dependent, except in the case where the limit model has $\xi = 0$.

Threshold selection is critical to any POT analysis. Too high a threshold could discard too much data leading to high variance of the estimate, but too low a threshold is likely to violate the asymptotic basis of the model, leading to bias. The standard practice is to adopt as low a threshold as possible, subject to the limit model providing a reasonable approximation. So, threshold selection is a commitment between choosing a high enough value for the asymptotic theorem can be considered accurate, and low enough to have the sufficient material to estimate the parameters ξ and β.

Other important assumption for the GPD is that the threshold exceedances are independent. Such an assumption is often unreasonable for weather and climate data because high values of meteorological and climatological quantities are often succeeded by high quantities (e.g. high rain day is likely to be followed by another high rain day). An approach frequently employed to handle such dependency is to decluster the data by identifying clusters and to utilize only a summary of each cluster; one of the simplest and most widely used methods for determining clusters is runs de-clustering (Gilleland & Katz, 2006).

5. Return levels (quantiles)

When considering extreme values of a random variable, one is interested in the return level of an extreme event, defined as the value z_p, such that there is a probability of p that z_p is exceeded in any given year, or alternatively, the level that is expected to be exceeded on average once every $1/p$ years ($1/p$ is often referred to as the return period); in extreme value terminology, z_p is the return level associated with the return period $1/p$. For example, if the 100-year return level for temperature is found to be 45°C, then the probability of temperature exceeding (return period) 45°C in any given year is $1/100 = 0.01$. In particular, a 20-year return value is the level that an annual extreme exceeds with probability $p = 5\%$. The quantity $1/p$ indicates the "rarity" of an extreme event and is usually referred to as the return period, or the waiting time for an extreme event.

The return level is derived from the distribution GEV or GPD by setting the cumulative distribution function equal to the desired probability/quantile, $1 - p$, and then solving for the return level. Estimates of extreme quantiles of the annual maximum distribution can be obtained by the equation 6.

$$f(x) = \begin{cases} \mu - \frac{\sigma}{\varepsilon} \left[1 - y_p^{-\varepsilon} \right], \text{ for } \varepsilon \neq 0 \\ \mu - \sigma \log y_p, \text{ for } \varepsilon \neq 0 \end{cases} \tag{6}$$

In the previous equation $y_p = -\log(1-p)$. If z_p is plotted against $\log y_p$, the plot is linear in the case $\xi = 0$. If $\xi < 0$ the plot is convex with asymptotic limit as $p \rightarrow 0$ at $\mu - \sigma/\xi$; if $\xi > 0$ the plot is concave and has no finite bound. This graph is a return level plot. Because of the simplicity of interpretation, and because the choice of scale compresses the tail of the distribution so that the effect of extrapolation is highlighted, return level plots are particularly convenient for both model presentation and validation.

6. Weather and climate extremes and its relationship with climate change

There is general agreement that changes in the frequency or intensity of extreme weather and climate events is increasing in many regions in response to global climate change and would have profound impacts on both human society and the natural environment. This has motivated many studies of extremes in the climate in the first decade of 2000's (Meehl et al., 2000; Easterling et al., 2000; Rusticucci & Barrucand, 2004; Kharin et al., 2007); Recent years have seen a number of weather and climate events cause large losses of life as well as a tremendous increase in economic losses from weather hazards. Examples of weather extremes in past decade (2001-2010) are showed in table 1 (WMO, 2011).

Year	Extreme weather events
2001	Extreme cold winter in Siberia and Mongolia. Minimum temperatures of near - 60°C across central and southern Siberia resulting in hundreds of deaths. Canada recorded the eighteenth straight warmer-than-average season.
2002	Exceptionally heavy rains in central Europe caused flooding of historic proportions, killing more than 100 people and forcing the evacuation of more than 450 000 people. Damage was estimated at US$ 9 billion in Germany alone.
2003	Europe recorded in August 2003 its worst heatwave. In many locations, temperatures rose above 40°C. In Belgium, France, Germany, Italy, the Netherlands, Portugal, Spain, Switzerland and the United Kingdom, 40 000 to 70 000 deaths were attributed to the heatwaves.
2004	Widespread winter storms in the Mediterranean region. Extreme hot conditions persisted in Japan during the summer, with record-breaking temperatures. A record number of 10 tropical cyclones made landfall in Japan, including. The first tropical cyclone since the start of satellite records made landfall on the southern coast of Brazil. In Afghanistan, drought conditions continued this year.
2005	This year was ranked in the top two warmest years along with 1998. Most active Atlantic hurricane season on record. In Central America and the Caribbean region, the most damage occurred from Hurricanes *Dennis, Emily, Stan, Wilma* and *Beta*. In the United States, Hurricane *Katrina* was the deadliest hurricane to hit the country since 1928, killing over 1 300 people. Australia officially recorded its warmest year on record.
2006	Heavy rains ended prolonged drought in the Greater Horn of Africa, leading to the worst flooding in October/November in 50 years. Disastrous tropical cyclones hit some south east Asian nations, including Typhoon *Durian* which killed nearly 1 200 people in the Philippines.
2007	Mexico suffered the worst flooding in five decades in November, causing the worst weather-related disaster in its history. Severe to exceptional drought continued in the south-east United States, with the driest spring on record and the second worst fire season after 2006.
2008	China witnessed the worst severe winter weather in five decades in January, with over 78 million people affected by the freezing temperatures and heavy snow. Tropical Cyclone *Nargis* with maximum winds of 215 km/hour was the most devastating cyclone to strike Asia since 1991, causing Myanmar's worst natural disaster ever.
2009	Australia was affected by exceptional heatwaves This was associated with disastrous bushfires that caused more than 170 fatalities. Victoria recorded its highest temperature with 48.8°C at Hopetoun, the highest temperature ever recorded so far south in the world.
2010	This year was ranked as the warmest year on record, along with 1998 and 2005. Hundreds of records for daily minimum temperatures were broken in the United States. Heavy snowfall disrupted air and road traffic in Europe, the United States and China. Australia faced its worst flooding in about 50 years.

Table 1. Some Weather Extremes in last decade (2001-2010), (WMO, 2011).

7. Weather and climate extremes: Review methods for their modeling

Quantifying and predicting changes in mean climate conditions and shifts in the frequency of extreme events is a daunting task that is vigorously pursued by many scientists, federal agencies, and private companies. Basically there are two essential tools for studying extreme weather and climate events: a) modeling using General Circulation Models (GCMs), and b) statistical modeling using Extreme Value Theory (EVT). Climate Models are derived from fundamental physical laws, which are then subjected to physical approximations appropriate for the large-scale climate system, and then further approximated through mathematical discretization (IPCC, 2007). Models show significant and increasing skill in representing many important *mean climate* features, such as the large-scale distributions of atmospheric temperature, precipitation, radiation and wind, and of oceanic temperatures, currents and sea ice cover. Simulations with global coupled ocean–atmosphere general circulation models (CGCMs) forced with projected greenhouse gas and aerosol emissions are the primary tools for studying possible future changes in climate mean, variability, and extremes. The ability of the generation of atmospheric general circulation models to simulate temperature and precipitation extremes was documented by Kharin et al. (2005, 2007). However, because the simulation of extremes pushes the limits of what GCMs are capable of, it is important to consider a number of advantages when using GCMs to study weather and climate extremes (Yin & Branstator, 2007). One advantage is the possibility of using a GCM to study extremes in climates different from that observed today, another important advantage is that GCM experiments can be designed to test hypotheses about the dynamics and other factors that influence extremes. A third advantage is that GCMs can be used to produce sample sizes of extremes much larger than found in the short observational record. Thus, while the observational record may be too short for robust statistics of extremes, a GCM dataset of hundreds or thousands of years of model time could produce robust extreme statistics. Accordingly by Tebaldi et al. (2006), GCMs are increasingly being used to study climate and weather extremes for societal and ecological impacts (Jentsch et al., 2007).

On the other hand, the complexity of the climate system always required the application of statistical methods to identify relationships among the climate quantities that were hidden in the multitude of spatial and temporal scales in play (Navarra, 1999). In the course of the years it has been shown that statistical techniques are a powerful tool that can lead to an understanding of nature as profound as other scientific devices. In particular, the use of EVT is a tool that seeks to provide an estimate of the tails of the original distribution using only extreme values of the data series. Many studies use EVT related to extreme weather and climate events and their impact: in ecology (Parmesan et al., 2000; Katz et al., 2002; Dixon et al., 2005); in disaster losses (Pielke, 2007; WMO, 2011); in heatwaves, extreme rainfall, snow events and droughts and related damages that affect to the community and stakeholders (Easterling et al., 2000; Meehl et al., 2000; Katz et al., 2005; Garcia-Cueto et al., 2010; Deguenon & Barbulescu, 2011).

8. Examples of application of EVT to extreme climate events in two cities of Mexico: Mexicali and Villahermosa

This part of the chapter focuses on the analysis of daily maximum temperature records from a local weather station in the northwest of Mexico, and daily rainfall records from a local weather station in southwest of Mexico. The city of northwest Mexico is named Mexicali, and the city of southeast Mexico is named Villahermosa. First we describe the climate of each city, and then Extreme Value Theory is applied. In the case of Mexicali City the summer maximum temperature data was fitted to a Generalized Extreme Value (GEV) distribution by using a block maxima approach. Furthermore, from 5-year to 500-year return level and shape parameter confidence limits were found, respectively. In the case of Villahermosa City, daily precipitation data over a threshold were fitted to Generalized Pareto Distribution (GPD). As in the case of Mexicali City, rainfall maximum return levels of several time periods for Villahermosa City are estimated.

8.1 Location and climate in Mexicali City

Mexicali is located in the Sonoran desert of northwestern Mexico, at 32.55°N, 115.47°'W and 4 meters above sea level; it borders Calexico, CA to the north and Sonora, Mexico to the west. Mexicali features a dry arid climate [Garcia's climate classification BW(h')(hs)(x')], with extremely hot summers and cold winters. Mexicali is one of the hottest cities of Mexico, with average July high temperatures of 42.2 °C. Average January highs are around 21 °C. Mexicali receives 90% of the maximum potential hours of daylight each year. On average Mexicali receives about 75 mm of rain annually. On July 28, 1995, Mexicali reached its all-time high of 52 °C.

8.2 EVT applied to summer maximum temperature at Mexicali City

Extreme Value Theory (EVT) was applied to find the probability of the highest summer temperatures, and quantify return levels, at Mexicali City. The analysis can be useful to help society prepare and protect itself from future dangerous temperatures. The data consists of the daily "summer" temperature records from a local weather station (Period 6/1/1951-9/15/2008, Station Mexicali 02033, Comision Nacional del Agua). Summer is defined in this research from 1 June to 15 September, the period for which maximum temperatures are very high. Data were fitted to a Generalized Extreme Value (GEV) distribution by using a block maxima approach. Furthermore, some predictions such as various return levels and shape parameter confidence limits were found. The Maximum Likelihood method for estimating the parameters (location, scale and shape) was used. The extremes package (Gilleland and Katz, 2006) of R (R Development Core Team, 2010) was used because it is an open source; it is particularly well oriented to climatic applications and has the ability to incorporate information about co-variables in order to estimate parameters. Figure 1 shows the plot of summer maximum temperature. It can be observed from figure 1 that the highest value is 52°C and lowest is 43.8°C. Maximum-likelihood fitted parameter values are presented in Table 2.

The figure 2 shows diagnostic plots from the proposed fitting. Combining the estimates and standard errors, the approximate 95% confidence intervals are (46.23, 47.01) for μ, (1.112, 1.645) for σ, and (-0.005, -0.287) for ξ.

When the probability and quantile graphs were examined (fig. 2), and since the plotted data approximately are near-linear, it suggested that the underlying suppositions for the GEV distribution are reasonable for the summer maximum temperature data. According to the shape parameter (ξ = -0.146), the most adequate distribution for modeling the summer maximum temperature occurrence is the Weibull distribution, which is bounded above, meaning that there are finite values which the maximum temperatures cannot exceed. Based on the results presented in Table 2, the estimate of upper limit is calculated as $\mu - \sigma/\xi$ = 46.623 - 1.379/-0.146 = 56.012 (°C). Testing the fit of the GEV probability density function to data was done using the chi-square goodness of fit test. The condition $\chi_n^2 \leq \chi_{n,0.05}^2$ is satisfied (3.04 < 3.84).

The cumulative distribution function (CDF) of block maximum z is calculated from equation (1) as: $G(z, \mu, \sigma, \xi) = \exp\{-[1-0.14695((z-46.62329)/1.3797)]^{(1/0.14695)}\}$. Figure 3 shows CDF for summer maximum temperature at Mexicali City.

Parameter	Estimate	Standard Error
Location (μ)	46.623	0.198
Scale (σ)	1.379	0.136
Shape (ξ)	-0.146	0.072
Negative log-likelihood	104.98	

Table 2. GEV parameter estimates from fitting summer maximum temperatures at Mexicali, Mexico.

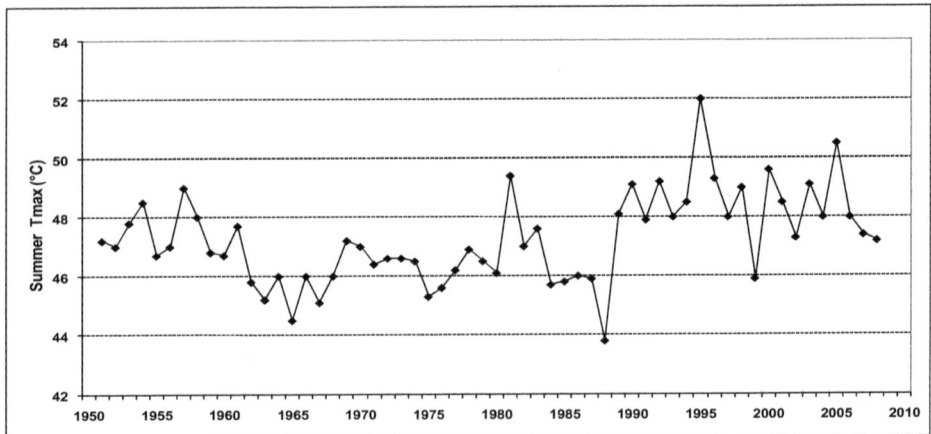

Fig. 1. Time plot of summer maximum temperature at Mexicali, Mexico (1951-2008).

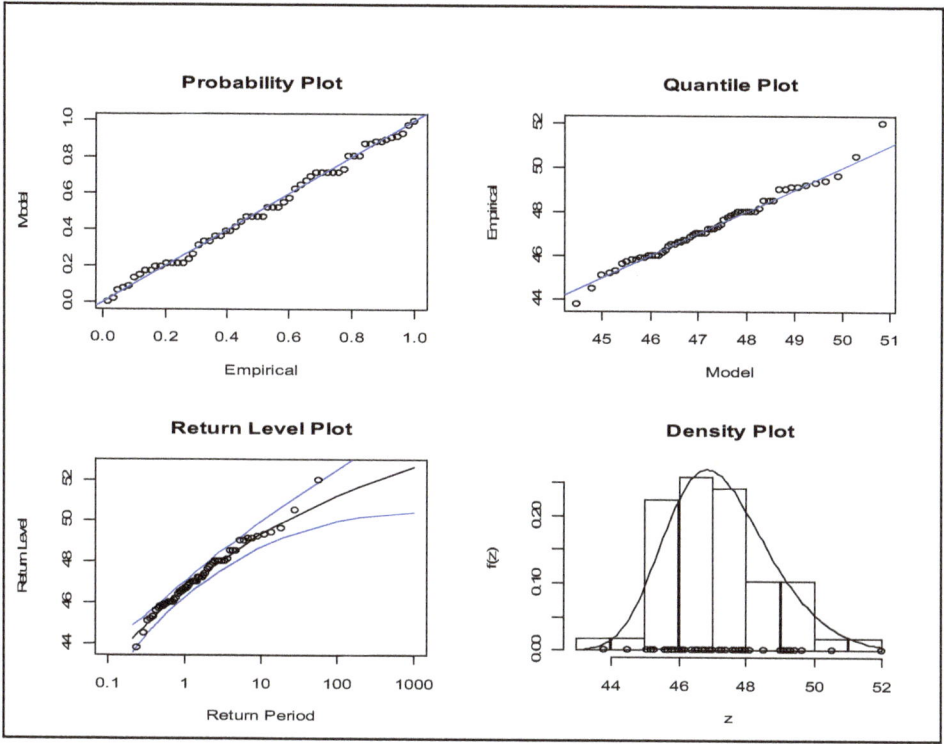

Fig. 2. GEV fit diagnostic plots for summer maximum temperature (°C) at Mexicali, Mexico (1951-2008).

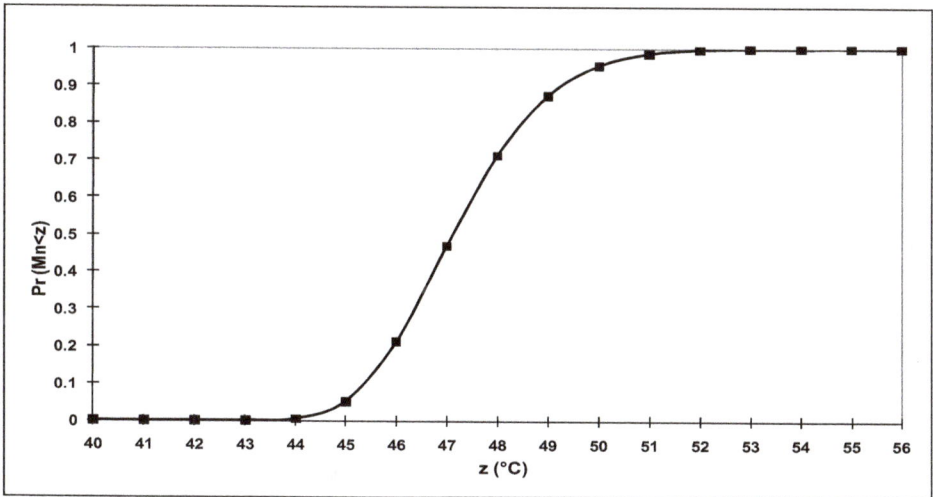

Fig. 3. GEV distribution fit to summer maximum temperature at Mexicali City, Mexico.

With the CDF, we can calculate the return levels of the summer maximum temperature. This plot is shown in bottom left of figure 2 along the point-wise 95% confidence bounds estimated by the delta method. The delta method assumes that the parameters estimates are symmetric, which is not always the case for the shape parameter or extreme return level (Gilleland & Katz, 2005). Greater accuracy of the confidence intervals can usually be achieved using the profile likelihood. Figure 4 show the profile log-likelihoods for the 100-year return level. The estimated return level is 51.2°C with 95% confidence interval of (50.4, 53.1). Similarly, the 95% confidence interval of the shape parameter ξ was (-0.269, -0.104). From the profile likelihood plots, the estimates were good because the dashed vertical lines intersected the likelihood at the same points as the lower horizontal line in both cases.

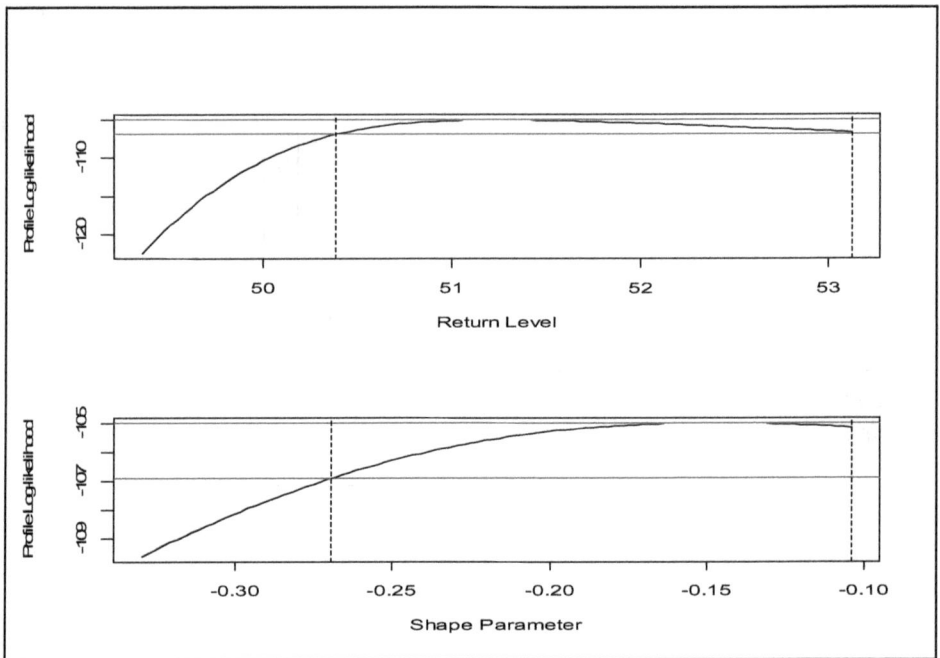

Fig. 4. 100-year profile likelihood plots for summer maximum temperature (°C) at Mexicali Mexico in period from 1951 to 2008.

Table 3 shows the results of estimated return levels and 95% confidence intervals.

It can be seen from Table 3 that the return levels for summer maximum temperature gradually increase for higher and higher return periods. Also the confidence intervals are increasingly wider as the return period is increasing. From the above results one would expect, per example, that summer maximum temperature (°C) at Mexicali Mexico will exceed about 49.3°C on average every 10 years, and will exceed about 51.2 on average every 100 years, and will exceed about 52.2°C every 500 years. The 95% confidence intervals, in (°C), were 48.7-50.0, 50.4-53.1, and 51.1-55.1, respectively.

Return period, yr	Return level (°C)	Lower bound (°C)	Upper bound (°C)
5	48.5	48.0	49.0
10	49.3	48.7	50.0
15	49.7	49.1	50.6
20	49.9	49.3	51.0
25	50.1	49.5	51.4
50	50.7	50.0	52.2
75	51.0	50.2	52.8
100	51.2	50.4	53.1
500	52.2	51.1	55.1

Table 3. Estimated return levels and 95% confidence intervals for several period returns from having fit summer maximum temperature (°C) at Mexicali, Mexico (summer) to the GEV distribution.

8.3 Location and climate in Villahermosa Mexico

Villahermosa, capital of the State of Tabasco, is located in southwest Mexico, at 17.98°N, 92.91°'W and 9 meters above sea level. Villahermosa features a tropical wet climate (Garcia's climate classification Am), with an annual average temperature of 26°C and an annual average precipitation of 1500 mm. the City of Villahermosa is characterized by its warm and humid climate. Temperatures during spring and summer seasons reach upward of 40°C, with humidity levels near around 30% during the same period. During its short "winter", Villahermosa's climate is very humid, but daytime temperatures decrease to around 28°C. In October 2007, Villahermosa suffered its worst flood in recorded history. Several hundred thousand people were displaced because of flooded homes.

8.4 EVT applied to daily maximum rainfall at Villahermosa City

Floods in Villahermosa City have recently become more destructive and projections show that this trend may become more pronounced. Part of this trend is linked to socio-economic factors, but a portion of the flood growth is linked to climate. As quantification of maximum rainfall is important for flood planning purposes, Generalized Pareto Distribution (GPD) by using Peaks Over Threshold (POT) was applied to the data base of daily rainfall records from a local weather station (period 10/1/1948-8/31/2010, Station 27054 Villahermosa, Comision Nacional del Agua of Mexico Country). Furthermore, predictions of several return levels and shape parameter confidence limits were determined. Also the maximum likelihood method for estimating the parameters of scale and shape was applied. Figure 5 shows the plot of daily rainfall in Villahermosa City.

In order to apply the POT, a threshold has to be fixed. The threshold has to be sufficiently large so that GPD is a suitable function for describing the tail of the cumulative distribution, and it has to be sufficiently small so that enough values are available to give an accurate estimation of the parameters of the GPD (Asensio, 2007). There is not any known, automatic procedure for the selection of the threshold. Here we choose a value of the threshold subjectively, and we verify the behavior of the parameters of the GPD for different values of the threshold. In our case, u has been chosen as the value that exceeds 98.5% (21517) of the data points of the time series, leaving only 1.5% of the data points (336) as extreme values. For the dataset shown in figure 5, we find u = 60 mm. So 336 points above the threshold are used to fit the GPD, neglecting any time dependence. The maximum rainfall value recorded since the date is 340 mm.

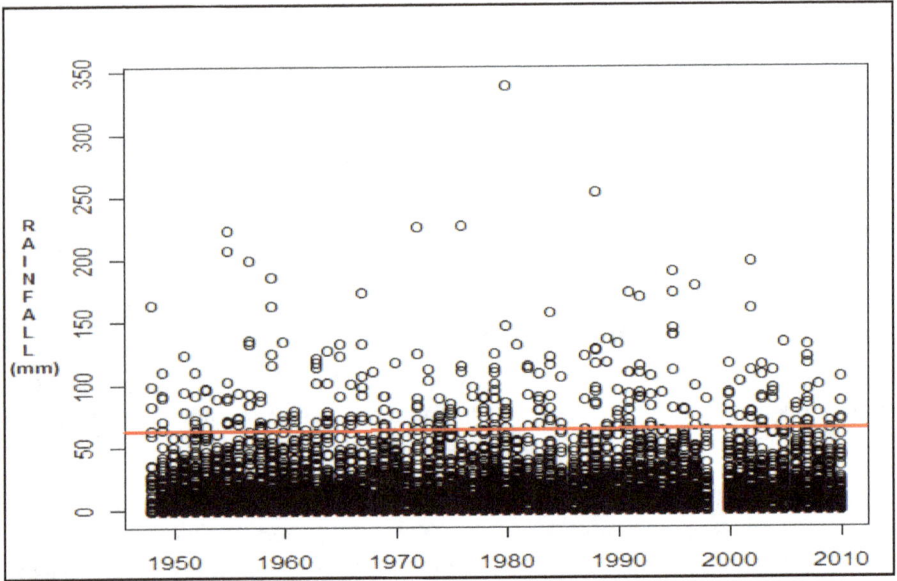

Fig. 5. Scatter plot of daily rainfall (mm) at Villahermosa City. Horizontal line is the threshold of 60 mm.

To show that the chosen threshold of 60 mm was adequate, the fit of the cumulative distribution with a range of thresholds was realized. The results are shown in figure 6 for different values of the threshold u, the upper panel showing the values obtained for σ (modified scale) and the lower panel the values for ξ (shape). If the GPD is a reasonable model for the exceedances above a certain threshold u_0, the estimates σ and ξ, should remain near-constant (Coles, 2001). It can be observed that selected threshold of 60 mm appears reasonable as the estimates of σ and ξ remain constant above level of 60 mm. It is interesting note that for u = 80 mm, approximately 0.8% of the data lie above the threshold, while for u = 100 mm, less than 0.4% of the data lie above. When there are less data available to estimate the parameters, it is obvious that the uncertainty increases.

The evaluation of significance of the model by likelihood ratio test was significant at the 5% confidence level. The empirical cumulative distribution function for points above the threshold is built and the values of σ and ξ that give the best fit were obtained. The parameters of the GPD were estimated by the maximum likelihood method using the statistical method for extreme values of R software; the values obtained are shown in table 4.

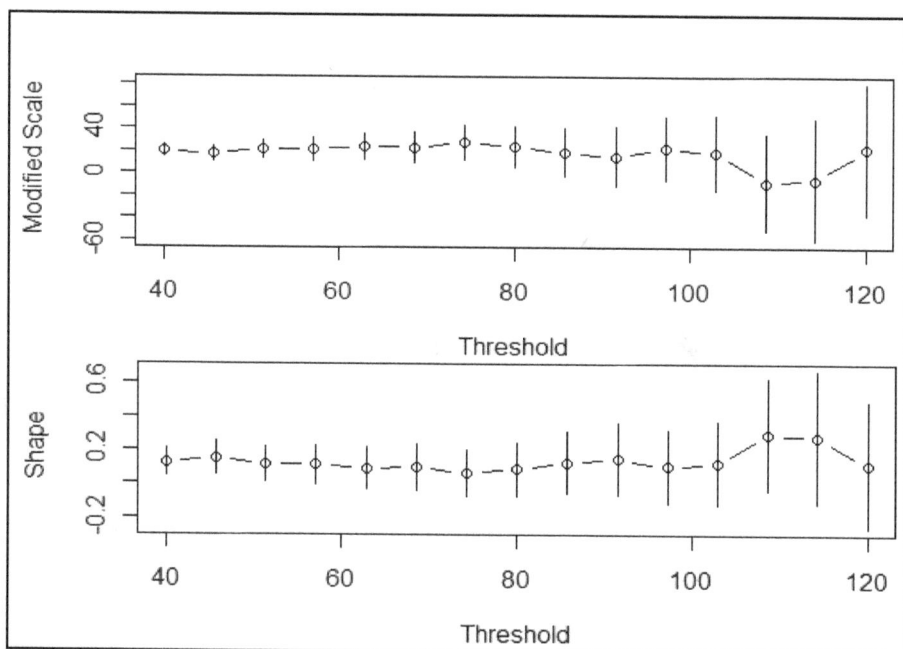

Fig. 6. Maximum likelihood estimates of σ (Modified Scale) and ξ (Shape) parameters obtained for different values of the threshold for the daily rainfall in Villahermosa during the period 1948-2010.

Parameter	Estimate	Standard error
Scale (σ)	27.15	2.27
Shape (ξ)	0.12	0.063

Table 4. Parameters of GPD rainfall data at Villahermosa City. The values are obtained with a threshold of 60 mm.

Figure 7 shows diagnostic plots from the proposed fitting, and suggest that probability and quantile graphs that the underlying assumption for the GPD are reasonably for these data. To assess the quality of fit between the GPD and this data set, the GPD density is superimposed on the histogram of the exceedances in figure 8. The estimated shape parameter is clearly positive ($\xi = 0.12$ with a standard error of 0.063) and it indicates that the distribution is heavy tailed.

With these values, the cumulative distribution function is shown in the figure 9, where we show the value over the threshold on the horizontal axis and the value of the GPD on the vertical axis.

Since $\xi > 0$, it is not useful to carry out a detailed inference of the upper limit, as was the case of summer maximum temperatures at Mexicali City. Instead, we focus on extreme return values. Return levels and shape parameter confidence limits (95%) for rainfall over threshold of 60 mm were estimated (Table 5).

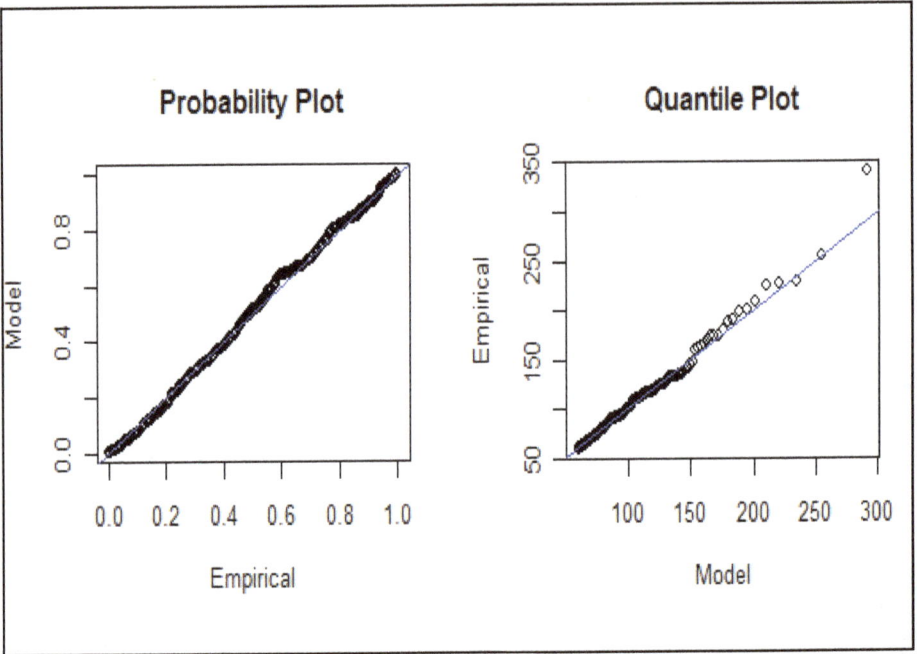

Fig. 7. Diagnostic plots for the threshold excess model fitted to daily rainfall data at Villahermosa, Mexico to a GPD.

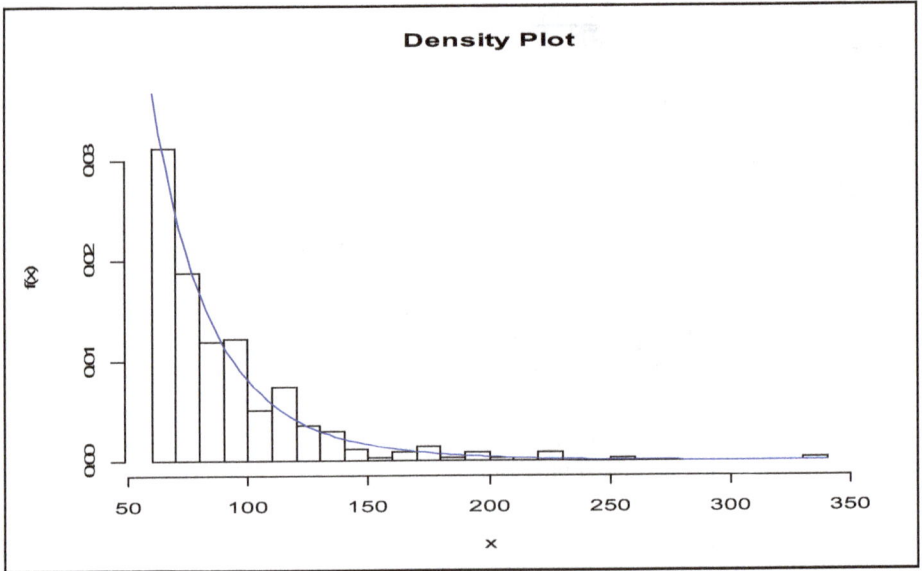

Fig. 8. Histogram and the fitted GPD density for the exceedances over threshold of 60 mm at Villahermosa, Mexico (1948-2010).

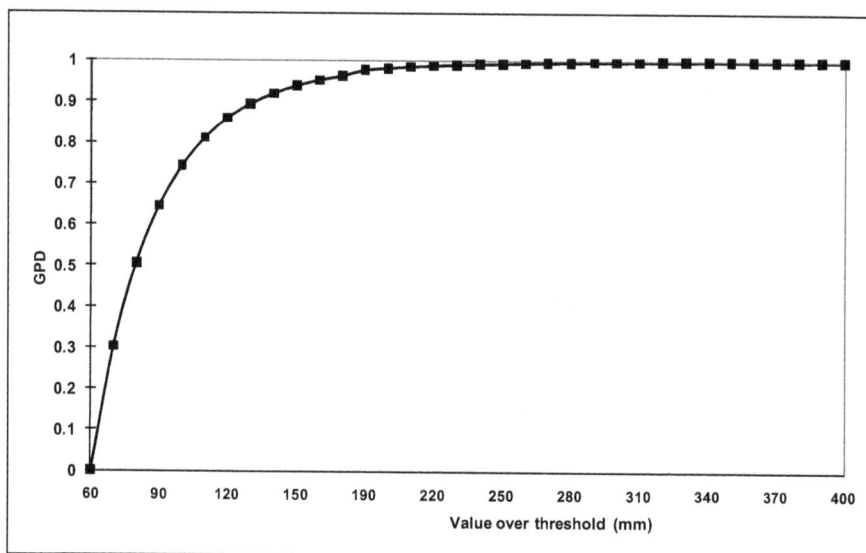

Fig. 9. GPD fit to rainfall over 60 mm at Villahermosa, Mexico (1948-2010).

Return period, yr	Return level (mm)	Lower bound (mm)	Upper bound (mm)
1	112.2	106.3	118.9
5	172.1	157.0	195.5
10	201.8	179.5	236.6
20	234.1	202.2	277.0
30	254.4	215.6	302.2
50	281.4	232.5	335.9
75	304.1	246.0	364.2
100	320.9	256.7	385.1
500	426.9	336.6	517.1

Table 5. Estimated return levels and 95% confidence intervals for several period returns from having fit rainfall daily data using a threshold of 60 mm at Villahermosa, Mexico to the GPD.

It can be seen from Table 5 that the return levels for rainfall over the threshold of 60 mm gradually increase for higher and higher return periods. Also the confidence intervals are increasingly wider as the return period increases. From the above results one would expect, per example, that rainfall over the threshold of 60 mm at Villahermosa Mexico will exceed about 201.8 mm on average every 10 years, and about 320.9 mm on average every 100 years, and about 426.9 mm every 500 years. The 95% confidence intervals (mm) were 179.5-236.6, 256.7-385.1, and 336.6-517.1, respectively.

9. Climate scenarios under climate forcing by anthropogenic effect

In a world where the climate remains "stationary", and that inputs have a periodic behavior, the return periods of climatic variables would be a function only of its history recorded in the database, which, as we have seen, can be estimated with some uncertainty under the theory of extreme values; however, if we refer to a dynamic climate in which climate change caused by greenhouse gas forcing is driving current climate going to another different climate regime, the estimate of future values of atmospheric variables, must be addressed with other techniques. These estimations that are called climate scenarios are a plausible representation of future climate that have been constructed for explicit use in investigating the potential impacts of anthropogenic climate change. Climate scenarios often make use of climate projections that are descriptions of the modeled response of the climate system to scenarios of greenhouse gas and aerosol concentrations, by manipulating model outputs and combining them with observed climate data.

Climate scenarios in our case studies were generated using a statistical-dynamic model, the Statistical Downscaling Model SDSM (Wilby et al., 2002). SDSM model is a combination of a stochastic weather generator and a transfer function method that needs two types of daily data. The first type corresponds to local predictands of interest (in our case maximum temperature and rainfall), and the second type corresponds to the data of large-scale predictors of a grid box closest to the study area. For model calibration the source of these predictors is the National Center for Environmental Prediction (NCEP) re-analysis data set and for downscaling future climate scenarios is the model HadCM3 with two emission scenarios, A2 and B2.

Previous to the scenarios generation, the local atmospheric predictands and NCEP predictors for the period of 1961-1990 were split into two parts. The first part of 1961-1975 was used for model's calibration while the remaining data of 1976-1990 were used for model's validation, as an independent set of data, which revealed satisfactory results. Next the corresponding large scale predictors of HadCM3 were used to downscale the future climate data of the selected cities under the A2 and B2 emission scenarios. These emissions scenarios were allowed which compare the results, since they belong to different evolutionary lines. While the A2 scenario assumes a very heterogeneous world, whose underlying theme is self-reliance and preservation of local identities, the B2 scenario describes a world in which the emphasis is on local solutions to economic, social and environmental sustainability. It is a world with continuously increasing global at a rate lower than A2. The scenarios projected of summer maximum temperatures at Mexicali City, and annual maximum daily rainfall at Villahermosa City, were built for periods 2010-2039, 2040-2069 and 2070-2099, called of the 2020's, 2050's and 2080's, respectively.

9.1 Scenarios of summer maximum temperature at Mexicali City

Scenarios of summer maximum temperatures at Mexicali City are shown in the Table 6.

Table 6 shows: 1) Each month shows a steady increase in summer maximum temperatures, both with medium emissions scenario (B2) and with high emissions (A2). With the same emissions scenario (A2 or B2) September is the one with the largest increase, indicating that the summer could be increasingly longer and more intense; 2) For the 2020's the largest increase in summer maximum temperature is with B2 emissions scenario, compared to A2. With respect to the baseline scenario (1961-1990) July presented the largest increase (2.5 °C), followed by the months of September, August and June, with 2.4 °C, 1.4 °C and 1.3 °C, respectively; 3) For the 2050's the largest increase in summer maximum temperature is with

A2 emissions scenario, except the month of September. The largest increase would be in September with 4.2 °C, followed by July, June and August, with 3.0 °C, 2.6 °C and 2.1 °C, respectively; and 4) For late XXI century, the largest increase in summer maximum temperature would have again with the A2 emissions scenario; the month of September would be the largest increase, up to 5.4 °C, followed by the months of July, August and June, with 4.1 °C, 3.7 °C and 3.5 °C, respectively.

Month	Summer Maximum Temperature (°C)						
	Observed Climate Baseline (1961-1990)	Future Scenarios					
		2020's		2050's		2080's	
		A2	B2	A2	B2	A2	B2
June	49.1	50.0	50.4	51.7	51.1	52.6	51.2
July	48.1	50.0	50.6	51.1	50.7	52.2	51.3
August	49.4	50.6	50.8	51.5	51.0	53.1	51.8
September	47.0	49.3	49.4	50.5	51.2	52.4	50.9

Table 6. Observed and Future Scenarios of summer maximum temperature for the 2020's, 2050's and 2080's at Mexicali City, Mexico with A2 and B2 emissions scenarios.

9.2 Scenarios of maximum daily rainfall at Villahermosa City

Figure 10 shows that neither with the high emissions scenario A2 would expect the maximum rainfall reaches 340 mm, recorded in October in baseline scenario of 1961-1990.

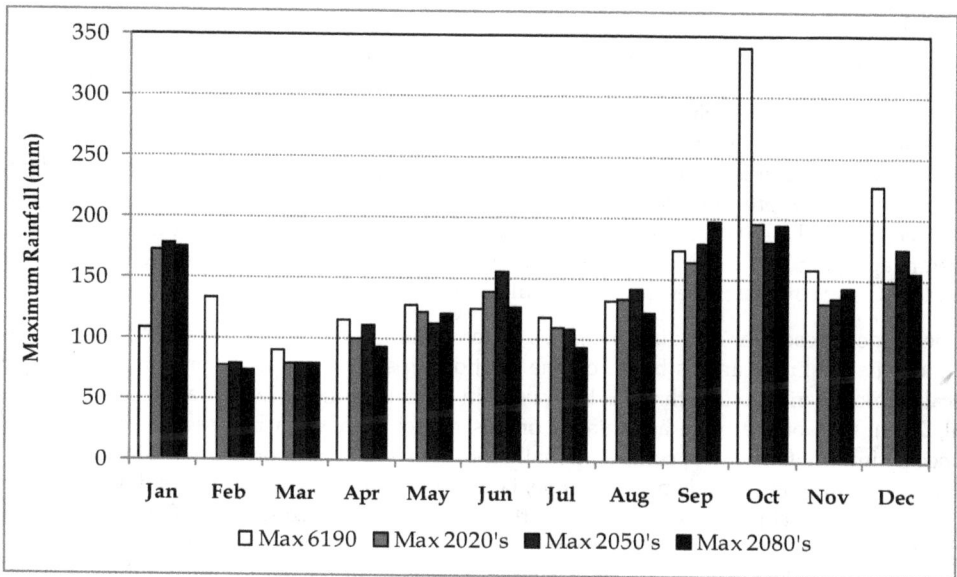

Fig. 10. Monthly daily maximum rainfall in a baseline scenario (1961-1990) and projected monthly maximum daily rainfall for the 2020's, 2050's and 2080's at Villahermosa City, with A2 emissions scenario.

Taking into consideration a measure of variability such as standard deviation, by the end of XXI century is expected that the maximum rainfall for the month of October is in the range from 140.5 mm to 250.3 mm. Also, by the end of the century in the month of January is expected a strong increase in the maximum rainfall, although the high variability determines an interval from 118.4 mm to 231.7 mm; in the months of February and December, noted that in contrast to that recorded in the baseline scenario, the maximum rainfall in the future scenarios are below that value, even considering that the standard deviation for the months of February and December are 16.6 mm and 54.9 mm, respectively.

10. Discussion

The results obtained in this chapter are based on the theory of extreme values that has a strong theoretical background and the realization of climate scenarios with a downscaling model, under A2 and B2 emissions scenarios. The basic premise relies on the assumption of asymptotic argument that converges to the three limiting distributions, as the size of the time series increases. Theoretically it is expected that for a time series infinitely large, the models correctly reproduce the tails of underlying distribution. In our case we have time series of limited size, so there always is an element of doubt when estimating the parameters of the chosen distribution, and when one is extrapolating values toward that, there is some uncertainty. However, EVT makes the best use of whatever data is available about extremes. In our case, the present results seem to support the fact that statistics of summer maximum temperature and daily maximum rainfall are well replicated under the background of extreme value theory. The fact that the theory explains the extreme events studied in this chapter supports the assertion of that EVT is able to estimate extreme quantiles for a short record of data.

In the case of modeling daily maximum rainfall in Villahermosa City, although the method applied is certainly very valuable, it has, however, some limitations: 1) the POT method was used without de-clustering techniques, and 2) the underlying distribution of analyzed stochastic variables should be stationary, but this condition is not always true in weather events. Therefore, it is clear that future studies will be needed to gain confidence regarding goodness of fit of proposed model (Smith, 2001). Accordingly: i) Excesses of rainfall over the chosen threshold must be studied in clusters, ii) The seasonal trend should be removed before applying POT technique, iii) The rainy and dry seasons should be studied separately, and iv) Other covariates, besides time should be included.

The model chosen for building scenarios of temperature and rainfall, the SDSM model, is conceptually simple and is based on the relationships between predictands and the large scale atmospheric forcing, that has been used in several studies (Hashmi et al., 2009; Chu et al, 2010). One of the main advantages of this model is that can be used to provide site-specific information and is computationally inexpensive. The major theoretical weakness is that its assumption basic is not verifiable, i.e., that the statistical relationship developed for the present day climate also hold under the different forcing conditions of possible future climates (Wilby & Harris, 2006). Although the model is not very good at projecting extreme values, the results found seem encouraging, especially in regard to the positive trend of summer temperatures at Mexicali City, as it is according to the results of General Circulation Models and Regional Circulation Models. With respect to scenarios of maximum daily rainfall at Villahermosa City does not show a definite trend in all months. Another feature of the projections is not exceeding historical amounts recorded in the months of October and December. Importantly, none of the currently available statistical downscaling techniques has been able to reproduce the rainfall with all characteristics very well, so it would be important to use another downscaling technique to compare the results obtained with SDSM model.

Respect to comparative analysis between return periods obtained with extreme value theory and climate scenarios obtained with SDSM model, with A2 and B2 emissions scenarios, we can observe that summer maximum temperatures at Mexicali City, with both methods yields similar values. Per example, by end of century, with application of extreme value theory, one would expect that summer maximum temperature at Mexicali City will exceed about 51.2°C, with a 95% confidence interval, in °C, of 50.4 to 53.1. Too, by end of century, but using SDSM model, the summer maximum temperature projected, with A2 emissions scenario, is about 53.1°C. Regarding the comparison of extreme rainfall events at Villahermosa City with both extreme value theory and downscaling model SDSM, it appears that the SDSM results are conservative with respect to the results obtained with extreme value theory. Per example, by 2050's, it is expected that return period of maximum rainfall is between 232.5 mm and 335.9, while the SDSM is between 132.2 mm and 230.0 mm; by the end of the century it is expected that return period of maximum rainfall is between 256.7 mm and 385.1 mm, while the SDSM is between 149.8 mm and 245.2 mm.

11. Conclusions

1. Extreme Value Theory was successfully applied to two climatic datasets.
2. GEV distribution was fitted by using a block maxima approach to summer maximum temperatures, using daily data collected at the meteorological station of Mexicali, Mexico. The diagnostic plots and χ^2 goodness-of-fit test confirm the adequacy of this distribution for the data analyzed.
3. The most adequate distribution for modeling the summer maximum temperature occurrence at Mexicali, Mexico was the Weibull distribution. The estimation of the upper limit was 56.012°C.
4. Return level confidence limits showed that the summer maximum temperature in Mexicali City was expect to exceed about 49.3°C on average every 10 years, about 50.7°C on average every 50 years, and about 51.2°C on average every 100 years. The 95% confidence intervals, in °C, were (48.7, 50.0), (50.0, 52.2), and (50.4, 53.1), respectively.
5. GPD model was fitted by using a peak over threshold approach for daily maximum rainfall, using daily data collected at the meteorological station at Villahermosa, Mexico. A threshold of 60 mm was chosen after a range of thresholds was demontrated. The evaluation of significance of the proposed model by likelihood ratio test was significant at the 95% confidence level. The diagnostic plots and quantile graphs suggest that the underlying assumption for the GPD is reasonable for the data analyzed.
6. Return levels confidence limits showed that daily maximum rainfall at Villahermosa City, with a threshold of 60 mm, was expect to exceed about 201.8 mm on average every 10 years, about 281.4 mm on average every 50 years, and about 320.9 mm on average every 100 years. The 95% confidence intervals, in mm, were (179.5, 236.6), (232.5, 335.9), and (256.7, 385.1), respectively.

7. There are shortcomings in the POT method proposed for modeling daily maximum rainfall including not using de-clustering techniques and not removing seasonal trend.

8. Scenarios for summer maximum temperature and maximum rainfall, with SDSM model, and emission scenarios A2 and B2, for the 2020's, 2050's and 2080's, were generated.

9. The comparison between the results obtained by extreme value theory (return periods) and those obtained by downscaling technique shows that for summer maximum temperatures at Mexicali City, are similar throughout this century. For extreme rainfall events SDSM model presents conservative results, i.e. lower than those found with respect to GPD model.

10. It would be important to use another downscaling technique, for estimate extreme events of rainfall at Villahermosa City, to compare the results obtained with SDSM model.

11. Estimating future return levels, and elaborate scenarios, in both temperature and rainfall extremes, at Mexicali City and Villahermosa City, respectively, provides essential input to urban adaptation and planning strategies through the establishment of, for example, heat watch-warning systems or flood prevention strategies. It is obvious that modeling work presented here reflects some uncertainties, as do all models of natural processes. However, inaction to protect lives and property may cost more if we dare not act because of unresolved questions.

12. References

Asensio, A. (2007). Extreme value theory and the solar cycle. *Astronomy and Astrophysics*, Vol. 472, No. 1, pp. 293-298, ISSN 0004-6361

Beirlant, J.; Goegebeur, Y.; Segers, J. & Tegels, J. (2004). *Statistics of extremes: theory and applications*, ISBN 0-471-97647-4, England

Berning, L. (2010). Improved estimation procedures for a positive extreme value index. *Thesis PhD*, Stellenbosch University, 259 pp.

Coles, S. (2001). *An Introduction to Statistical Modeling of Extreme Values*, ISBN 1-85233-459-2, London, Great Britain

Chu, J.; Xia, J.; Xu, C. & Singh, V. (2010). Statistical downscaling of daily mean temperature, pan evaporation and precipitation for climate change scenarios in Haihe, China. *Theoretical and Applied Climatology*, Vol. 99, pp. 149-161, ISSN 0177-798X

Deguenon, J. & Barbulescu, A. (2011). Study of Extreme Rainfall using GPD Model. *International Journal of Mathematics and Computation*, Vol. 11, No. 11, pp. 28-37, ISSN 0974-5718

De Haan, L. & Ferreira, A. (2006). *Extreme Value Theory: An Introduction*. ISBN 978-0387239460, New York

Dixon, P.; Ellison, A. & Gotelli, N. (2005). Improving the precision of estimates of the frequency of rare events. *Ecology*, Vol. 86, No. 5, pp. 1114-1123, ISSN 0012-9658

Easterling, R. (2000). Climate extremes: Observations, Modeling and Impacts. *Science*, Vol. 289, No. 5847, pp. 2068-2074, ISSN 0036-8075

Easterling, D.; Evans, J.; Groissman, P.; Karl, R.; Kunkel, E. & Ambenje, P. (2000). Observed variability and trends in extreme climate events: a brief review. *Bulletin of the American Meteorological Society*, Vol. 81, No. 3, pp. 417-425, ISSN 0003-0007

Embrechts, P., Klüppelberg, C. & Mikosch, T. 1997. *Modelling extremal events for insurance and finance*. Berlin: Springer

Finkenstädt, B. & Rootzén, H. 2004. Extreme values in finance, telecommunications and the environment. ISBN 1-58488-411-8

Fisher, R.A. & Tippett, L.H.C. 1928. On the estimation of the frequency distributions of the largest or smallest member of a sample. *Proceedings of the Cambridge Philosophical Society*, vol.24, pp.180-190

García-Cueto, R.; Tejeda-Martinez, A. & Jauregui, E. (2010). Heat waves and heat days in an arid city in the northwest of Mexico: current trends and in climate change scenarios. *Journal International of Biometeorology*, Vol. 54, No. 4, pp. 335-345

Gilleland, E. & Katz, R. (eds) (2005). Tutorial for the extremes toolkit: weather and climate applications of extreme value statistics, http://www.assessement.ucar.edu/toolkit

Gilleland, E. & Katz, R. (2006). Analyzing seasonal to interannual extreme weather and climate variability with the extremes toolkit (extRemes), *18th Conference on Climate Variability and Change, 86th American Meteorological Society (AMS) Annual Meeting, 29 January - 2 February, 2006, Atlanta, Georgia.*

Gnedenko, B. (1943). Sur la distribution limite du terme maximum d'une série aléatoire. *Annals of Mathematics*, vol.44, no.3, pp. 423-453

Gumbel, E. (1958). *Statistics of Extremes*, New York, ISBN 0-486-43604-7

Hadživuković, S. & Emilija, N. (2005). R. A. Fisher and Modern Statistics. International Statistical Institute, 55th Session.

Hashmi, M.; Shamseldin, A. & Melville, B. (2009). Statistical downscaling of precipitation: state-of-the-art and application of Bayesian multi-model approach for uncertainty assessment. *Hydrology and Earth Systems Science*, Vol. 6, pp. 6535-6579.

IPCC (2007). *Climate Change 2007: The Physical Science Basis. Contribution of Working Group I to the Fourth Assessment Report of the Intergovernmental Panel on Climate Change* [Solomon S., D. Qin, M. Manning, Z. Chen, M. Marquis, K.B. Averyt, M. Tignor & H.L. Miller (eds.)]. Cambridge University Press, UK and NY, USA, 996 pp.

Jentsch, A.; Kreyling, J. & Beieirkuhnlein, C. (2007). A new generation of climate-change experiments: events, not trends. *Frontiers in Ecology and the Environment*, Vol. 5, No. 7, pp. 365-374, ISSN 1540-9295

Katz, R. & Brown, B. (1992). Extreme Events in a Changing Climate: Variability is More Important than Averages. *Climatic Change*, Vol. 21, No. 3, pp. 289–302, ISSN 0165-0009

Katz, R.; Brush, G. & Parlange, M. (2005). Statistics of extremes: modeling ecological disturbances. *Ecology*, Vol. 86, No. 5, pp. 1124-1134, ISSN 0012-9658

Katz, R.; Parlange, M, & Naveau, P. (2002). Statistics of extremes in hydrology. *Advances in Water Resources* Vol. 25, No. x, pp. 1287-1304, ISSN 0309-1708

Kharin, V. & Zwiers, F. (2005). Estimating extremes in transient climate change simulations, *Journal of Climate* 18, pp. 1156-1173

Kharin, V.; Zwiers, F.; Zhang, J. & Hegerl, G. (2007). Changes in temperature and precipitation extremes in the IPCC ensemble of global coupled model simulations. *Journal of Climate*, Vol. 20, No. x, pp. 1419-1444, ISSN 0894-8755

Kotz, S. & Nadarajah, S. (2000). *Extreme Value Distributions: Theory and Applications*. ISBN 978-1860942242 London

Leadbetter, M.; Lindgren, G. & Rootzén H. (1983). *Extremes and related properties of stationary sequences and processes*. ISBN 0387907319 New York

Meehl,G;, Zwiers, F.; Evans, J.; Knutson, T.; Mearns, L. & Whetton, P. (2000). Trends in extreme weather and climate events: issues related to modeling extremes in projections of future climate change. *Bulletin of the American Meteorological Society*, Vol. 81, No. 3, pp. 427-436, ISSN 0003-0007

Navarra, A. (1999). The Development of Climate Research, In *Analysis of Climate Variability*, H. von Storch & A. Navarra (Eds), 3-10, Springer, ISBN 3-540-66315-0, Heidelberg, Alemania

Naveau, P.; Nogaj, M.; Amman, C.; Yiou, P.; Cooley, D. & Jomelli, V. (2005). Statistical methods for the analysis of climate extremes. *Comptes Rendus Geoscience*, Vol. 337, No. x, pp. 1013-1022, ISSN 1631-0713

Palmer, T. & Räisänen J. (2002). Quantifying the risk of extreme seasonal precipitation events in a changing climate. *Nature*, No. 415, pp. 512-514, ISSN 0028-0836

Parmesan, C.; Root, T. & Willig, M. (2000). Impacts of extreme weather and climate on terrestrial biota. *Bulletin of the American Meteorological Society*, Vol. 81, No. 3, pp. 443-450, ISSN 0003-0007

Pickands, J. (1975). Statistical inference using extreme order statistics. *Annals of Statistics*, vol.3, pp. 119-131

Pielke, R. (2007). Extremes in global climate models. *Proceedings Winter Workshop Extreme Events*, pp. 131-140, Hawai, USA, January 23-26, 2010

R Development Core Team, (2010). *A language and environment for statistical computing*. R Foundation for Statistical Computing, ISBN 3-900051-07-0, *http://www.R-project.org*¸

Reiss, R. & Thomas, M. (2007). *Statistical analysis of extreme values: with Applications to Insurance, Finance, Hydrology and Other Fields*, ISBN 978-3764372309, New York

Rusticucci, M. & Barrucand, M. (2004). Observed trends and changes in temperature extremes over Argentina. *Journal of Climate*, Vol. 17, No. X, pp. 4099-4107, ISSN 0894-8755

Smith, R. (2001). *Environmental Statistics*, Department of Statistics, University of North Carolina, USA pp. 377, Available from http://www.stat.unc.edu/postscript/rs/envnotes.ps

Tebaldi, C.; Hayhoe, K; Arblaster, J. & Meehl, G. (2006). Going to the extremes. *Climatic Change*, Vol. 79, pp. 185-211, ISSN 0165-0009

Wilby, R.; Dawson, C. & Barrow, E. (2002). SDSM- a decision support tool for the assessment of regional climate change impacts. *Environmental Modeling Software*, Vol. 17, pp. 145-157.

Wilby, R. & Harris, I. (2006). A framework for assessing uncertainties in climate change impacts: low flow scenarios for the River Thames, UK. *Water Resources Research*, Vol. 42, W02419, doi10.1029/2005WR004065

WMO, (2011). Weather extremes in a changing climate, ISBN 978-92-63-11075-6, Geneva

Yin, J. & Branstator, G. (2007). Extremes in global climate models. *Proceedings Winter Workshop Extreme Events*, pp. 67-72, Hawai, USA January 23-26, 2010.

Part 3

Ocean, Biosphere and Soil Interactions

8

Modeling Freezing and Thawing of Subsurface Soil Layers for Global Climate Models

Kazuyuki Saito
[1]*University of Alaska Fairbanks*
[2]*Japan Agency for Marine-Earth Science and Technology*
[1]*USA*
[2]*Japan*

1. Introduction

Freezing ground, including permafrost and seasonally-frozen ground, is found in more than half of the Northern Hemisphere (NH). The changes in soil freezing/thawing, or the subsurface thermal conditions, are a local phenomenon, but can have large impacts on climate and life in cold regions. Permafrost (soil is at or below 0°C for more than two consecutive years), occupying about 20% of the exposed NH land, is an essential component in the hydro-eco-climate system in the Arctic. It defines and controls biogeochemical activities, types of dominant vegetation, hydrological states and conditions (e.g., water holding and storage capacity, river and channel routing, limnology) in the regions. However, the influences of local subsurface hydrological-thermal regime changes are not confined to the high-latitude regions. Multiple pathways, which are physical, ecological and/or biogeochemical, can transfer the subsurface changes to other regions, and affect the regional to global climate with different consequences. The carbon and nitrogen cycles are closely controlled by the presence of permafrost, degradation of which may cause an irreversible change in the cycle and accelerate the emission of greenhouse gases, such as carbon dioxide and methane. Some types of vegetation may not survive when the underlying permafrost degrades, leading to migration of other vegetation types. Consequent changes in albedo and evapotranspiration functionality will shift the budget of energy and water on large scales. Precipitation in the high-latitude, continental summer owes strongly to water recycling capability. Water availability in summer, therefore, may be altered following such changes in the surface and subsurface hydrological conditions.

The influences of frozen ground are not limited to nature, but also affect the socio-economy. Non-negligible consequences may occur at local and remote areas. Changes in the location and capacity of water holding (for short term) and storage (for long term) can affect patterns and cycles of vegetation, precipitation and near-surface hydrology, leading to water resource issues. Freezing, either permanent or seasonal, is one of the essential limitations determining the range and capability of agriculture. For example, seedling timing and viability of plant are determined largely by start date of soil thawing and the thickness of seasonal thawing (the active layer in permafrost regions) or freezing (in case of seasonally-frozen regions). Roads, railways, and buildings, as well as residential zones, are also affected by changes in occurrence, frequency, and magnitudes of ground freezing and thawing. It is important to have a good assessment capability of freeze/thaw occurrence for social, economic, and political planning to adapt to, mitigate or counteract such changes.

Land process models, like other process models incorporated in large-scale climate models, need to fulfill two contradicting demands: good reproducibility of the detailed actual states and processes, and conciseness to run smoothly and economically in the conglomerate climate models. The former requires implementation of realistic physics to simulate the complex reality, while the latter is important for balanced evaluation of the total behavior of the climate system where a system component, like land, is interwoven through interactions and feedback with other system components. Offline simulation of land process models forced by historical surface meteorology offers a good opportunity to evaluate the reproducibility of the model, and, moreover, to examine the sensitivity of the focused processes (in this study's case, freeze/thaw dynamics) under direct control that is not possible in the case of coupled (online) simulations with the atmospheric (and additionally, oceanic) component.

Development and examination of permafrost dynamics in numerical models started in the early days of climate modeling. The performance and the sensitivity of land process models with the freeze/thaw process have been examined in the context of climate studies since the 1990s within the global climate modeling community (e.g., Riseborough et al., 2008), through both offline and in coupled simulations. Saito (2008b) examined, in a series of one-dimensional offline experiments for a tundra site, the sensitivity of the subsurface thermal and hydrological regimes to the complexity of the resolved physics under present-day conditions and under a warming scenario. The importance of the physical refinement was demonstrated in terms of the effect of freezing on the evaluation of hydrological-thermal property values, presence of the top organic layers, and presence of unfrozen water under the freezing point. Studies by coupled simulations evaluated the comprehensive performance of the climate models, as a whole, that incorporate the freeze/thaw process (e.g., Lawrence et al., 2008 and Saito et al., in preparation). Such evaluations, however, would not make sense if the used land process models are unable to reproduce the historical subsurface regimes when forced by historical data.

In this chapter, the author aims to evaluate the performance of land process models with different complexities of soil freeze/thaw dynamics, regarding reproducibility of surface and subsurface hydrological-thermal conditions during the latter half of the last century, and its sensitivity to the implemented physics. However, there are not so much observational data for the subsurface hydrological or thermal states that cover the spatial and temporal extent matching large-scale climate model outputs. Therefore, the comparison and validation tend to be limited in time or space, or indirect.

2. Model, data and experimental design

Two sets of global-scale offline simulations were conducted using different types of forcing data. One set of simulations is a hindcast where the models were forced by a compiled dataset of the historical, observed surface meteorology. Another set is a climatology forced by the climatological values derived from historical values. The land process model employed in this study is called MATSIRO (originally developed by Takata et al. 2003, and improved in terms of freeze/thaw dynamics by Saito 2008a, 2008b), a component of the global climate model MIROC (K-1 developers, 2004) developed in Japan. The refinement of MATSIRO for surface and subsurface hydrological-thermal regimes was documented by Saito (2008). The design of the sensitivity examinations is described in 2.1. The forcing data and other data used for evaluations and validations are explained in 2.2.

2.1 Experimental design and boundary factors

Four different experiments were conducted to examine the difference and sensitivity of the relevant configurations and factors (Table 1), i.e., a) total depth, total number of layers and thickness of layers in the soil column, b) presence or absence of top organic layers (TOL; e.g. a litter layer and a humus layer), c) basic value of the soil characteristics, such as thermal conductivity and heat capacity for different soil types, and d) freeze/thaw dynamics parameterizations. The last factor includes evaluation of thermal and hydraulic properties such as thermal and hydraulic conductivity, heat capacity, and presence of unfrozen water under the freezing point.

Exp.	Soil column depth and thickness	Top organic layers	Soil property values	thermal-hydrological parameterization
CNV	14 meters (6 layers: 5, 20, 75, 100, 200 cm, 10 m)	No	Default	Conventional
noTOL14			Adapted	Improved
TOL14		Yes		
TOL100	100 meters (12 layers: 1, 2, 5, 12, 25, 55, 100, 150, 250, 400 cm, 10, 80 m)			

Table 1. Summary of the experiments.

Figure 1 illustrates the distribution of prescribed vegetation (left) and soil characteristics (right). Both distributions were basically the same as the ones used in Saito et al. (2007), but are adapted to the 1948 cropland condition (Ramankutty & Foley 1999) and fixed through the simulations. They were common to all the experiments, except that top organic layers were neglected for CNV and noTOL14 runs. Consult Takata et al. (2003) for details of each category of vegetation and soil characteristics. Figure 2 shows the annual range of soil temperature for the bottom layer (i.e., for 4-14 m for the 6-layer runs and 20-100 m for the 12-layer run). It demonstrates that the soil column needs at least 20 m or more to be consistent with the specified lower boundary condition of a constant (often set to zero) heat flux.

Vertical profiles of soil thermal conductivity (lines with symbols) and soil moisture (simple lines) under different frozen ground conditions are shown in Figure 3. Figures 3a and 3c show results for grid points close to the southern permafrost region in North America and Eurasia, respectively. Figure 3b is for the northern part of permafrost region, and Figure 3d is for a seasonally-frozen region. The figure illustrates that the improved parameterization (i.e. TOL14 and TOL100 runs shown in green and blue, respectively) capture the feature well, for example, the summer thermal conductivity values are generally lower than those for winter because solid ice has higher conductivity than liquid water. The conventional model, the CNV run (shown in orange), shows variations following the total moisture but is less sensitive to the ratio of soil moisture and ice (difference between values in summer and in winter). The vertical structure of thermal conductivity is not clear for the cases of mineral layers only (i.e., CNV and noTOL14). The cases with top organic layers (i.e., TOL14 and TOL100), however, reproduce the vertical structure of the soil property that is comparable to the observations (shown in black in Figure 3a).

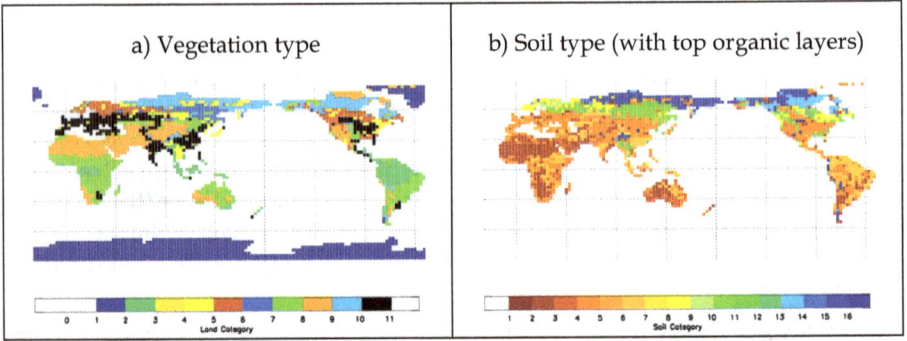

Fig. 1. a) Classification map of vegetation used in the simulation. Categories are classified as 12 types. 0: ocean, 1: continental ice, 2: broadleaf evergreen forest, 3: broadleaf deciduous forest & woodland, 4: mixed coniferous & broadleaf deciduous forest & woodland, 5: coniferous forest & woodland, 6: high latitude deciduous forest & woodland, 7: wooded C4 grassland, 8: shrubs & bare ground, 9: tundra, and 10: cultivation. b) Same as a) except for soil characteristics used in the simulation. 1: coarse, 2: medium/coarse, 3: medium, 4: fine/medium, 5: fine, 6: ice, 7: coarse in Taiga, 8: medium/coarse in Taiga, 9: medium in Taiga, 10: fine/medium in Taiga, 11: fine in Taiga, 12: coarse in Tundra, 13: medium/coarse in Tundra, 14: medium in Tundra, 15: fine/medium in Tundra, 16: fine in Tundra. For the mineral-only cases (i.e. no top organic layers such as CNV and noTOL runs), the categories 7 to 11 and 12 to 16 were replaced by categories 1 to 5.

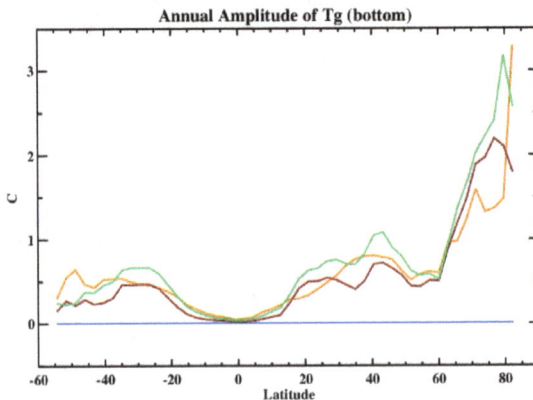

Fig. 2. Zonal average of annual range of soil temperature at the bottom layer, i.e. the 4-14 m for the 6-layer soil column, and 20-100 m for the 12-layer column. The averaging period is 1948 to 2000. The model CNV is shown in orange, noTOL14 in red, TOL14 in green, and TOL100 in blue.

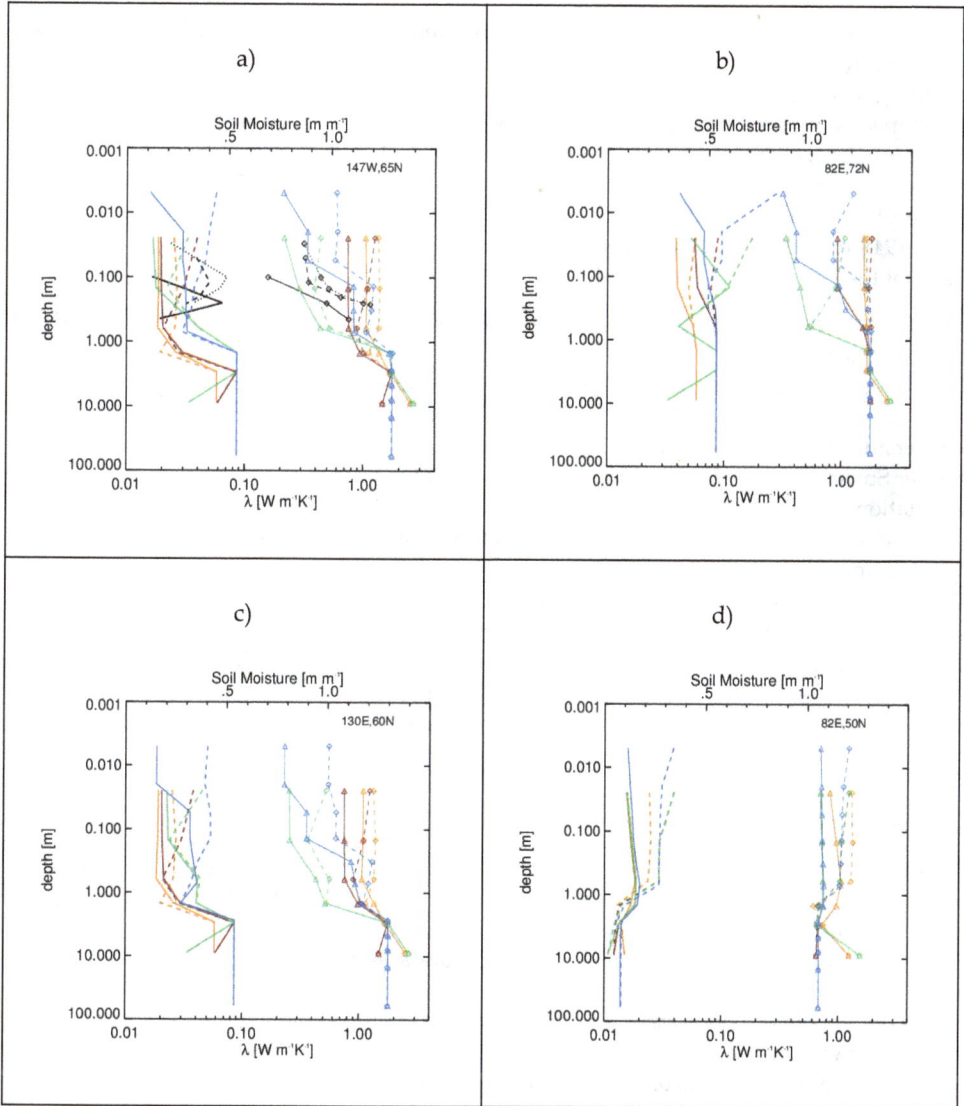

Fig. 3. Profile of calculated soil thermal conductivity (scales at the bottom axis) for summer (triangle with solid lines) and winter (diamond with dashed lines) for different simulation runs. The models are shown in the same color scheme as in Figure 2. Also shown is the profile of total soil moisture content (scales at the top axis) for summer (solid) and winter (dashed). Black lines in a) are taken from summer observations measured at a Seward Peninsula site in Alaska. Different line types denote different pits at the same site. The regions for each panel are identified in the text.

2.2 Data used for forcing and evaluations

The historical atmospheric data covering different periods in the 20th century were collected from different data sources and used to evaluate statistical uncertainties resulting from the forcing data. A global dataset of meteorological forcing that was developed, compiled and provided by the Land Surface Hydrology Research Group, Princeton University was used (Sheffield et al. 2006). The 3-hourly data at 1.0-degree resolution was interpolated into 1-hourly, T42 horizontal grids (c. 2.85x2.85 degree grids in the direction of longitude and latitude) for the experiments for the simulation period from 1948 to 2000. The climatology of the forcing data was derived from the first thirty years of the period.

The permafrost map, compiled by the International Permafrost Association (IPA) (Brown et al. 1997) and referred to as IPA map hereafter, is consulted to validate the modeled distribution of permafrost and the ground ice volume. River discharge of the selected large arctic basins was compared with the R-Arctic Net dataset collected and compiled by the University of New Hampshire (R-ArcticNET v4.0). Ground temperature distribution in the former Soviet Union was also used; it was interpolated from the observations at more than 200 stations onto the T42 grid at the National Snow and Ice Data Center (NSIDC) (described in Saito et al. 2007). Surface air temperature data was taken from the 0.5 degree latitude/longitude dataset of monthly surface temperature CRU05 produced by the Climate Research Unit, East Anglia University (New et al. 2000). Observational values of active layer depth were taken from the Circumpolar Active Layer Monitoring (CALM) dataset (Brown et al. 2003). The vertical profile of soil thermal conductivity and soil moisture was measured at a coastal tundra site on the Seward Peninsula in summer of 2010 by the chapter author. The Decagon KD2 Pro thermal analyzer and Campbell Scientific Hydrosense were used to measure the respective aforementioned quantities.

3. Results and discussions

In 3.1, the hindcast runs, which used historical values of the forcing data compiled by Sheffield et al. (2006), were analyzed to highlight how the modeled surface heat budget (3.1.1), subsurface thermal regimes (3.1.2) and hydrological budget (3.1.3) in the cold regions are dependent on the different modeling configurations and different implementations of physics. In addition, 3.2 shows results of the climatological runs, forced by 1948-1978 climatology derived from the above historical data. The climatological runs produced a repeating two-year cycle in which snow accumulation is markedly larger in one year and smaller in the other. One arbitrary two year sequence (after sufficient spin-up integrations) was used for analysis to demonstrate the effects of snow on the subsurface thermal and hydrological regimes, and how sensitive those effects are to the complexity of the resolved land processes.

3.1 Historical hindcast from 1948 to 2000

A historical hindcast from 1948 to 2000 (53 years) was integrated after a 500 year spin-up under 1949 conditions, which was used instead of 1948 to avoid the leap year. All soil layers reached equilibrium in all the runs after the spin-up. No ensemble was made for this study.

3.1.1 Influences on energy budget at the surface

From the viewpoint of large-scale climate model development, if the subsurface model performance has little or no influence on the atmosphere, the complexity of the physics implemented for the land process models does not matter. Impacts on surface budgets of either energy or water, or both, are evidence that subsurface models do matter to the total climate model.

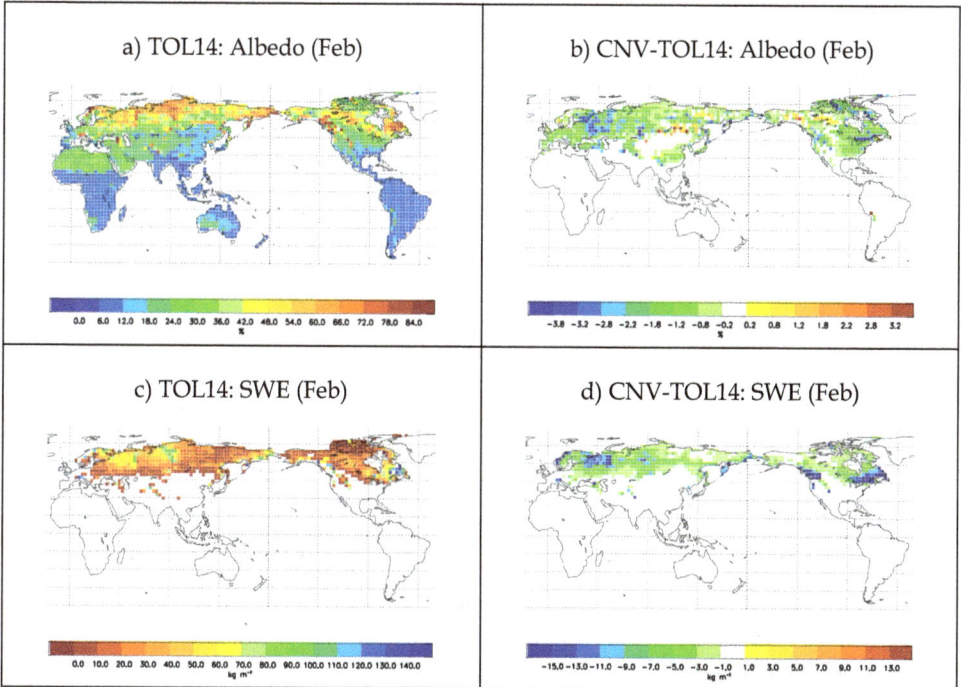

Fig. 4. a) Distribution of surface albedo in February averaged for the period 1948 to 2000 for the TOL14 run, and b) the difference from the CNV run. c), d) Same as a), b) except for snow water equivalent (SWE; kg m-2).

The marked difference and sensitivity found in winter between the experiments is in the albedo and, consequently, the shortwave radiation budget due to different snow accumulations. Figure 4 shows the surface albedo (%) and snow water equivalent (kg m-2) in February. In most of the NH land snow accumulation (or for the sake of radiation budget, snow cover extent) reaches its maximum in January to February, and the solar radiation is stronger and spread wider in February than in January. In order to emphasize the comparison and due to space limitations, the figures in this subsection show only the results for the TOL14 run and the deviation of the conventional model (CNV–TOL14), but the presence of the top organic layers was the major factor of the difference (not shown). Snow accumulation and, hence, diagnosed snow cover area were larger in the northern high-latitude land for the TOL runs, due to cooler soil temperature below the surface at the beginning of snow accumulation season. A larger snow area led to a higher albedo in winter that potentially cools the near-surface atmosphere if the atmosphere is interactive through reduced absorption of the solar energy at the surface. The areas of large difference for the TOL runs are northern Europe, the European part of Russia, Kamchatka, the Rocky Mountains and northeastern Canada for the higher albedo, and southern Siberia and British Columbia for the lower albedo.

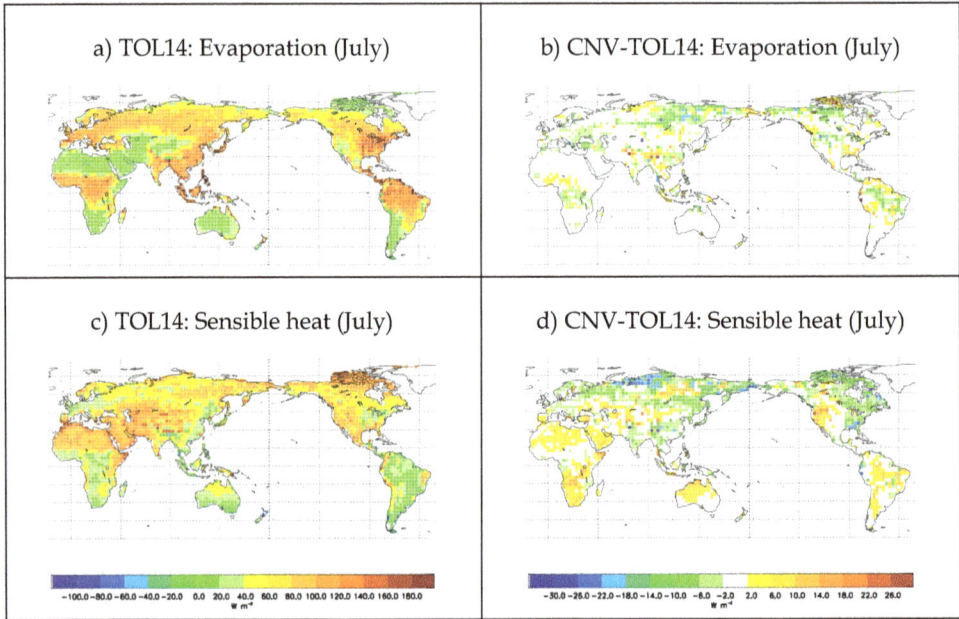

Fig. 5. a), b) Same as Figure 4 a), b) except for latent heat flux (W m-2) in July. c), d) Same as a), b) except for sensible heat flux (W m-2).

In warmer seasons, the eddy heat fluxes (i.e., sensible and latent heat flux) become more important and enhanced than in cold seasons. Figure 5 shows a result for the summer conditions in which the surface heat balance was different between the experiments because the amount and the partition of surface available energy was changed: evapotranspiration (latent heat flux) was enhanced in permafrost regions, i.e., northeastern Eurasia and northern North America (Figure 5a-b), and sensible heat flux increased in large areas in the northern high-latitude land areas for the refined models (Figure 5c-d). One of the important factors that caused this change was the reduction of geothermal heat to the ground. In spring to summer the geothermal heat flux is negative (downward, i.e., from the atmosphere to ground; Figure 6a). The lower thermal conductivity value for the top organic layers hinders heat conduction to the ground in comparison to the only mineral soils (Figure 6b). Convergent heat at the surface can increase sensible heat fluxes. Another reason is availability of water in upper soil layers in permafrost regions (see profiles of thermal conductivity and soil moisture for summer shown by solid lines, in Figures 3a-c). Layers of mosses or dead organics can hold water more effectively to make it available for evaporation from the ground surface, or for plants to transpire.

Geothermal heat flux showed a marked difference in two transitional seasons, early summer and early winter (Figure 6). Early in the cold season, the geothermal heat is generally positive (upward, i.e., from the ground to the atmosphere; Figure 6c). The conventional scheme transferred more heat from the atmosphere in summer and to the atmosphere in winter so that the annual amplitude of ground temperature was large. The refined parameterizations take the amount of ground ice into account in computing the working thermal conductivity, which makes wintertime thermal conductivity larger even when the total soil moisture (both in solid and liquid) is the same between summer and winter. This makes seasonal asymmetry in conduction of ground heat leading to different vertical profiles and seasonality of modeled annual ground temperature, as discussed in Saito (2008b). The changes in the surface heat balance and heat conduction influence the subsurface thermal regimes differently in different regions, and modify the distribution of frozen ground (permafrost, seasonally frozen, and no freeze regions) as shown in 3.1.2.

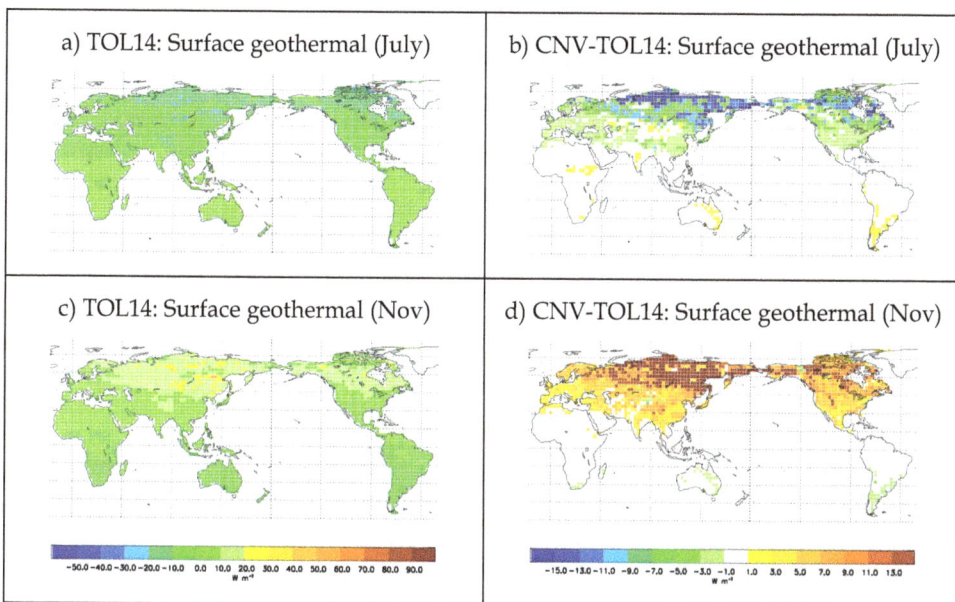

Fig. 6. a), b) Same as Figure 4 a), b) except for geothermal heat flux at ground surface (W m^{-2}) in July. c), d) Same as a), b) except for November.

3.1.2 Evaluations of subsurface thermal regimes

The current thermal state of permafrost from the last century was widely investigated and documented during the International Polar Year (IPY 2007-2009). Permafrost in Eurasia (Romanovsky et al., 2010) and North America (Smith et al., 2010) is stable in the northern cold regions, but shows a warming trend in the southern warmer regions.

Thermal state means a seasonal cycle and distribution of temperature. Figure 7 shows an example of modeled subsurface temperature (T_g) distribution and a comparison with an observation-based estimate for the regions of the former Soviet Union. Notice that the distribution of modeled Tg is very close to that of the forcing data (surface air temperature at 2m, $T2$; Figure 7c). The difference between the prescribed $T2$ and the modeled Tg was mostly in the range between 2 to 4°C, and Tg was warmer. This difference is due to several agents, such as vegetation, snow cover, and seasonality of water storage. Estimates of subsurface temperature distribution, such as the one shown in Figure 7b, are neither common nor easy to produce. It is a valuable map, but has weaknesses too. As discussed in Saito et al. (2007) the regions of permafrost, e.g., eastern Siberia, feature a less dense network of observation stations so that the estimated values have large uncertainty. In the seasonally frozen regions (the southwestern part of the map) the correspondence between the modeled and estimates are good, but the difference is large in the northeastern part. Judging together with other observational evidence (e.g. Figure 9f in this chapter) the estimates are too warm in the region.

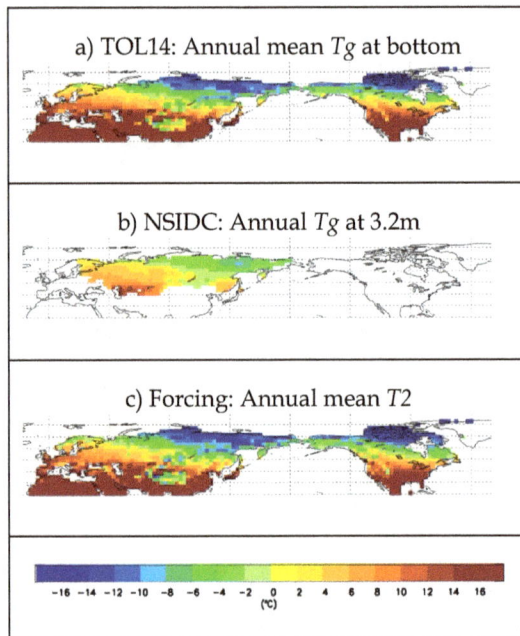

Fig. 7. Comparison of the annual average of a) modeled soil temperature (Tg) at the bottom layer (4-14 m) for the TOL14 run, b) estimated soil temperature at 3.2m derived from the former Soviet Union observations, and c) surface air temperature at 2m ($T2$) used as the forcing. Color scheme is common to all the plots.

The active layer is "the layer of ground that is subject to annual thawing and freezing in areas underlain by permafrost" (van Everdingen (ed), 1998). The maximum depth of the active layer is usually reached in the late summer, and referred hereafter as maximum ALD or mALD. Change of mALD occurred due to several reasons, both natural and anthropogenic. Temperature change, as well as surface disturbances such as wildfire, construction, and cultivation, can modify mALD. Figure 8 shows examples of the evolution of mALD, both modeled and observed, at two different sites characterized as permafrost.

Although the average and the interannual variations of the forcing $T2$ (thin black lines) do not differ from the amplitude of the interannual variations in the observed (thick black lines. only shown as anomaly from the average) and modeled mALD (colored lines) are very different. In Figure 8a the CNV run showed a larger interannual amplitude than the TOL14 run. This is because the large and rapid conduction of heat to ground for the CNV run reached to a deeper layer in some years but not in other years. In the TOL runs, the heat conduction is slower and damped, and the depth of mALD is more stable. In Figure 8b the maximum active layer depth is deeper than in Figure 8a, and, since the CNV run allows no unfrozen water, once a layer is frozen it stays at 0°C until all water freezes to ice or ice melts away completely, and the table of permafrost stays at the same depth. The refined parameterization, however, allows unfrozen water and the ground temperature can be below 0°C while some water stays in liquid phase, so that the depth of the maximum ALD can fluctuate over years. At both points, the mALD was shallower in the 1960s and the early 1970s, and deepened after the mid 1980s. All the runs with the refined physics showed this similar trend, but the TOL14 run computed an abrupt deepening of ALD in Figure 8b, the cause of which was not clear.

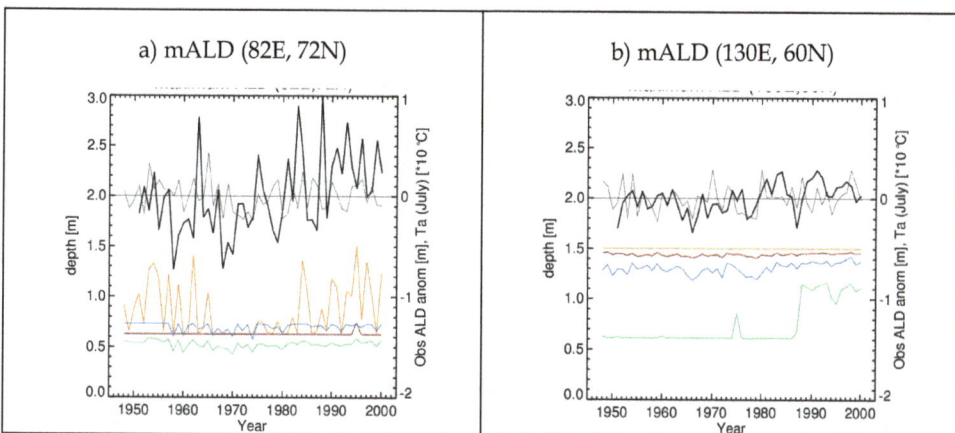

Fig. 8. Evolution of maximum active layer depth (mALD) simulated (shown in same colors as in Figure 2. Scale is shown on the left axis), and estimated anomaly (in thick black line. Scale is shown on the right axis) at a) 82E, 72N, and b) 130E, 60N in the continuous permafrost region. Also shown is the forcing T2 at the points (Scales shown on the right axis. Note the values need to be multiplied by ten).

Figure 9 shows the geographical distribution of mean maximum ALD for the integration period. The CNV run computed the deepest ALD of the four, ranging from 1m along the northern coast and more than 3m in some southern regions. The noTOL14 run simulated somewhat shallower ALD, but had a wide zone of ALD deeper than 3m compared to the CNV run. This is again because of the presence of unfrozen water (or continuous transition at and below the freezing point), which can allow a gradual change of ALD. The experiments with top organic layers modeled much shallower ALD (Figures 9c-d). Figure 9e is derived from the NSIDC soil temperature data for the former Soviet Union, and 9f is taken from the CALM dataset. Since the CALM dataset is directly taken from the observations (no spatial or temporal interpolation) it is one of the most trustable sources. The result from CALM, in comparison with Figure 9e, provides evidence that the estimate of the interpolated soil temperature has a warm bias in permafrost regions.

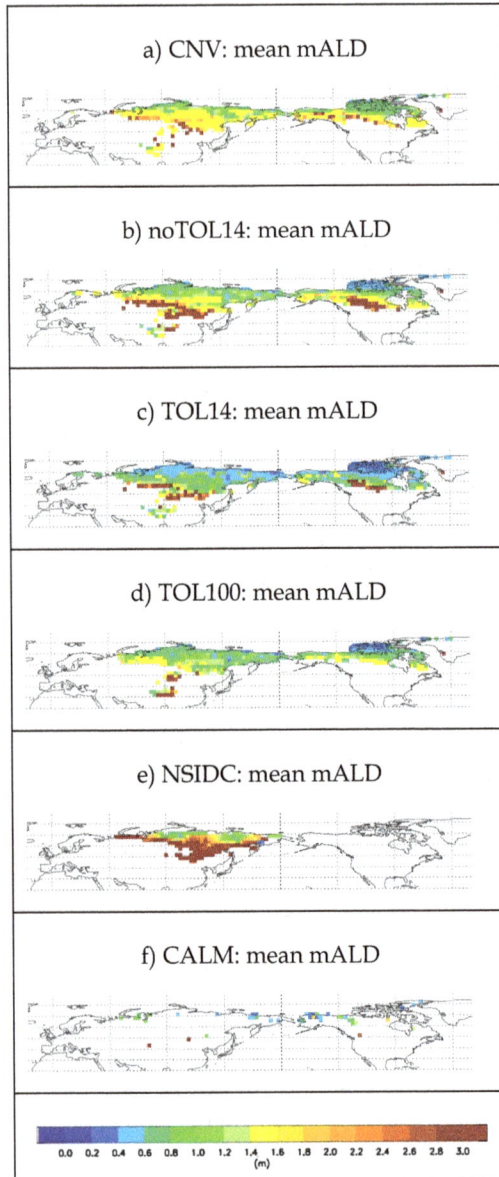

Fig. 9. Comparison of maximum active layer depth among simulations in a) CNV run, b) noTOL14 run, c) TOL14 run, and d) TOL100 run, e) estimations from the former Soviet Union observations, and f) direct observations by CALM.

Those differences in the simulated thermal conditions (temperature and heat flow) led to a difference in the modeled distribution of the thermal regimes, and frozen ground (Figure 10 and Table 2). The regions of modeled permafrost were defined by whether any of the subsurface layer stays frozen for two or more consecutive years. Regions were classified as seasonally-frozen if some soil layer freezes at any time of a year but excluded permafrost.

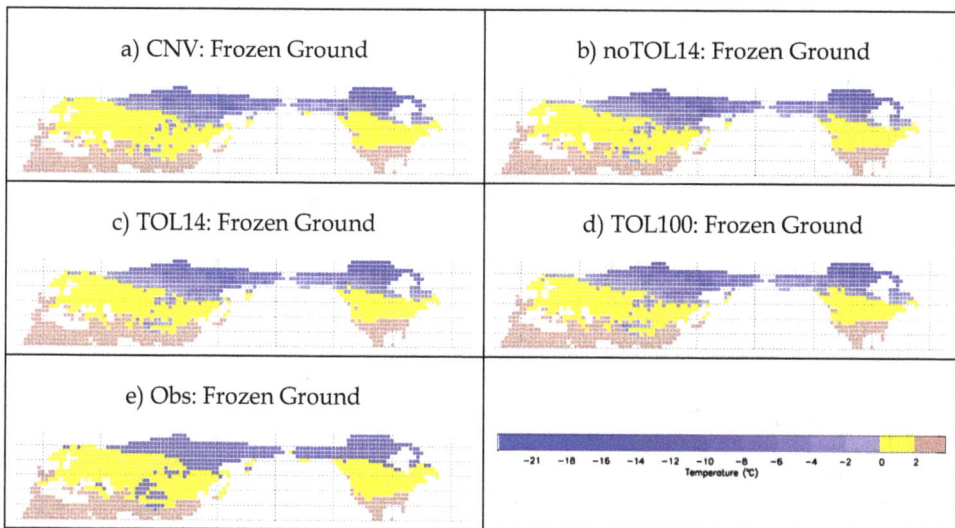

a) CNV: Frozen Ground	b) noTOL14: Frozen Ground
c) TOL14: Frozen Ground	d) TOL100: Frozen Ground
e) Obs: Frozen Ground	

Fig. 10. Modeled and observed distribution of frozen ground classified as "no freezing" (pink), seasonal freezing (yellow), and permafrost (blue). Mean annual permafrost temperature at the permafrost table is shown for the modeled results.

	Total	Seasonally frozen	Permafrost
CNV	48.5	30.2	18.3
noTOL14	48.5	27.6	20.9
TOL14	48.7	27.6	21.2
TOL100	48.7	26.9	21.8
Obs	51.0	33.4	17.6

Table 2. Comparison of the areal extent of frozen ground (million km²). Observational distribution is based on IPA map (Brown et al., 1997) for permafrost, and derived from observed monthly surface air temperature for seasonally frozen ground (cf. Saito et al., 2007), and mapped on to the T42 grids.

An observation-based map (Figure 10e) was derived from the IPA map for the permafrost region by choosing the continuous (90-100%) and discontinuous (50-90%) permafrost regions. Distribution of seasonally-frozen ground was diagnosed from the CRU temperature data following the methodology of Zhang et al. (2003) as described in Saito et al. (2007). In general, the modeled permafrost extent is larger than the values from the observation-based one, while the area of seasonally frozen ground is smaller for the simulations. Since the surface air temperature was prescribed (as forcing), the southern limit of seasonally frozen ground was mostly fixed for the modeled case in these offline simulations. It is of interest to investigate whether the area of seasonally frozen ground would change when the land process models are coupled to the atmospheric (and even to the oceanic) climate models. The results of the coupled simulations are compiled in another paper.

3.1.3 Hydrological evaluations in the cold regions

Discharge of the river basins to the Arctic Ocean shows a significantly increasing trend since the last century (Peterson et al., 2002; White, D., et al., 2007). The reason of the increase is not clear yet, but the possible agents include changes in atmospheric moisture conversion and precipitation, dam construction, permafrost thaw, and wildfire (Yang et al., 2003; McClelland et al., 2004). Possible attribution of the permafrost thaw is from changes in the water content held in permafrost through changes in the area of permafrost and the depth of the active layer. Figure 11 shows the difference of the averaged water content between experiments (Figures 11a, 11c and 11e). The CNV run showed smaller water content in the high latitudes, partly because the specified porosity (volume not occupied by soil) for mineral soil was smaller and because infiltration of water at the beginning of the cooling season was small in the conventional model. Figure 11 also shows the distribution of ground ice. The model difference is most prominent between the conventional and the refined one (11a and 11b), ground ice content tended to larger for the CNV run despite smaller water content as a whole in permafrost. This is because some percent (up to 20% or more, depending on the temperature and soil characteristics) of water remains unfrozen in the refined models.

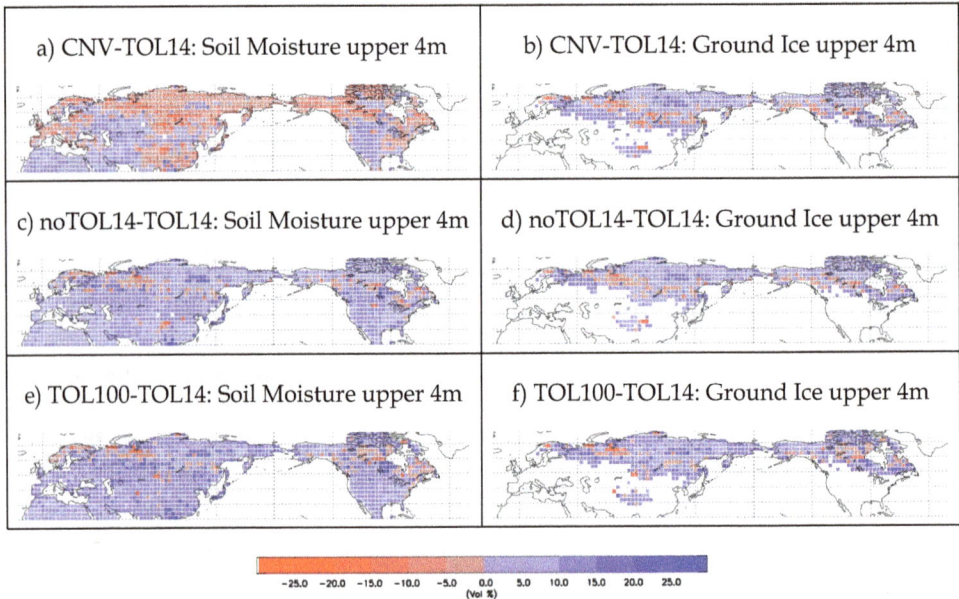

Fig. 11. Difference of annual mean total soil water content (volumetric %; both liquid water and solid ice) in the upper 4m-layer from the TOL14 run for a) CNV run, c) noTOL14 run, and e) TOL100 run. b), d) and f) Same as a), c), and e) except for ground ice content (volumetric %).

The meridional profile of the modeled ground ice was compared with the observation-based estimate in Figure 12. The experiments successfully captured the general characteristics of the latitudinal distribution in the northern hemisphere, but the absolute value diverts roughly by a factor of three. Some parts of the discrepancy (i.e., overestimation) arguably resulted from neglect of sub-grid scale heterogeneity of land profile. For example, a grid categorized as grassland may have small desert, lakes or rocky hills that tend to have limited ground-ice content compared to a grassland. In turn, the classification of ground-ice content in the IPA map is susceptive to be more qualitative than quantitative in areas with sparse observational evidence.

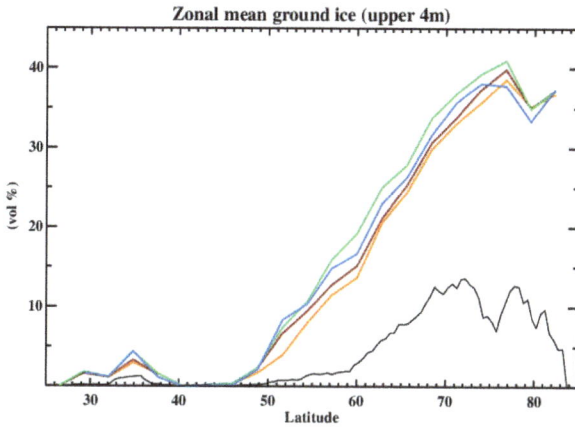

Fig. 12. Zonal average of ground ice content (volumetric %) in the upper 4m soil for northern hemisphere. Modeled results averaged for 1948-2000 are shown in the same color scheme as in Figure 2. Estimates from the observations (the IPA map) are shown in black.

Historical changes in the modeled and observed annual river discharge, together with those in the forcing precipitation, are shown in Figure 13 for the selected arctic basins, and the result of analyses for the attribution of the changes are summarized in Table 3. The observational data are taken from the R-Arctic Net database (R-ArcticNet v4.0). For those river basins underlain in parts by permafrost, such as the Lena, Mackenzie and Anadyr Kolyma, the correlation of the modeled and observed river discharge and precipitation was significant, although the averaged level of discharge tended to differ between the experiments. Those river basins mostly in the seasonally frozen region showed little or no correspondence between the modeled and the observed.

Since precipitation (water input from the atmosphere) is common to all the experiments and basins, there should be other reasons (or mechanisms), e.g., evapotranspiration, or water storage, that determine the discharge from those basins. The lower row of the analysis results in Table 3 shows the correlation between the modeled and observed discharge after removing precipitation. It serves as an index for possible sources of fluctuation other than precipitation. Significant results were obtained only for Anadyr Kolyma and Nelson, for which other factors can explain the discharge fluctuation (although it is not known what exact factors do explain it, based on this method only). The discrepancy between the model and observations for the Ob, Yenisei and Yukon rivers remains unexplained by these analyses.

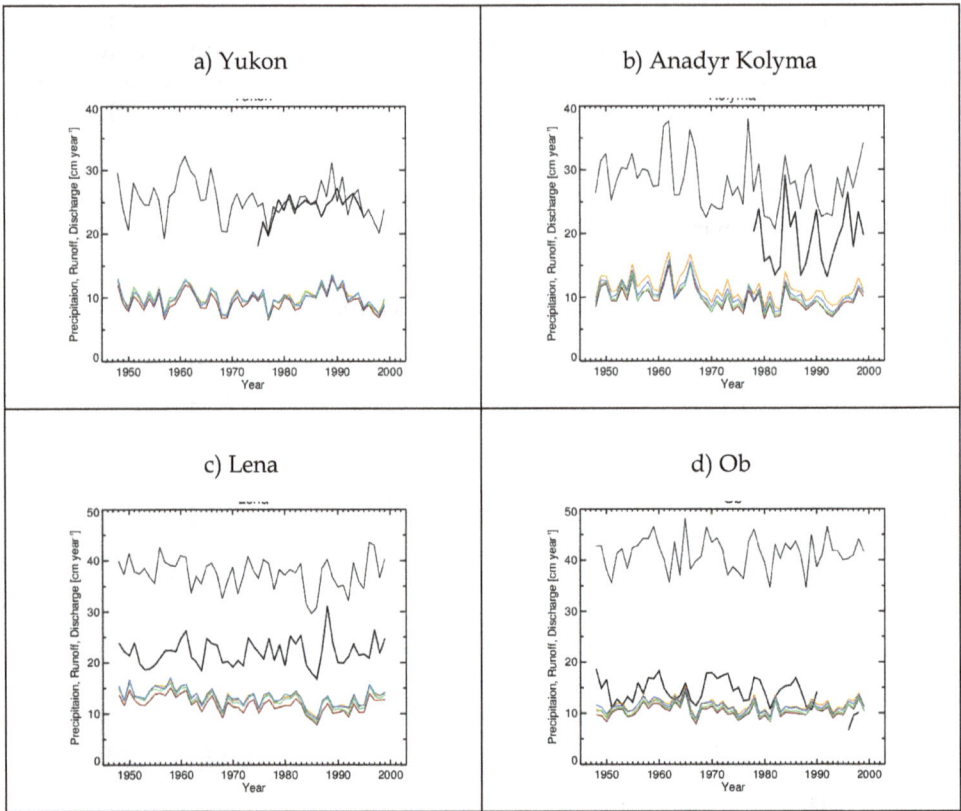

Fig. 13. Time series of annual precipitation (forcing. Shown in thin black), and observed (thick black) and modeled (in colors) discharge for the selected river basin. a) Yukon, b) Anadyr Kolyma, c) Lena, and d) Ob. A year cycle is from September to August.

3.2 Snow effects on hydrological-thermal regimes

Seasonal snow cover is one of the major players in cryosphere-climate interactions. It influences the surface energy balance between the atmosphere and land through its high albedo. It also influences the state of the subsurface hydrological and thermal regimes, and its seasonal change though its insulation effect and storage of water during winter. The climatological simulations all showed a two-year cycle with a high snow winter and a low snow year. Snow accumulation (denoted by snow water equivalent, SWE) is a prognostic variable in the model, i.e., computed within the model instead of externally prescribed or forced. Reproduction of the amount and distribution of snow accumulation is one of the areas in which current global climate models are limited. The produced two-year cycle is an artificial phenomenon of the land process model, flip-flopping between two stable states, when forced by climatological data (i.e., no interannual variations). Nevertheless, this gives a unique opportunity to investigate the effects of snow reflected on the subsurface hydrology and thermal regimes.

	Severnaya Dvina	Ob	Yenisey	Lena	Anadyr Kolyma	Yukon	Nelson	Mackenzie
Discharge area at the station (km²)	415943.9	2765830.0	1878807.8	2510086.8	449626.0	911706.8	1128703.9	1705077.9
Modeled discharge area (km²)	348000.0 (96.8%)	2160000.0 (98.5%)	1760000.0 (89.5%)	2430000.0 (91.2%)	526000.0 (117.0%)	831386.2 (93.7%)	1010000.0 (78.1%)	1680000.0 (83.7%)
Years	49	45	44	50	20	19	38	26
CNV	**0.61** 0.16	-0.01 *-0.30*	0.24 0.14	**0.48** 0.02	**0.72** *0.55*	0.38 0.16	**0.38** *0.32*	**0.55** 0.27
noTOL14	**0.59** 0.12	-0.00 *-0.30*	0.25 0.16	**0.49** 0.06	**0.83** **0.66**	0.33 0.07	**0.38** *0.32*	**0.58** 0.29
TOL14	**0.56** 0.11	0.09 -0.21	*0.28* 0.20	**0.50** -0.01	**0.87** **0.73**	0.35 0.07	**0.36** 0.29	**0.58** 0.24
TOL100	**0.61** 0.15	0.11 -0.19	*0.27* 0.19	**0.51** 0.04	**0.77** *0.59*	0.33 0.07	**0.39** *0.33*	**0.57** 0.28

Table 3. Pearson's correlation between the observed and simulated annual discharge (upper row) and the partial correlation with the effect of precipitation removed (lower row) in the selected arctic basins. Statistically significant correlation coefficients at the 99% confidence level are shown by bold fonts, and those at the 95% level are shown by italic fonts. Numbers in parentheses are the percentage of the modeled discharge area relative to the discharge area at the corresponding station.

Table 4 and Figure 14 summarize the effects of snow on the subsurface hydrological and thermal regimes by the four experiments summarized in Table 1. The grids 14a, 14b and 14c are from permafrost regions, and the grid 14d is from a seasonally frozen region, at which the refined model showed no large difference among experiments. Note that the scales for snow are different among the figures.

Snow accumulation (snow water equivalent) is largest at grid 14b (northwestern Siberia) with the monthly maximum reaching more than 100 kg m⁻² even in a low year, whereas grid 14a (Interior Alaska) has only 8 kg m⁻², even in a high snow year. Variations in snow between the experiments are small at all the points, as well as ground temperature for the top soil layer. In contrast, geothermal fluxes and soil water content are highly variable among the experiments. As for the effect of snow, water content and ground temperature show a large difference in summer, while geothermal fluxes have greater sensitivity in spring and autumn.

An observational fact is that the maximum active layer depth (mALD) is more correlated with the summer temperature than the snow amount from the previous winter. The maximum depth of the active layer was larger and soil temperature was higher in summer following the high snow year at all permafrost grids (Table 4a), so it could not be examined here. mALD is largest by a factor of two or more for the CNV run among the experiments, and the difference in snow conditions was also the largest.

The average amount of annual local runoff was proportional to the average SWE, but interannual variations were not proportional to the snow fluctuation (Table 4b), which is partly explained by the differing amount of soil water content after snowmelt (more water content after high snow winter, Figure 14), suggesting the buffering (or smoothing) functionality of soil.

The conventional model showed the largest amount of cumulative geothermal heat conducted both from and to the atmosphere, while it is least sensitive to the effects of snow (Table 4c). Upward flux is larger after high snow accumulation at all grids, but the downward flux varies among the grids and among the experiments.

a) ALD (cm)	147W, 65N	130E, 60N	82E, 72N	82E, 50N
CNV	272.1 (0.55)	237.5 (0.63)	91.6 (0.68)	n/a
noTOL14	150.0 (0.99)	150.0 (0.95)	62.5 (1.00)	n/a
TOL14	62.5 (0.97)	112.1 (0.54)	55.4 (0.96)	n/a
TOL100	150.0 (0.95)	145.0 (0.82)	72.5 (0.97)	n/a

b) Runoff (cm/year)	147W, 65N	130E, 60N	82E, 72N	82E, 50N
CNV	3.17 (0.92)	10.1 (0.98)	25.7 (1.07)	11.5 (1.01)
noTOL14	2.45 (1.07)	8.64 (1.07)	25.8 (1.02)	9.85 (1.07)
TOL14	4.47 (0.96)	9.95 (0.94)	29.1 (1.04)	10.6 (1.00)
TOL100	2.86 (0.94)	9.75 (1.03)	30.3 (1.03)	10.6 (1.02)

c) Geo- thermal flux at surface (W/m2)	147W, 65N	130E, 60N	82E, 72N	82E, 50N
CNV	230.2 (0.85) -233.1 (0.95)	293.7 (0.79) -285.1 (0.97)	245.4 (0.85) -235.3 (0.97)	180.0 (0.94) -199.4 (0.99)
noTOL14	169.9 (0.76) -160.4 (1.10)	224.6 (0.74) -212.5 (1.04)	219.4 (0.83) -203.2 (0.99)	143.1 (0.96) -177.2 (0.98)
TOL14	142.7 (0.69) -132.2 (1.13)	175.0 (0.62) -159.4 (1.14)	125.6 (0.85) -128.0 (0.99)	142.4 (0.96) -174.1 (0.99)
TOL100	221.1 (0.76) -210.1 (1.04)	263.7 (0.74) -251.4 (0.99)	204.9 (0.87) -194.3 (0.98)	147.6 (0.93) -179.8 (0.99)

Table 4. Comparison of a) maximum active layer depth (mALD), b) local annual runoff (cm year^{-1}), and c) cumulative geothermal heat fluxes to/from ground (W m^{-2}) for the large snow years. A positive value in heat flux means upward, i.e. from ground to the atmosphere. Ratio of the low snow year value relative to the high snow year value is shown in the parentheses.

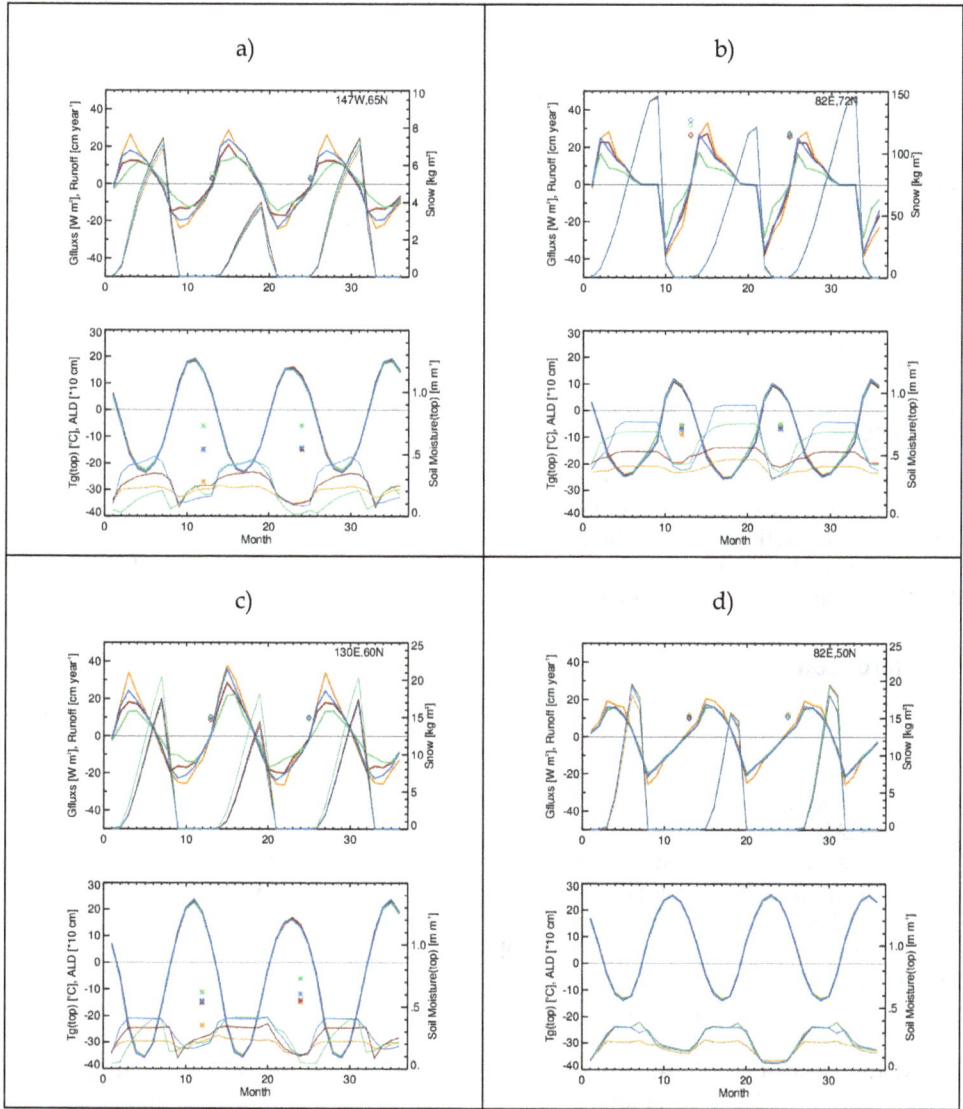

Fig. 14. Effects of snow on the subsurface states at locations with different frozen ground conditions. [Upper panel] The geothermal heat flux at the surface (W m-2; thick lines) and the snow water equivalent (kg m-2; thin lines). Scales are on the left and right axes, respectively. Local annual runoff (cm year-1) is shown by diamonds. [Lower panel] Soil temperature (oC; thick lines), and soil moisture content (m m-1; thin lines) for the top layer (0-5 cm for the 6-layer runs and 0-1 cm for the 12-layer run). Scales are on left and right axes, respectively. The maximum active layer depth (x10 cm) is shown by an asterisk. A year cycle begins in September and is repeated three times (high snow-low snow-high snow years). The color scheme is the same as in Figure 2.

4. Conclusion

Two sets of global-scale offline experiments were run with land process models of different complexity of freeze/thaw dynamics and related subsurface conditions. One set is a hindcast with historical forcing data, and the other was a climatological run. The simulated surface and subsurface variables were compared to each other for the sensitivity of model conditions, and with observation-based data for validation. It was demonstrated that subsurface physics can affect the atmosphere through the surface energy budget, that the subsurface regime is simulated more correctly using top organic layers and the refined freeze/thaw dynamics. The hydrological regime and snow effects showed significant sensitivity to the model complexity, although the results remained descriptive and qualitative, rather than quantitative, partly due to an immature observation-model comparison framework.

5. Acknowledgment

All the numerical simulations were run at the Arctic Region Supercomputing Center at the University of Alaska Fairbanks. Field measurement of soil properties on the Seward Peninsula was partly supported by the IARC-JAXA (Japan Aerospace Exploration Agency) Information System collaborative studies. The manuscript was proofread and edited by Ned Rozell, Geophysical Institute, University of Alaska Fairbanks.

6. References

Brown, J., Ferrins Jr., O. J., Heginbotom, J. A., & Melnikov, E. S. (1997). Circum-Arctic map of permafrost and ground-ice conditions, Geological Survey for the International Permafrost Association, 1:10,000,000, U. S. Geol. Surv. Circum-Pac. Map, CP-45.

Brown, J., Hinkel, K. & Nelson, F. (comp.) 2003. Circumpolar Active Layer Monitoring (CALM) Program Network: Description and data. In International Permafrost Association Standing Committee on Data Information and Communication (comp.). 2003. Circumpolar Active-Layer Permafrost System, Version 2.0. Edited by M. Parsons and T. Zhang. Boulder, CO: National Snow and Ice Data Center/World Data Center for Glaciology. CD-ROM.

Frauenfeld, O. W., Zhang, T., Barry, R. G. & Gilichinsky, D. (2004), Interdecadal changes in seasonal freeze and thaw depths in Russia, Journal of Geophysical Research, 109, D05101, doi:10.1029/2003JD004245.

K-1 Model Developers (2004). K-1 coupled GCM (MIROC) description, Technical Report 1, edited by H. Hasumi and S. Emori, 34 pp., Center for Climate System Research, University of Tokyo, Kashiwa, Japan.

Lawrence, D. M., Slater, A. G., Romanovsky, V. E., & Nicolsky, D. J. (2008). Sensitivity of a model projection of near-surface permafrost degradation to soil column depth and representation of soil organic matter, Journal of Geophysical Research, 113, F02011, doi:10.1029/2007JF000883.

McClelland, J. W., Holmes, R. M., Peterson, B. J. & Stieglitz, M. (2004). Increasing river discharge in the Eurasian Arctic: Consideration of dams, permafrost thaw, and fires as potential agents of change, Journal of Geophysical Research, 109, D18102, doi:10.1029/2004JD004583.

New, M. G., Hulme, M. & Jones, P. D. (2000). Representing twentieth-century space-time climate variability. Part II: Development of 1901-1996 monthly grids of terrestrial surface climate. *Journal of Climate*, 13, pp. 2217-2238.

Peterson, B. J., Holmes, R. M., McClelland, J. W., Vörösmarty, C. J., Lammers, R. B., Shiklomanov, A. I., Shiklomanov, I. A., & Rahmstorf, S. (2002). Increasing river discharge to the Arctic Ocean, *Science*, 298, pp. 2171-2173.

Ramankutty, N., & Foley, J. A. (1999). Estimating historical changes in global land cover: Croplands from 1700 and 1992, *Global Biogeochemical Cycles*, 13, pp. 997-1027.

R-ArcticNET v4.0. (n.d.) A Regional, Hydrometeorological Data Network For the pan-Arctic Region. September 2011, Available from* <http://www.R-ArcticNET.sr.unh.edu>

Riseborough, D., Shiklomanov, N., Etzelmüller, B., Gruber, S., & Marchenko, S. (2008). Recent Advances in Permafrost Modelling, *Permafrost and Periglacial Processes*, 19, pp. 137-156.

Romanovsky, V. E., Drozdov, D. S., Oberman, N. G., Malkova, G. V., Kholodov, A. L., Marchenko, S. S., Moskalenko, N. G., Sergeev, D. O., Ukraintseva, N. G., Abramov, A. A., Gilichinsky, D. A. & Vasiliev, A. A. (2010). Thermal State of Permafrost in Russia, *Permafrost and Periglacial Processes*, 21, pp. 136-155.

Saito, K., Kimoto, M., Zhang, T., Takata, K., & Emori, S. (2007), Evaluating a high-resolution climate model: Simulated hydrothermal regimes in frozen ground regions and their change under the global warming scenario, *Journal of Geophysical Research*, 112, F02S11, doi:10.1029/2006JF000577.

Saito, K. (2008a). Refinement of Physical Land Scheme for Cold-region Subsurface Hydro-thermal Processes and its Impact on Global Hydro-climate. *Proceedings of the 9th International Conference on Permafrost*, Fairbanks, Alaska, USA., June 27-July 3, pp. 1555-1560.

Saito, K. (2008b). Arctic land hydrothermal sensitivity under warming: Idealized off-line evaluation of a physical terrestrial scheme in a global climate model, *Journal of Geophysical Research*, 113, D21106, doi:10.1029/2008JD009880.

Sheffield, J., Goteti, G., & Wood, E. F. (2006). Development of a 50-yr high-resolution global dataset of meteorological forcings for land surface modeling, *Journal of Climate*, 19, pp. 3088-3111.

Smith, S.L., Romanovsky, V.E., Lewkowicz, A.G., Burn, C.R., Allard, M., Clow, G.D., Yoshikawa, K., & Throop, J. (2010). Thermal State of Permafrost in North America: A Contribution to the International Polar Year, *Permafrost and Periglacial Processes*, 21, pp. 117-135.

Takata, K., Emori, S., & Watanabe, T. (2003). Development of the minimal advanced treatments of surface interaction and runoff, *Global and Planetary Change*, 38, pp. 209- 222.

van Everdingen, Robert, ed. (1998, revised May 2005). Multi-language glossary of permafrost and related ground-ice terms. Boulder, CO: National Snow and Ice Data Center/World Data Center for Glaciology.

White, D., Hinzman, L., Alessa, L., Cassano, J., Chambers, M., Falkner, K., Francis, J., Gutowski Jr., W. J., Holland, M., Max Holmes, R., Huntington, H., Kane, D., Kliskey, A., Lee, C., McClelland, J., Peterson, B., Scott Rupp, T., Straneo, F., Steele, M., Woodgate, R., Yang, D., Yoshikawa, K., & Zhang, T. (2007). The arctic freshwater system: Changes and impacts, *Journal of Geophysical Research*, 112, G04S54, doi:10.1029/2006JG000353.

Yang, D., Kane, D. L., Hinzman, H. D., Zhang, X., Zhang, T. & Ye, H. (2003). Siberian River Hydrologic Regime and Recent Change, *Journal of Geophysical Research*, 107(D23), 4694, doi:10.1029/2002JD002542.

Zhang, T., Barry, R. G., Knowles, K., Ling, F. & Armstrong, R. L. (2003). Distribution of seasonally and perennially frozen ground in the Northern Hemisphere, in *Proceedings of the 8th International Conference on Permafrost*, Zurich, Switzerland, 21–25 July, 2003, pp. 1289–1294.

9

The Role of Stochastic Forcing in Climate Models: The Case of Thermohaline Circulation

M.N. Lorenzo, J.J. Taboada and I. Iglesias
University of Vigo / Ephyslab
Spain

1. Introduction

At present both daily and seasonal weather forecasts as well as long-term climate predictions are the result of the implementation of multiple and complex numerical models. The history of numerical models to predict the weather goes back to the 1950s and 1960s and since then computational breakthroughs have enabled us to go from simple box models to complex models that take into account a multitude of phenomena and processes occurring in the whole climate system. However despite the great strides made in recent years and great experience in the development of such models we must interpret the results taking into account the limitations and uncertainties inherent in any numerical model.

Among the most challenging major constraints in developing a climate model, we can identify three potentially difficult to solve: The first is that the climate system by its nature presents a nonlinear chaotic behaviour, which makes a weather forecast of more than seven or nine days unpredictable. This behaviour is alluded to in the famous butterfly effect discovered by Edward Lorenz in 1963. The second problem facing the modelers is the number of simplifications and assumptions introduced in the models through what is known as parameterizations. The parameterizations simulate the behaviour of processes that due to either the temporal or spatial scales of the model cannot be solved. Finally the third problem that arises is that despite progress in recent years models are always limited to represent part of the climate system but not all. Many fluctuations due to complex processes or biological or chemical components are omitted. In this way, we can find processes that are not considered but have a relevant role in the outcome. An example of this has been the recent inclusion of sulfur aerosols in climate models. This improvement has allowed us to propose an explanation for the asymmetry of the climatic evolution in the two hemispheres in the context of climate change experienced in recent decades. The improvement of models are continuous, but there are many important processes of the global system not present in the actual models: sources and sinks of oceanic and continental carbon, cycle of methane, increment of tropospheric ozone, the role of organic and mineral aerosols, synoptic variability, etc. These improvements involve a greater cost both computationally and in mathematical development. In recent years, modelers have begun working with the ideas proposed by Hasselmann and Epstein which propose to mimic some processes not considered in the models with different stochastic forcings (Hasselmann, 1976). At the moment, the list of success of results is large: introducing random noise has demonstrated considerable skill in improving weather forecasts through the ensemble technique (Buizza et al. 1999), in modelling El Niño events (Zavala-Garay et al. 2003), in the study of the atmospheric quasi-biennial oscillation (Piani et al. 2004), in modelling atmospheric convection (Lin & Neelin 2002), in enhancing ocean sea-surface temperature predictability (Scott 2003) and in modelling the impacts of ocean eddies (Berloff 2005).

The purpose of this chapter is to present the problem of unresolved scales in climate models using stochastic climate models. In particular we will present the study that considers the behaviour of the ocean thermohaline circulation (THC). The choice of this dynamical system is because in the last decades it has been established (Rahmstorf, 2000) that THC plays a main role regulating North Atlantic climate and also because this system can experience sudden changes. At present we do not completely understand THC dynamics but we know that it is a global phenomenon with interhemispheric and inter-ocean exchanges influenced by the topography of the ocean floor, the global distribution of wind stress, surface heat and freshwater fluxes. Despite not knowing completely the dynamics of THC, the conceptual picture is very well known, consisting of convection areas that 'push' the THC, in addition to a broad upwelling branch that occurs over the worlds' oceans, 'pulling' the THC (Manabe & Stouffer, 1993). Therefore, among other factors, the behaviour of THC depends on north Atlantic surface water being sufficiently cold and salty to destabilize the water column and produce deep water formation. In this way, THC formation is very sensitive to air-sea heat exchange and freshwater input in the North Atlantic. On the other hand, these two parameters are expected to vary due to climate change produced by greenhouse gases accumulation in the atmosphere and eventually these changes can produce a complete shutdown of the ocean overturning in the North Atlantic. The fate of THC under those new conditions has been the subject of a scientific debate (Broecker, 1987). THC was nominated the Achilles heel of actual climate, highlighting the possibility that minor changes in parameters can cause a sudden change in climate conditions (Broecker, 1991; Broecker, 1997). Those abrupt changes are not new in the climate system and can be produced by natural variability. Specifically, paleoclimatic data suggests that past abrupt changes of THC dynamics have taken place (Stocker et al., 2001; Rahmstorf, 2002) with THC oscillating between a warm mode similar to the present-day Atlantic, a cold mode with North Atlantic deep water (NADW) forming south of Iceland, and a switched-off mode. Those changes have provoked abrupt changes in surface climate (Clark et al., 2002). Therefore the possibility of a sudden change is not a theoretical one but a real possibility.

The influence of THC on climate, the possibility of an abrupt change in its mode of operation and the evidence in paleoclimate data of abrupt changes in the past have resulted in detailed investigations of the dynamical behaviour of the THC. Many modelling studies have been done in the last decades on this topic, but lack of predictability near thresholds implies that abrupt climate changes will always have more uncertainty than gradual climate change (Knutti & Stocker, 2002). Moreover, there are other underlying uncertainties added to nonlinearity. First of all, to know what the exact situation is now is not an easy question. In this regard, the initial condition is somewhat undefined. We must recall that in nonlinear systems this is a critical question, because depending on how close to the threshold we are, the behaviour of the whole system facing the same perturbation will be very different.

WOCE (World Ocean Circulation Experiment) project has obtained a value of 15 +/-2 Sv that can be used as a reference value (Ganachaud & Wunsch, 2000). From direct observations made using the deployed array of moored instruments in the context of the Rapid Climate Change program (RAPID) (Cunningham et al., 2007) we can see variability around this value. The second source of uncertainty is that the increasing rate of freshwater input (Stocker & Schmittner, 1997) and the location of this input are also important and unknown variables. Moreover in 3D models it has been demonstrated that THC behaviour depends greatly on the vertical mixing parameterization (Knutti et al, 2000). Moreover, it is necessary to know the "hydrological sensibility" (Rahmstorf and Ganopolsky, 1999), namely, the rate of change of hydrological cycle in a warming atmosphere. How much is the warming going to increase runoff from glaciers, rivers and marine ice? Also, the problem of climate change itself is not yet well established and some quite surprising feedbacks can appear. In this way, Delworth & Dixon, 2000 have proposed that, in a scenario of global heating, the NAO (North Atlantic Oscillation) would be preferentially in the positive phase, enhancing North Atlantic winds. These enhanced winds will cause a cooling of surface water that can cancel the effects of freshening. On the other hand, Latif et al. 2000 has proposed a similar mechanism but related to the ENSO (El Niño-Southern Oscillation) pattern. But, superimposed on those uncertainties is the fact that internal or natural variability on different scales may also play a main role in determining the THC stability (Knutti & Stocker, 2002; Alley et al., 2001; Monahan, 2002; Taboada & Lorenzo, 2005; Lorenzo et al., 2009). The influence of noise on the fate of THC in the context of climate change has not been as extensively studied as atmosphere-ocean global climate models (AOGCMs) and intermediate models without noise. As we have mentioned above, the main aim of this chapter is to show the effects of adding noise to the equations that drive the dynamics of the THC, in order to better estimate unresolved scales.

2. The role of stochastic forcings

As mentioned in the introduction section, modern climate models integrate basic fluid dynamical equations. They simulate the time-dependent three-dimensional flow fields and associated transports of mass, heat, and other fluid properties at a resolution of typically a few hundred kilometres. However, processes below this resolution (such as clouds and ocean eddies) cannot be represented explicitly and must be parameterized, that is, expressed in terms of the resolved larger scale motions. This is the one of biggest sources of uncertainty of GCMs. On the other hand, in most climate models, short time scale weather forcing is not represented or is smoothed.

These simplifications do not always ensure a correct result in the long-term behaviour of a system as complex as the climate system. Remember that the chaotic nature of this system allows small changes or disturbances in certain conditions or parameters to create large variations in the medium-term prediction, with even larger errors possibly accumulated in the long term. In meteorological studies, medium-and long-term solutions to these problems were found by developing ensemble prediction techniques. In this case, different models are run with different initial or boundary conditions or diverse parameterizations in order to create a wider range of dispersion in the model results, thus covering all possible values that could develop naturally in the atmosphere.

With the ensemble prediction technique, some of the shortcomings of the medium-term predictions have been improved, but in climate predictions we need more. As mentioned previously, many processes that relate to small spatial and temporal scales are omitted or approximated in climate prediction studies and this may taint some results. In recent decades the idea of Hasselmann and Epstein to simulate these processes with random noise has gained strength. The role of these noisy sources in the models would play a similar role to the noise on a particle moving in a potential well. Without noise, the system has a tendency to remain in a state. When noise is added, the random perturbations increase the likelihood of the particle to overcome the potential barrier and move to the other state. This phenomenon is known as the paradigm of stochastic resonance (figure 1) and plays an important role in the transition probabilities which are known to depend on sensitivity to noise levels (Gammaitoni et al., 1998; García-Ojalvo & Sancho, 1999; Pérez-Muñuzuri et al., 2003; Lorenzo et al., 2003).

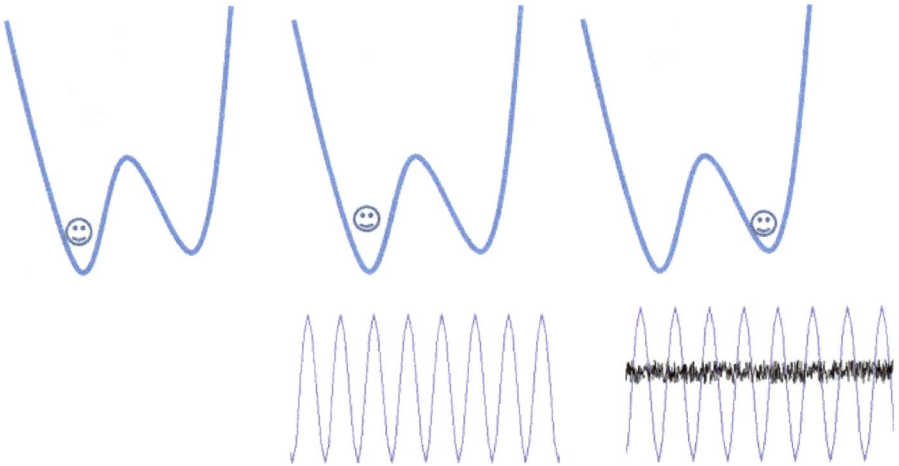

Fig. 1. Paradigm of stochastic resonance: In a symmetric double-well potential the presence of an optimal dose of noise will induce hopping events between the two wells.

As an example, most GCMs cannot be used to understand the long- term evolution of ENSO. An analysis of several GCMs suggests that it is very rare that an atmospheric GCM reproduces the correct spatial distribution, phase and seasonal behaviour of the observed subannual variability (Zhang 2005). However this kind of variability exists and when the models don't consider it, one could expect these inaccuracies to propagate to other time-scales. Zavala et al., 2008 have studied the possibility to improve our understanding of ENSO irregularity, adding the unresolved variability from observations to the GCMs. In this particular case, the irregular behaviour of ENSO has been suggested to result from nonlinear resonances of the ENSO mode with the annual cycle, from small changes in the background state or from the sporadic forcing of subannual variability on the ENSO mode (Zavala-Garay et al., 2008 and references in there). In other studies simple linear models were used to estimate the decadal response of the extratropical ocean to wind stress forcing, assuming a flat bottom, a mean state at rest, and no dissipation (Frankignoul et al., 1997; Cessi & Louazel, 2001).

In particular, we will present in this chapter the results found in previous works of the authors where stochastic forcing has been added to numerical models to analyze the behaviour of the Thermohaline Circulation (THC). As we mentioned above the choice of this example is because THC is a bistable process which implies the existence of critical points, where predictability becomes dramatically low. A shut down of the THC might be an irreversible process because of its multiple equilibriums, which have been reported by ocean and climate models of different complexity (Stommel, 1961; Bryan, 1986; Marotzke & Willebrand, 1991; Stocker & Wright, 1991). Previous works on simple models have studied the effect of these transitions induced by noise on the behaviour of the THC. Timmerman and Lohman in (Timmermann & Lohmann 2000) used a simple box model to study the stability of the THC as a function of the meridional temperature noise level. This stochastic temperature can either be interpreted as the representation of unresolved physical processes or as the expression of a thermohaline time scale separation. Monahan et al. (2008) study how the Meridional Overturning Circulation (MOC), the branch of the North Atlantic THC, varies over a wide range of spatial and temporal scales in response to fluctuating 'weather' perturbations that may be modelled as stochastic forcing. In that study, the authors analyzed the effects of noise on the variability of the MOC. In particular, they saw the role of noise on the transitions between the states of MOC and its role in the appearance of the Dansgaard-Oeschger events that are characteristic of glacial periods (Monahan et al., 2008). Other authors (Monhanan, 2001, Kuhlbrodt & Monahan, 2003) have studied a somewhat counterintuitive effect of adding noise to the equations that drive the dynamics of THC. This effect is the stochastic stabilization. In the first case with a simple box model, the effect of adding noise is that THC remains in a more stable state than would occur in the deterministic case. The second work focuses on deep convection and also arrives at the conclusion that adding noise changes the dynamical characteristic of the system. Such research proves that adding noise to simulate climate variability in different temporal or spatial scales influences the stability of the climate system in general and of the THC in particular.

The aforementioned studies illustrate the possibility of noise-induced transitions between different regimes of THC when random noise is considered in climate models, in this case related to a hypothesis of a massive sudden freshwater input.

3. Experiments and results

Two experiments are presented in this paper, both showing the possibility of a shutdown or weakening of the THC when a stochastic discharge of freshwater input is added around Greenland. In the first experiment a low complexity ocean-atmosphere coupled model was used to investigate the possibility of a collapse of the THC, taking into account the synoptic atmospheric variability. This kind of model can only give qualitative results but, they it is very useful for understanding parameter space and for building hypotheses. In a second experiment, the qualitative results found in the simplified model were reproduced in a model of intermediate complexity that is more realistic and represents a prelude to general circulation models.

3.1 Low complexity ocean-atmosphere coupled model

The "toy" model considered in the first part of our study is an atmosphere-ocean model which has been taken from a previously published work by Roebber (1995). The atmospheric part of the model is represented by a low-order model introduced by Edward Lorenz (1984) and defined by three equations:

$$\frac{dX}{dt} = -Y^2 - Z^2 - aX + aF$$

$$\frac{dY}{dt} = XY - bXZ - Y + G \qquad (1)$$

$$\frac{dZ}{dt} = bXY + XZ - Z$$

Where X, Y and Z represent the meridional temperature gradient and the amplitudes of the cosine and sine phases of a chain of superposed large scale eddies, respectively. F characterizes the meridional gradient of diabatic heating and G is the asymmetric thermal forcing, representing the longitudinal heating contrast between land and sea. a=0.25 and b=0.4 are two constants.

The oceanic part of the model is a box model (Figure 2) representing THC in the North Atlantic Ocean. In this model, Qs denotes the equivalent salt flux, T_{a1} and T_{a2} correspond to restoring air temperatures, and q is the Thermohaline circulation. The explicit equations of the model and the ocean model constants can be found in Roebber 1995 and Taboada & Lorenzo (2005).

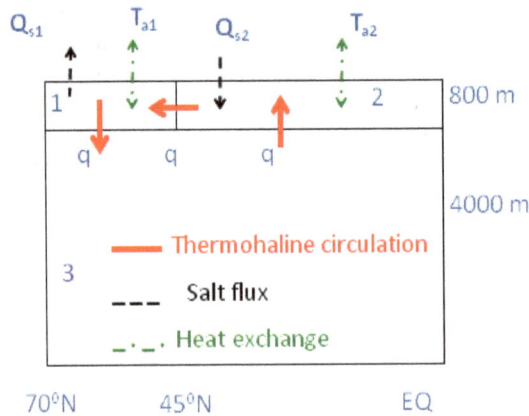

Fig. 2. Geometry of the North Atlantic through a box model. q represents the thermohaline circulation and the sense of the arrows indicates a positive circulation.

The influence of the atmosphere on the ocean model is through the variables F and G, which present a seasonal variation and whose equations are:

$$F(t) = F_0 + F_1 \cos \omega t + F_2(T_2 - T_1)$$
$$G(t) = G_0 + G_1 \cos \omega t + G_2 T_1 \qquad (2)$$

Where ω is the annual frequency and t=0 at winter solstice. The values chosen are, F_0=4.65, F_1=1, F_2=47.9, G_0=-3.60, G_1=1.0 and G_2=4.0254. These suppositions insure that F and G remain bounded by qualitatively plausible constraints for collapse flows (Latif et al., 2006). On the other hand, the ocean is coupled to the atmospheric model through the restoring temperatures T_{a1} and T_{a2} and the equivalent salt flux Q_s with the next expressions:

$$T_{a1}(t) = T_{a2} - \gamma X(t)$$
$$Q_s(t) = 0.00166 + 0.00022(Y^2 + Z^2)$$
(3)

where γ=0.06364 and T_{a2}=25°C are constants and the parameterization of $Q_s(t)$ is due to the assumption that the eddy water vapour transport is directly proportional to the eddy sensible heat flux given by Y^2+Z^2.

In this toy model the synoptic scale variability is not considered and following the ideas of Hasselmann and Epstein, random fluctuations were added in an additive way in parameters F and G to introduce the synoptic variability inside this model. These fluctuations were specified by white Gaussian noise with zero mean and whose correlation function is given by Equation 4. This kind of noise parameterization suitably simulates the processes that lay below the resolution of the model. In other previous work (Clark et al., 2002), the effect of atmospheric processes on the ocean has been taken into account as white noise.

$$\langle \xi_w(t)\xi_w(t') \rangle = 2A\delta(t - t')$$
(4)

where δ is the delta function, A is the intensity of the noise, 2A its variance, and $\langle \rangle$ denotes an average over the probability distribution of the random field.

In our experiment it was analyzed whether high-frequency variations in the meridional gradient of diabatic heating or in the longitudinal heating contrast between land and sea could induce significant changes in the ocean circulation with a transition to an off state of the THC. To carry out this study the model was run under present-day conditions, with an approximated value of 15Sv (1 Sv $\cong 10^6$ m^{3}s^{-1}). Q_{s1}=Q_{s2}=$Q_s(t)$ and stochastic perturbations of white Gaussian noise were introduced in the parameter F or G of the model.

The addition of noise in F, produces a greater variability in the THC but a transition from "on" to "off" state is not observed. This result is understandable because the effect of variations in F is translated as noise in the difference between temperature T_{a1} and T_{a2} and increments or decrements between northern and southern latitudes in North Hemisphere could strengthen or weaken the THC, but a collapse would be very difficult because the basic differences between north and south are maintained (Broecker, 1987).

Different is the case of a forcing in the parameter G (Figure 3) because this parameter is introduced in salt flux Q_s and in this case noise affects the difference between the temperature in the land and the temperature in the ocean. An increment in the temperature of the land can provoke a net freshwater flux into the northern part of the North Atlantic with the consequent loss of salinity and the collapse of THC (Broecker, 1987). In the real climate, that means that it is easier to collapse THC changing the budget evaporation-precipitation, than changing differences in temperatures.

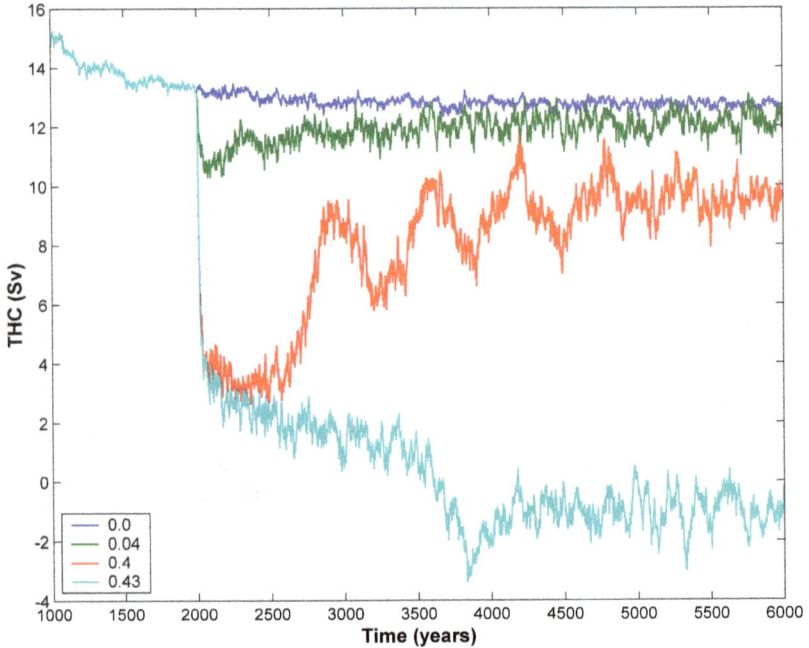

Fig. 3. Behaviour of the THC when a stochastic forcing on the zonal gradient in diabatic heating, G, is applied. The different values correspond to different values of the intensity of the applied noise.

With this toy model, it was also observed that the threshold necessary to collapse the THC is very sensitive to atmospheric greenhouse gases concentration. In Figure 4 it is shown that when decreasing the value of γ, related with the effects of global warming, the difference in temperature between the equator and the poles will diminish (Rahmstorf, 1996) and the high-frequency synoptic variations could provoke a collapse in THC easier than in actual conditions with lower values of concentration of CO_2.

Therefore, with this toy model it is possible to observe that the noise added to a model in order to mimic the variability not present in the model is able to modify the expected behaviour. Bistable systems such as the THC change from one state to another depending on their proximity to critical transition points when some fluctuations are considered. Although in actual conditions a collapse of the THC is not probable, changes such as global warming may bring the system nearer to critical points where the transition to a state of collapse will be made possible.

3.2 Earth model of intermediate complexity

The results discussed in the previous sections from a toy model have been reproduced in an analogous way in an earth model of intermediate complexity (EMIC). This section shows some results obtained from the model ECBilt-Clio after adding noise from freshwater input around Greenland (Lorenzo et al., 2008). This work corroborates the importance of considering the most likely processes in the development of a climate model to achieve more realistic results.

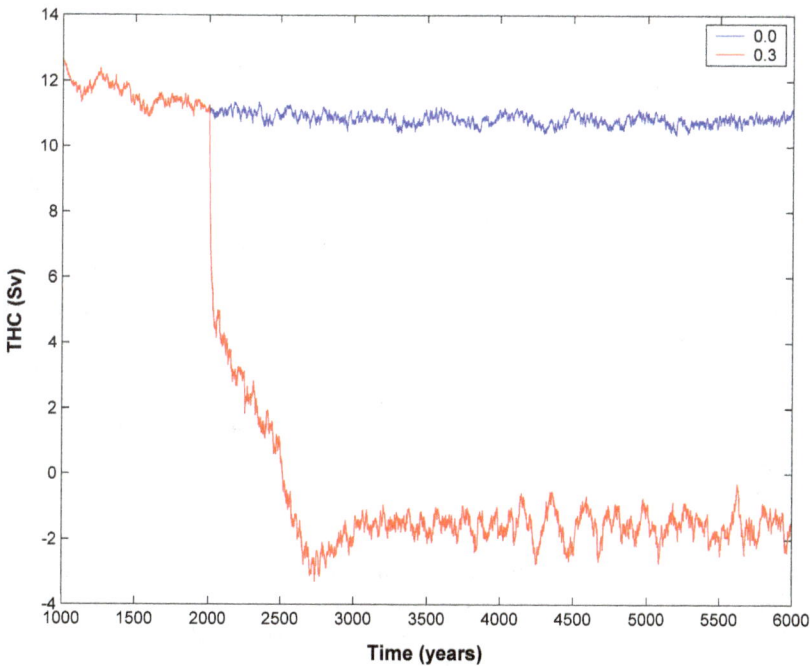

Fig. 4. Effects of a stochastic perturbation on the zonal gradient in diabatic heating G after a weakening of the THC due to a increasing of the CO_2 concentration, $\gamma=0.045$, for two values of the intensity of the applied noise A.

The model chosen is an earth system model of the intermediate complexity. Such models are realistic in the sense that they contain the minimum amount of physics that is necessary to simulate the midlatitude planetary and synoptic-scale circulations in the atmosphere as well as its variability on various timescales. We chose this model, developed at the Koninklijk Nederlands Meteorologisch Instituut, because it is simple to execute and it is one of the models of intermediate complexity used in the elaboration of the Intergovernmental Panel on Climate Change (IPCC) report (IPCC, 2001; IPCC, 2007). The atmospheric component uses the ECBilt model, a spectral T21 global three-level quasi-geostrophic model that uses simple parameterizations to simulate the diabatic processes (Opsteegh et al., 1998). The oceanic component is simulated by the Clio model which comprises a primitive equation, free-surface ocean general circulation model coupled to a thermodynamic-dynamic sea-ice model (Goosse & Fichefet, 1999). The oceanic component includes a relatively sophisticated parameterisation of vertical mixing and a three-layer sea-ice model, which takes into account sensible and latent heat storage in the snow-ice system. It simulates the changes of snow and ice thickness in response to changes in surface and bottom heat fluxes. In the computation of ice dynamics, sea ice is considered to behave as a viscous-plastic continuum. The horizontal resolution of the Clio model is 3° in both latitude and longitude and there are 20 unevenly spaced vertical layers in the ocean. The Clio model has a rotated grid over the North Atlantic Ocean in order to circumvent the singularity at the pole.

In our study, we stabilize the model ECBilt-Clio in the actual state and then we add stochastic forcing to introduce natural decadal and multidecadal atmospheric variability not simulated by the model. This forcing was introduced in the model through the addition of random Gaussian noise in the freshwater discharge around Greenland with a variability of 10 and 70 years (Lorenzo et al., 2008). The length of the experiments was 800 years, using initial conditions that were generated using a 1200 year spin-up run during which no stochastic forcing was applied. The intensity of the freshwater input varied, following a white Gaussian noise distribution that had zero mean and a standard deviation around 0.16 Sv. This value was chosen because previous studies showed that the strength of the forcing freshwater flux should lie between 0.1 Sv, which is the magnitude predicted for a large CO_2 induced climate change (4 x CO_2), and 1 Sv, which is within the range envisaged for events driven by meltwater release during the last glacial era and the deglaciation (Stouffer et al., 2006; Clarke et al., 2003). Moreover, previous work showed that the difference in freshwater export between La Niña and El Niño years is on the order of 0.1 Sv for the Atlantic (Schmittner et al., 2000; Schmittner & Clement, 2002).

The results obtained with the intermediate complexity model confirm the hypothesis suggested with the toy model. In the actual context a complete collapse of the THC is not probable, but a weakening of the circulation is observed when the synoptic and decadal natural variability of some climate processes is introduced, and results show that it is enough to observe significant changes in the climate system.

Figure 5 shows the temporal evolution of the THC in 3 different cases: without introducing a random discharge of freshwater around Greenland, introducing a random input decadal variability and introducing a random multidecadal discharge through that region.

In any of the considered cases the THC does not collapse, but a weakening is experienced. In that case the signal does have a significant impact on the behaviour of variables such as air surface temperature, sea surface temperature, precipitation or streamfunction. In Figure 6 the anomalies observed in these variables are shown.

We can see that although the THC is not collapsed several effects are considered when natural variability is introduced in the model using a stochastic forcing. In this case, the results correspond to a multidecadal variability related with the existence of an oscillatory mode of variability with a period of ~70 years involving fluctuations in the intensity of the THC in the North Atlantic (Delworth & Mann, 2000). The images show a cooling in the North Hemisphere with changes in the global precipitation patterns, mainly in the Intertropical Convergence Zone, ITCZ, and changes in the jet stream (Lorenzo et al., 2009).

Other variables, such as the sea surface salinity, geopotential or the NAO index (Lorenzo et al., 2008) have been analyzed and notable changes have been measured when a decadal or multi-decadal variability were considered in the freshwater flux of the GIN Sea. This should alert us to the need to add to the climate models many of the scales of natural variability that are not considered in most models in the name of simplicity.

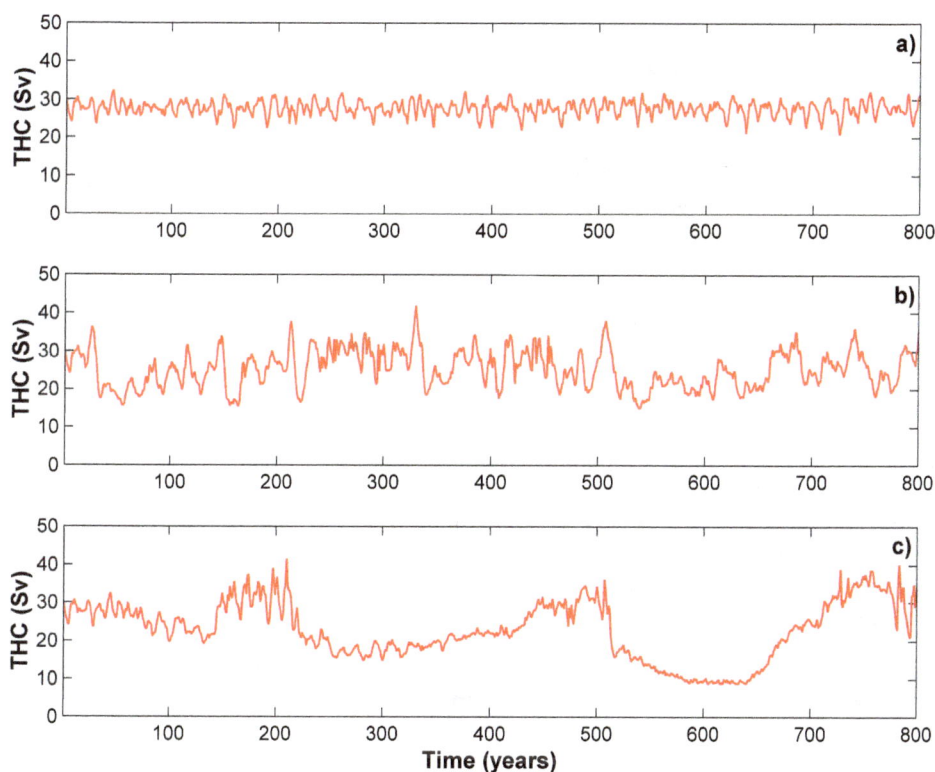

Fig. 5. Behaviour of the THC three simulations. (a) without forcing; (b) with random freshwater input with decadal variability added in the ocean basin of the Greenland/Iceland/Norwegian Sea (GIN Sea); and c) the same b) but with multidecadal (70 years) variability.

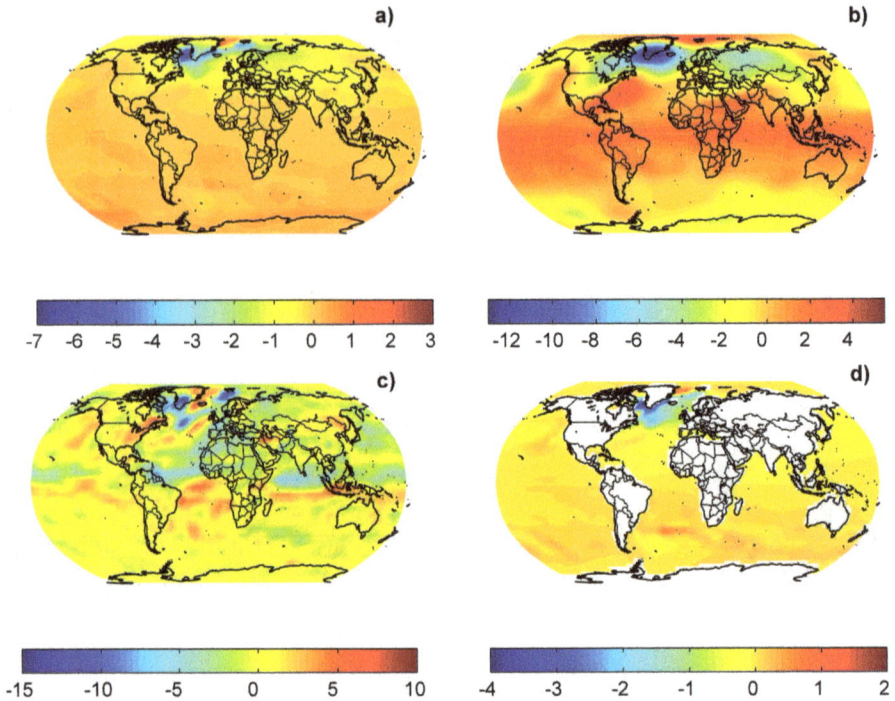

Fig. 6. a) Anomaly of mean surface temperature (°C). b) Anomaly of mean streamfunction (m² s⁻¹). c) Anomaly of mean precipitation (cm year⁻¹) and d) Anomaly of mean sea surface temperature SST (°C) all them during the 800-year simulation for a variability of 70 years in the flux of freshwater. The averaging period for the anomalies is computed relative to an unperturbed control run.

4. Conclusions

In this chapter we have shown the effect of the climate's natural variability, simulated through white Gaussian noise at different temporal scales, on the behaviour of THC. This noise has been added to two different models: a low complexity ocean-atmosphere coupled model and an earth model of intermediate complexity previously used in the reports of the IPCC. In these experiments the relevance of the variability of high frequency in the climate was observed. Here, the attention was focused on the case of thermohaline circulation and the addition of white Gaussian noise but in the bibliography there are numerous examples in which the addition of random noise, white or colored, can improve the understanding of certain climate processes like El Niño events, the quasi-biennial oscillation, the atmospheric convection or ocean eddies, amongst others. In our particular case, it was observed that a critical noise level exists which induces an abrupt transition of the salinity statistics. This abrupt transition induces a sudden change in the THC behaviour, which provokes a significant change in the climate. Even if adding noise does not induce a complete shutdown, climate in the North Atlantic would be affected. The concept of noise-induced transitions explains this qualitative behaviour. Furthermore, changing noise amplitudes could lead to rapid changes in the mean state and, in the context of climate variability, to abrupt climate transitions.

In this chapter we intend to call attention to those unresolved scales and processes in the models. We have shown the example of the variability in the discharge of freshwater in the GIN Sea, but there are many other unresolved processes models: gravity waves, clouds, ocean eddies, small-scale turbulence... The complexity and computational cost of a conventional resolution of them still seems far away, so the alternative stochastic approach, based on noise, seems to be the more suitable solution (Williams, 2005). In agreement with Williams (2005) we think that the inclusion of stochastic forcing in the next generation of climate models is necessary to give a more realistic view of climatic processes that occur around us.

5. Acknowledgment

This work is supported by the Xunta de Galicia and FEDER under Research Grant No: 10PXIB383169PR. J.J. Taboada acknowledges the financial support from the Department of Environment of the Galician Government (Xunta de Galicia).

6. References

Alley, R.B., Anandakrishnan, S., & Jung, P. (2001). Stochastic resonance in the North Atlantic. *Paleoceanography*, 16, pp. 190-198.

Berloff, P. S. (2005). Random-forcing model of the mesoscale oceanic eddies. *J. Fluid Mech.*, 529, pp. 71-95 (doi:10.1017/S0022112005003393).

Broecker, W.S. (1987). Unpleasant surprises in the greenhouse. *Nature*, 328, pp. 123-126.

Broecker, W. S. (1991). The great ocean conveyor. *Oceanography*, 4, pp. 79-89.

Broecker, W. S. (1997). Thermohaline circulation, the Achilles' heel of Our Climate System: Will Man-Made CO2 Upset the Current Balance? *Science*, 278, pp. 1582–1588.

Bryan, F. (1986). High-latitude salinity effects and interhemispheric thermohaline circulations. *Nature*, 323, pp. 301-304.

Buizza, R., Miller, M., & Palmer T.N. (1999). Stochastic representation of model uncertainties in the ECMWF ensemble prediction scheme. *Q. J. R. Meteorol. Soc.*, 125, pp. 2887-2908 (doi:10.1256/smsqj.56005.).

Cessi, P., & Louazel S. (2001). Decadal Oceanic Response to Stochastic Wind Forcing. *J. Phys. Oceanogr.*, 31, pp. 3020–3029. doi: 10.1175/1520-0485(2001)031<3020: DORTSW>2.0.CO;2

Clark, P.U., Pisias, N.G., Stocker, T.F., & Weaver, A.J. (2002). The role of the thermohaline circulation in abrupt climate change. *Nature*, 415, pp. 863-870.

Clarke, G.K.C., Leverington, D.W., Teller, J.T., & Dyke, A.S. 2003. Superlakes, megafloods, and abrupt climate change. *Science*, 301, pp. 922-923.

Cunningham, S.A., Kanzow, T., Rayner, D., Baringer, M.O., Johns, W.E., Marotzke, J., Longworth, H.R., Grant, E.M., Hirschi, Joël J.-M., Beal, L.M., Meinen, C.S., & Bryden, H. L. (2007). Temporal Variability of the Atlantic Meridional Overturning Circulation at 26.5°N. *Science*, 317, pp. 935-938.

Delworth, T.L., & Dixon, K.W. (2000). Implications of the recent trend in the Arctic/North Atlantic Oscillation for the North Atlantic thermohaline circulation. J. Clim., 13, pp. 3721-3727.

Delworth, T.L., & Mann, M.E. (2000). Observed and simulated multidecadal variability in the Northern Hemisphere. Climate Dymanics, 16, pp. 661-676.

Frankignoul, C., Müller, P., & Zorita, E. (1997). A Simple Model of the Decadal Response of the Ocean to Stochastic Wind Forcing. J. Phys. Oceanogr., 27, pp. 1533-1546. doi: 10.1175/1520-0485(1997)027<1533:ASMOTD>2.0.CO;2

Gammaitoni, L., Hänggi, P., Jung, P., & Marchesoni, F. 1998. Stochastic resonance. Rev. Mod. Phys., 70, pp. 223-287.

Ganachaud, A., & Wunsch, C. (2000). Improved estimates of global ocean circulation, heat transport and mixing from hydrographic data. Nature, 408, pp. 453-457.

García-Ojalvo, J. & Sancho, J.M. (1999). Noise in spatially extended systems. Institute for Nonlinear Science, Springer-Verlag, ISBN-13: 978-0387988559, New York

Goosse, H., & Fichefet, T. (1999). Importance of ice-ocean interactions for the global ocean circulation: a model study. J. Geophys. Res., 104, C10, pp. 23337-23355.

Hasselmann, K. (1976). Stochastic climate models. Part I. Theory. Tellus, 28, pp. 473-485.

IPCC. in Climate Change (2001). The Scientific Basis. Contribution of Working Group I to the Third Assessment Report of the Intergovernmental Panel on Climate Change, edited by Houghton, J. T., Y. Ding, D. J. Griggs, M. Noguer, P. J. v. d. Linden, X. Dai, et al., pp. 881 pp, Cambridge University Press, Cambridge, United Kingdom and New York, NY, USA, 2001.

IPCC. Climate change (2007). the scientific basis. In: Contribution of Working Group I to the Fourth Assessment Report of the Intergovernmental Panel on Climate Change, Cambridge University Press, Cambridge, UK, 2007.

Kuhlbrodt, T., & Monahan A.H. (2003). Stochastic stability of open-ocean deep convection. Journal of Physical Oceanography, 33, pp. 2764-2780.

Knutti, R., Stocker, T. F., & Wright D. G. (2000). The effects of sub-grid-scale parameterizations in a zonally averaged ocean model. Journal of Physical Oceanography, 30, pp. 2738-2752.

Knutti, R., & Stocker, T.F. (2002). Limited predictability of the future thermohaline circulation close to an instability threshold. J. Climate, 15, pp. 179-186.

Latif, M., Roeckner, E., Mikolajewicz, U., & Voss, R. (2000). Tropical Stabilization of the Thermohaline circulation in a Greenhouse Warming Simulation. J. of Climate, 13, pp. 1809-1813.

Latif, M., Böning, C., Willebrand, J., Biastoch, A., Dengg, J., N. Keenlyside, Madec, G., & Schweckendiek, U. (2006). Is the thermohaline circulation changing? J. Cim. 19, pp. 4631-4637.

Lorenz, E. N. (1963). Deterministic nonperiodic flow. J. Atmos. Sci., 20, 2, pp. 130-141.

Lorenz, E. N. (1984). Irregularity. A fundamentel property of the atmosphere. Tellus, 36A, pp. 98-110.

Lin, J. W. B., & Neelin, J. D. (2002). Considerations for stochastic convective parameterization. J. Atmos. Sci., 59, pp. 959-975.

Lorenzo, M. N., Santos, M. A., & Pérez-Muñuzuri, V. (2003). Spatiotemporal stochastic forcings effects in an ensemble consisting of arrays of diffusively coupled Lorenz cells. Chaos, 13, pp. 913-920.

Lorenzo, M. N, Taboada, J. J., Iglesias I., & Álvarez, I. (2008). The role of stochastic forcing on the behaviour of the Thermohaline circulation. *Annals of the New York Academy of Sciences, Special Issue 'Trends and Directions in Climate Research'*, 1146, pp. 60-86.

Lorenzo, M. N., Taboada, J.J., & Iglesias, I. (2009). Sensitivity of thermohaline circulation to decadal and multidecadal variability. *ICES Journal of Marine Science*, 66, pp. 1439-1447, doi:10.1093/icesjms/fsp061.

Manabe, S., & Stouffer, R. J. (1993). Century-scale effects of increased atmospheric $CO2$ on the ocean-atmosphere system. *Nature*, 364, pp. 215-218.

Marotzke, J., & Willebrand J. (1991). Multiple equilibria of the global thermohaline circulation. *Journal of Physical Oceanography*, 21, pp. 1372-1385.

Monahan, A.H., (2001). Nonlinear principal component analysis: Tropical Indo-Pacific sea surface temperature and sea level pressure. *J. Climate*, 14, pp. 219-233.

Monahan, A. H. (2002). Stabilization of climate regimes by noise in a simple model of the thermohaline circulation. *J. Phys. Oceanogr.*, 32, pp. 2072-2085. (doi:10.1175/ 1520-0485(2002)032!2072:SOCRBNO2.0.CO ;2).

Monahan, A, Alexander, J., & Weaver, A. J. (2008). Stochastic models of the meridional overturning circulation: time scales and patterns of variability. *Phil. Trans. R. Soc. A* 366, 2527–2544. doi:10.1098/rsta.2008.0045

Pérez-Muñuzuri, V, Lorenzo, M. N., Montero, P., Fraedrich, K., Kirk, E., & Lunkeit F. (2003). Response of a global atmospheric circualtion modelo n spatiotemporal stochastic forcing: Ensemble statistic. *Nonlinear Processes in Geophysics*, 10, pp. 453-461.

Piani, C., Norton, W.A., & Stainforth, D.A. (2004). Equatorial stratospheric response to variations in deterministic and stochastic gravity wave parameterizations. *J. Geophys. Res.*, 109, pp. D14 101 (doi:10.1029/2004JD 004656.).

Opsteegh, J.D., Haarsma, R.J., Selten, F.M., & Kattenberg A. (1998). ECBILT: A dynamic alternative to mixed boundary conditions in ocean models. *Tellus*, 50A, pp. 348-367.

Rahmstorf, S., 1996. On the freshwater forcing and transport of the Atlantin thermohaline circulation. *Clim. Dynam.*, 12, pp. 799-811.

Rahmstorf, S., & Ganopolski, A. (1999). Simple theoretical model may explain apparent climate instability. *J. Clim.*, 12, pp. 1349-1352.

Rahmstorf, S. (2000). The thermohaline ocean circulation – a system with dangerous thresholds? *Clim. Change*, 46, pp. 247–256.

Rahmstorf, S. (2002). Ocean circulation and climate during the past 120,000 years. *Nature*, 419, pp. 207-214.

Roebber, P. J. (1995). Climate variability in a low-order coupled atmosphere-ocean model. *Tellus*, 47A, pp. 473–494.

Scott, R. B. (2003). Predictability of SST in an idealized, one-dimensional, coupled atmosphere-ocean climate model with stochastic forcing and advection. *J. Clim.*, 16, pp. 323-335 (doi:10.1175/1520-0442(2003) 016!0323:POSIAIO2.0.CO;2).

Schmittner, A., Appenzeller, C., & Stocker, T.F. (2000). Enhanced Atlantic freshwater export during El Niño. *Geophysical Research Letters*, 27, 8, pp. 1163-1166.

Schmittner, A., & Clement, A.C. (2002). Sensitivity of the thermohaline circulation to tropical and high latitude freshwater forcing during the last glacial-interglacial cycle. *Paleoceanography*, 17, Doi 10.1029/2000PA000 591.

Stocker, T. F., & Wright, D. G. (1991). Rapid transitions of the ocean's deep circulation induced by changes in surface water fluxes. *Nature*, 351, pp. 729-732.

Stocker, T. F., & Schmittner, A. (1997). Influence of CO_2 emission rates on the stability of the thermohaline circulation. *Nature*, 388, pp. 862-865.

Stocker, T. F., Knutti, R., & Plattner, G. K. (2001). The oceans and rapid climate change: past, present, and future. *Geophysical Monograph*, 126, AGU, Washington D.C., USA.

Stommel, H. (1961). Thermohaline convection with two stable regimes of flow, *Tellus*, 13, pp. 224-230.

Stouffer, R. J., Gregory, J. M. Yin J., Dixon, K. W., Spelman, M. J., Hurlin, W., Weaver, A. J., Eby, M., Flato, G. M., Hasumi, H., Hu, A., Jungclaus, J. H., Kamenkovich, I. V., Levermann, A., Montoya, M., Murakami, S., Nawrath, S., Oka, A., Peltier, W. R., Robitaille, D. Y., Sokolov, A., Vettoretti, G., & Weber, S. L. (2006). Investigating the causes of the response of the thermohaline circulation to past and future climate changes. *J. Cim.*, 19, pp. 1365-1387.

Taboada, J. J., & Lorenzo, M. N. (2005). Effects of the synoptic scale variability on the thermohaline circulation. *Nonlinear Process Geophys*, 12, pp. 435-439.

Timmermann, A., & Lohmann, G. (2000). Noise-Induced Transitions in a Simplified Model of the Thermohaline Circulation. *Journal of Physical Oceanography*, 30, 8, pp. 1891-1900. Doi 10.1175/1520-0485(2000)030<1891:NITIAS>2.0.CO;2

Williams, P.D. (2005). Modelling climate change: The role of unresolved processes. *Phil. Trans. R. Soc. A*, 363, pp. 2931-2946.

Zavala-Garay, J., Moore, A.M., & Perez, C.L. (2003). The response of a coupled model of ENSO to observed estimates of stochastic forcing. *J. Clim.*, 16, pp. 2827-2842 doi:10.1175/1520-0442(2003)016!2827: TROACMO2.0.CO;2.

Zavala-Garay, J., Zhang, C., Moore, A. M., Wittenberg, A. T., Harrison, M. J., Rosati, A., Vialard J., & Kleeman, R. (2008). Sensitivity Of Hybrid Enso Models To Unresolved Atmospheric Variability. *Journal of climate*, 21, pp. 3704-3721.

Zhang, C., (2005). Madden-Julian Oscillation. *Rev. Geophys*, 43, RG2003, doi:10.1029/2004RG000158, 2005.

10

Numerical Investigation of the Interaction Between Land Surface Processes and Climate

Kazuo Mabuchi
Meteorological Research Institute
Japan

1. Introduction

The close connection between land surface ecosystem and climate is an important element in the Earth's environmental system, and land surface is very important to regional and global climates (Dickinson & Henderson-Sellers, 1988; Avissar & Pielke, 1989; Sato et al., 1989; Shukla et al., 1990; Bonan et al., 1992; Lofgren, 1995). The land surface is the lower boundary for the atmosphere, and energy and greenhouse gases are exchanged between the land surface and the atmosphere. Vegetation on the land surface is an important determinant of these fluxes. Snow on the ground also acts as a very important boundary condition for the atmosphere, and influences the hydrological cycle both directly and indirectly (Barnett et al., 1989; Yasunari et al., 1991; Vernekar et al., 1995).

In this chapter, examples of land biosphere model and numerical studies concerning the interactions between land surface processes and the climate are described. Several results of simulations using a global and a regional climate models are introduced with brief review of pioneering research works. These results indicate that the understanding of energy and carbon cycle mechanisms related to the land surface processes is very important to improve estimates for future situations of the Earth.

Section 2 describes the land biosphere model connected on-line to the climate models. Section 3 presents the investigation results using the global climate model, while Section 4 presents those using the regional climate model. Finally, concluding remarks are given in Section 5.

2. Land surface model for use within climate models

2.1 Introduction

Since the middle 1980s, various kinds of land surface models, such as the Biosphere Atmosphere Transfer Scheme (BATS) (Dickinson et al., 1986) or the Simple Biosphere Model (SiB) (Sellers et al., 1986), which include realistic representations of vegetation and soil, have been developed and used in climate models. These models can simulate energy fluxes between the land surface and the atmosphere, but cannot simulate the carbon cycles. Subsequently, to investigate the global carbon cycle related to global warming, the land surface model (SiB2) that includes parameterizations of carbon dioxide uptake and emission by vegetation was developed (Sellers et al., 1992, 1996).

In this section, a terrestrial ecosystem model for use within global and regional climate models is introduced.

2.2 Biosphere-atmosphere interaction model

A land surface model, termed the Biosphere-Atmosphere Interaction Model Version 1 (BAIM), was developed (Mabuchi et al., 1997; Mabuchi, 1997; Mabuchi, 1999). In BAIM, the morphological and physiological properties of vegetation are explicitly parameterized, and realistic treatments of snow on the ground and water/ice in the soil are included. The model uses three groups of parameters: morphological (e.g., leaf area index, canopy height, and leaf angle distribution), physiological (e.g., green leaf fraction and Rubisco capacity), and physical parameters (e.g., transmittance and reflectance of the leaf, soil reflectance, and hydraulic conductivity of soil). Some of the morphological parameter values, for example the canopy roughness length or the zero-plane displacement height of canopy, are derived from some fundamental parameters in BAIM. Therefore, BAIM has flexibility for the treatment of the morphological structure of vegetation.

The ecosystem modeled in BAIM consists of up to two vegetation layers and three soil layers. The temperatures and stored moistures for vegetation, snow, and soil layers are predicted by the equations describing the heat and water budget in each layer. In BAIM, the carbon dioxide flux caused by the ecosystem's absorption by photosynthesis and emission through respiration is explicitly simulated. Use of BAIM can result in estimates of not only the energy fluxes, but also the carbon dioxide flux between terrestrial ecosystems and the atmosphere. Accumulation and melting processes of snow on the leaves and on the ground are also simulated. In the presence of snow cover on the ground, the snow layer is divided into a maximum of three layers, with the temperature and amounts of snow and water stored in each layer predicted. The model can also predict the freezing and melting of water in the soil. The prediction of temperature and moisture is divided into three categories according to the amount of snow on the ground. The model adopts realistic descriptions of photosynthesis for C_3 and C_4 plants. The bulk stomatal resistances of canopy and ground cover, which are closely related to the water vapor and carbon dioxide fluxes between the ecosystem and the atmosphere, are obtained from the integration of the leaf-level stomatal resistance, calculated from the consideration of the enzyme kinetics and electron transport properties of chloroplasts and the ambient environmental parameters.

The upper boundary conditions for BAIM are the atmospheric variables at the reference height within the atmospheric boundary layer, i.e., air temperature, T_m (K), water vapor pressure, e_m (Pa), carbon dioxide concentration, C_m (mol mol^{-1}), and wind speed, U_m (m s^{-1}); also downward short-wave radiation, R_s (W m^{-2}), and long-wave radiation, R_l (W m^{-2}), at the bottom of the atmosphere, and precipitation rate, P_m (m s^{-1}). The lower boundary condition is the deep soil temperature, T_4 (K). The downward short-wave radiation used in BAIM, R_s, consists of four components, i.e., visible direct beam radiation, visible diffuse radiation, near infrared direct beam radiation, and near infrared diffuse radiation. The visible radiation is photosynthetically active radiation (PAR), which is absorbed strongly by the leaves of vegetation. The downward long-wave radiation is treated as diffuse radiation.

Model sensitivity tests for the values of parameters used in the model were performed using point micrometeorological data observed at a grassland site (Mabuchi et al., 1997). By changing the values of the parameters by $\pm$ 50 %, the maximum variations of the net radiation flux, the sensible heat flux, the latent heat flux, and the soil heat flux were about $\pm$ 15 W m^{-2}, $\pm$ 8 W m^{-2}, $\pm$ 9 W m^{-2}, and $\pm$ 1 W m^{-2}, respectively. The maximum variations in the net carbon dioxide flux were about $\pm$ 5 μmol m^{-2} s^{-1} for C$_3$ parameters and $\pm$ 7 μmol m^{-2} s^{-1} for C$_4$ parameters. These maximum variation values are comparable to observation errors.

Several terrestrial ecosystem models that can simulate the biomass carbon cycle have been developed (e.g., Ito & Oikawa, 2002; Schwalm & Ek, 2004; Garcia-Quijano & Barros, 2005; Matala et al., 2005). The Biosphere-Atmosphere Interaction Model Version 2 (BAIM2) is an improved land surface model, based on BAIM. In BAIM2, the carbon storage of the vegetation is divided into five components, i.e., leaves, trunk, root, litter, and soil. When BAIM2 is connected with a climate model, the carbon exchanges among the components of vegetation and the atmosphere, and the carbon dioxide concentration in the atmosphere, are calculated with other physical processes at each time step of the fully coupled model integration. The values of some of the morphological parameters used in the model (e.g., leaf area index and canopy height) are derived from the carbon storage values of the components, and the phenological changes of vegetation are reproduced by the seasonal climate conditions simulated by the model.

3. Simulation using the global climate model

3.1 Introduction

The relationship between the land-surface ecosystem and the climate is an important element in the Earth's environmental system. In the past, numerous studies have focused on land surface processes and interactions between the land surface and the atmosphere. In particular, tropical vegetation controls the physical and biogeochemical interactions in climatically influential areas over the Earth, and plays important roles in forming regional and global climates (Henderson-Sellers et al., 1993; Zhang et al., 1996a, 1996b; Hahmann & Dickinson, 1997; Lean & Rowntree, 1997; Xue, 1997; Clark et al., 2001; Hales et al., 2004; Snyder et al., 2004).

Werth & Avissar (2002, 2005a, 2005b) investigated the impacts of deforestations in Amazonia, Africa, and Southeast Asia. They found that deforestation induces the reduction of precipitation in each local area. Moreover, they suggested that the effects of deforestation in the tropics spread beyond the deforested area and reach into the mid-latitudes. Hasler et al. (2009) performed an impact study of tropical deforestation with multi-model ensemble simulations using three different global climate models. Results indicated a strong decrease in precipitation in the deforested areas and precipitation changes outside those areas. They also pointed out that the effect in the northern mid-latitudes is weaker, but some evidence of a wave train forced by the tropical changes could be seen. Nobre et al. (2009) studied the role of ocean-atmosphere interactions on climate change related to Amazon deforestation. Their simulations utilized an atmospheric general circulation model (AGCM) forced with climatological global sea surface temperature (SST), and a coupled ocean-atmosphere general circulation model (CGCM) coupled over the global tropics. Both model simulations indicated local surface warming and rainfall reduction, with larger impact suggested by the CGCM than by the AGCM. They also emphasized that the remote response detected with the CGCM experiments was an increased El Niño-Southern Oscillation (ENSO)-like variability over the Pacific, as a result of ocean-atmosphere interactions originated by the Amazon deforestation.

Among the vegetation changes in the tropics, forest removal in the Asian tropical region has recently become a severe problem. Jang et al. (1996) assessed changes in global forest conditions between 1986 and 1993, using satellite data converted to terrestrial net primary production (NPP). They observed that Indonesia, Papua New Guinea, and Burma accounted for 10 % of the world regions where the NPP had decreased by more than 800 g m $^{-2}$ between 1986 and 1993, due to degradation of the tropical forest. A recent FAO report (2007) addressed annual deforestation and net forest area changes from 1990 to 2000 and from 2000 to 2005. The report indicated that the annual deforestation in Southeast Asia was -2.8 million ha year $^{-1}$, and the annual rates of net forest area change were - 1.20 % year $^{-1}$ for 1990 to 2000 and - 1.30 % year $^{-1}$ for 2000 to 2005. These rates of net forest area change in Southeast Asia were the most severe of those for all the tropical areas worldwide.

Kanae et al. (2001) investigated the impact of deforestation on regional precipitation over the Indochina peninsula by analyzing precipitation data observed at meteorological stations in Thailand. Significant decreases in precipitation over Thailand were detected in September. Sen et al. (2004) investigated the effects of Indochina deforestation on the East Asian summer monsoon. Comparison of changes in the model results with observed rainfall trends pointed to deforestation in the Indochina peninsula as a major factor for climate changes in the region. Werth & Avissar (2005b) examined the effects of Southeast Asian deforestation, and found a strong local effect, with a reduction in Asian precipitation that persisted throughout the year.

With the physical aspects, the carbon cycle and the climate system are closely connected. Terrestrial ecosystems take up 2 to 4 PgC of carbon per year, which amounts to 20 to 30 % of annual total anthropogenic carbon dioxide emissions into the atmosphere, and exhibit strong interannual variability (Prentice, 2001). To understand the origin of the variations in the atmosphere-biosphere carbon exchange, it is necessary to understand the mechanisms of carbon dynamics between the atmosphere and the terrestrial biosphere.

The studies of Betts et al. (1997) and Mabuchi et al. (2000) focused on vegetation physiology and the carbon circulation associated with vegetation activity and climate. Betts et al. (1997) used a general circulation model iteratively coupled with an equilibrium vegetation model to quantify the effects of both physiological and structural vegetation feedbacks on a doubled carbon dioxide climate. It was shown that changes in vegetation structure partially offset physiological vegetation–climate feedback on a global scale over the long term, but overall vegetation feedback provided significant regional scale effects. Mabuchi et al. (2000) investigated the relationship between climate and the carbon dioxide cycle around the Japanese Islands. Interannual variations of carbon dioxide concentrations in the lower troposphere were found to be related to vegetation activity, while downward short-wave radiation was determined to be the most important element for vegetation activity around the Japanese Islands. Ito (2003) and Cao et al. (2005) used ecosystem models to simulate the global scale carbon dioxide exchange between the atmosphere and the terrestrial biosphere over long time periods in the past. Matthews et al. (2005) also examined the behavior of the terrestrial carbon cycle under both historical and future climate changes, using a global climate model coupled with a dynamic terrestrial vegetation and carbon cycle model.

Mabuchi et al. (2009) recently used a regional climate model that includes a terrestrial biosphere model to investigate the impact of climate factors on the carbon cycle in the East Asian terrestrial ecosystem. They concluded that a typical relationship exists between variations of the carbon cycle over land areas and those of climate factors on a regional scale in East Asia.

In this section, several results of numerical studies using a global climate model that includes a realistic biological land surface model are introduced. The numerical simulations were performed to investigate the impacts of Asian tropical vegetation change on the climate and carbon cycle (Mabuchi et al., 2005a, 2005b; Mabuchi, 2011).

3.2 Global climate model

The atmospheric model used in the experiments is the general circulation spectral model developed by the Japan Meteorological Agency (JMA). This general circulation model has a triangular truncation at wave number 63 (T63), and employs hybrid vertical coordinates at 21 levels. The horizontal resolution is 1.875° (192 × 96 grid points). The basic equations adopted for the model are the primitive equations. The model's atmospheric prognostic variables are the temperature, specific humidity, divergence and vorticity of the wind, the carbon dioxide concentration in each atmospheric layer, and surface pressure. The time step interval of the integration is about 20 minutes. The model includes short-wave and long-wave radiation processes (Sugi et al., 1990; Lacis & Hansen, 1974). Large scale precipitation and convective precipitation are estimated separately, with convective precipitation calculated by the Kuo scheme (Kuo, 1974). Vertical diffusion is calculated by the turbulent closure model (level 2.0) proposed by Mellor & Yamada (1974).

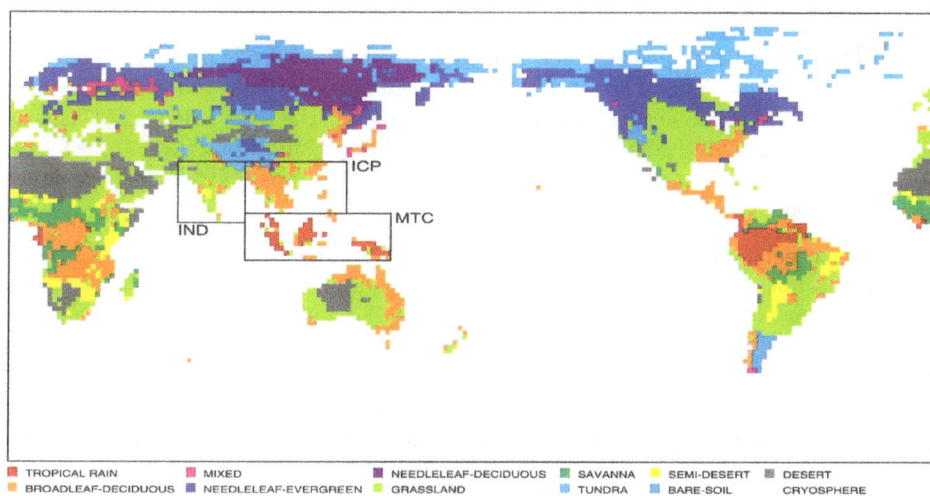

Fig. 1. Distribution of the original vegetation used in the model and the experiment regions. The vegetation is indicated by the color legend below the figure. The experiment regions are defined by the boxed areas, and are the Indian subcontinent area (IND), the Indochina Peninsula area (ICP), and the maritime continent area (MTC).

BAIM2 was integrated into this general circulation model, resulting in a climate model that can simulate the effects of vegetation on climate. The type of vegetation at each model land-area grid-point was specified, and the interactions between the land surface vegetation and the atmosphere were estimated by BAIM2 at each grid-point.

Figure 1 indicates the distribution of the vegetation used in the model and the experiment areas. The vegetation type of each model land-area grid-point was derived from the Major World Ecosystem Complexes Ranked by Carbon in Live Vegetation dataset (Olson et al., 1983). This vegetation data set has 47 types of vegetation. Fundamentally, these types of vegetation cover are divided into a number of groups consisting of forest, grassland, crop, shrub, taiga, savanna, wetland, semi-desert, desert, tundra, and cryosphere. The actual vegetation of a given global land surface grid was classified into one of 12 types (Fig. 1), including desert and cryosphere. Shrub and wetland were classified as grassland, with forest and taiga in east Siberia regarded as needle-leaf deciduous-forest-type vegetation. In the experiments, crop-type vegetation was classified as C_3 grassland vegetation.

3.3 Climatic impact of vegetation change in the Asian tropical region

3.3.1 Experiment design

The purpose of this study is not only to examine the impact of deforestation, but to also investigate the role of vegetation in the formation of the climate through numerical simulations. The experiments were performed under conditions that the land surface vegetation was changed morphologically, physiologically, and physically. Through these experiments, the mechanisms of the interactions between the land surface vegetation and the atmosphere can be understood. And more importantly, the influence of deforestation on the climate under various conditions can be estimated more accurately.

To investigate the impact of vegetation changes on climate in the Asian tropical region, three experiment areas were defined: the Indian subcontinent area (IND), the Indochina Peninsula area (ICP), and the maritime continent area (MTC) (Fig. 1). The IND, ICP, and MTC areas are mainly covered by the grassland, tropical seasonal forest, and tropical rain forest types of vegetation, respectively. The photosynthesis process for the grass of the actual vegetation in the experiment areas was assumed as the C_3 type.

Prior to the vegetation change impact experiments, a control time integration (CN) was performed. In this control integration, the actual global vegetation (see Fig. 1) and climatic SST values were used. The sea surface temperatures and sea ice values were taken from the GISST2.2 dataset (Rayner et al., 1996). The monthly climatic values of these data were assigned to each model ocean-area grid point. A 10-year spin-up calculation was carried out in order to estimate the initial values of soil water content, including the ice content in the soil and soil temperature. Using the soil values obtained from the spin-up calculation, the control integration was continued for 20 years.

After the control integration, three vegetation change impact experiments were performed: a bare soil experiment (BS), a C_4 grass experiment (C4), and a green-less experiment (GR). In the BS experiment, it was assumed that the vegetation on the ground was almost removed. In the C4 experiment, while the morphological and physical parameters were set as C_3 grass type, the physiological parameters associated with the photosynthesis processes for C_4

plants were used (see Mabuchi et al., 1997). In the GR experiment, the types of vegetation in the experiment areas were the same as those in the control, but the greenness values of the vegetation in the areas were all set to zero. Namely, it was assumed that the morphological character of vegetation was not changed, but all leaves were considered dead. The purpose of the GR experiment was to purely simulate the effect of physiological activity of vegetation on climate. In the BS and C4 impact experiments, the actual vegetation types in the experiment areas (the IND, ICP, and MTC areas), were changed to a single vegetation type for each of the impact experiments. In each impact experiment, a 10-year spin-up integration was first performed, starting from the soil conditions at the end of the control run, and then the main experiment impact time integration was continued for 20 years under the changed vegetation conditions. The results of these three 20-year impact time integrations were compared with the results of the 20-year control integration. In this study, the analysis is generally performed on the 20-year mean of the seasonal mean values for June-July-August (JJA) and December-January-February (DJF).

3.3.2 Results for JJA

3.3.2.1 Verification of the results of the model control integration

The comparison of the results of the model control integration with the analysis data of the June-July-August (JJA) mean was performed. The analysis data used in the verification is the global objective analysis data compiled by the JMA. The grid resolution of the analysis data is 1.875°.

Although the contrast of the sea surface pressure values between the model ocean and continent was clearer than those in the analysis data, the pressure distribution pattern of the model was rather consistent with that of the analysis data. As for the surface wind vectors, the differences between the model results and the analysis were relatively small. The model heights at the 500-hPa level exhibited slightly higher values than the analysis data. The distribution patterns of the model results, however, indicated good agreement with those of the analysis data. The model wind vector patterns at 500-hPa also coincided with those of the analysis data. Although the precipitation distribution pattern of the model roughly agrees with the CPC Merged Analysis of Precipitation (CMAP) data (Xie & Arkin, 1997), the values of the model precipitation along the intertropical convergence zone (ITCZ), especially at 150° W and the surrounding area, and over the western equatorial Pacific are less than those of the CMAP data. These differences in the atmospheric elements between the results of the model control integration and the analysis data were considered during the examination of the results of the impact experiments.

3.3.2.2 Changes of the surface albedo and roughness length

Table 1 indicates the comparison of experiment area means of the calculated physical values for each simulation. The results of the three impact experiments, BS, C4, and GR, are compared with those of the control experiment.

When changing the vegetation, some physical characteristics of the land surface are altered by the physiological and morphological features of the vegetation. Among them, the surface albedo and the roughness length are clearly altered by the vegetation change.

	IND (88)	Diff	ICP (107)	Diff	MTC (66)	Diff
(ALB)						
CN	0.215	-	0.165	-	0.154	-
BS	0.161	-0.054	0.157	-0.008	0.157	+0.003
C4	0.225	+0.010	0.224	+0.059	0.256	+0.102
GR	0.281	+0.066	0.173	+0.008	0.124	-0.030
(Z0)						
CN	0.34	-	1.84	-	3.62	-
BS	0.0007	-0.34	0.0007	-1.84	0.0007	-3.62
C4	0.11	-0.23	0.11	-1.73	0.09	-3.53
GR	0.34	0.	1.84	0.	3.62	0.
(RNET)						
CN	9.63	-	10.62	-	9.15	-
BS	9.83	+0.20	10.42	-0.20	8.67	-0.48
C4	9.68	+0.05	9.91	-0.71	8.04	-1.11
GR	8.89	-0.74	10.30	-0.32	8.78	-0.37
(E)						
CN	6.06	-	8.13	-	7.32	-
BS	6.49	+0.43	8.77	+0.64	7.83	+0.51
C4	6.85	+0.79	7.98	-0.15	6.95	-0.37
GR	5.26	-0.80	5.78	-2.35	4.84	-2.48
(H)						
CN	3.49	-	2.18	-	1.81	-
BS	3.18	-0.31	1.26	-0.92	0.75	-1.06
C4	2.76	-0.73	1.64	-0.54	1.05	-0.76
GR	3.55	+0.06	4.20	+2.02	3.94	+2.13
(TC)						
CN	26.54	-	23.24	-	24.31	-
BS	-	-	-	-	-	-
C4	26.09	-0.45	23.28	+0.04	24.46	+0.15
GR	26.36	-0.18	23.95	+0.71	25.08	+0.77
(TG)						
CN	26.85	-	23.43	-	23.92	-
BS	28.45	+1.60	24.46	+1.03	25.38	+1.46
C4	26.57	-0.28	23.12	-0.31	24.36	+0.44
GR	26.60	-0.25	23.75	+0.32	24.20	+0.28

(WA)						
CN	75.32	-	103.75	-	98.39	-
BS	78.80	+3.48	104.74	+0.99	97.82	-0.57
C4	70.91	-4.41	103.95	+0.20	97.62	-0.77
GR	80.08	+4.76	103.37	-0.38	104.39	+6.00
(P)						
CN	7.26	-	8.20	-	6.89	-
BS	7.10	-0.16	8.38	+0.18	6.26	-0.63
C4	7.15	-0.11	7.76	-0.44	5.47	-1.42
GR	7.03	-0.23	6.69	-1.51	8.94	+2.05

Table 1. Comparison of the calculated physical values at the land surface in the experiment areas. Values are JJA means, and are listed for the Indian subcontinent area mean (IND), the Indochina Peninsula area mean (ICP), and the maritime continent area mean (MTC). Numbers in parentheses of the area are grid point numbers located in each experiment area. The labels are also for the surface albedo (ALB), the roughness length (Z0) (m), the net radiation (RNET) (MJ m $^{-2}$ day $^{-1}$), the latent heat flux (E) (MJ m $^{-2}$ day $^{-1}$), the sensible heat flux (H) (MJ m $^{-2}$ day $^{-1}$), the canopy temperature (TC) (°C), the soil surface temperature (TG) (°C), the soil water content (WA) (cm), and the precipitation (P) (mm day $^{-1}$). The experiments are the control integration (CN), bare soil experiment (BS), C$_4$ grass experiment (C4), and green-less experiment (GR). The difference values from the CN are also indicated (Diff.).

The general features when undergoing a change of vegetation from the actual vegetation (CN) to bare soil (BS) is that the surface albedo values significantly increase in the area where the actual vegetation was the forest type, and significantly decrease in the area where the vegetation was grassland. In the ICP area, the albedo values increase in the areas where the vegetation was the seasonal rain forest, and decrease in the areas where the vegetation was grassland. The albedo value decreases in the mean of the overall area. In this study, the value of soil surface reflectance for the BS experiment was set to the value for soil in the forest (not for desert). Therefore, the albedo values were decreased by the change from the grassland type to bare soil.

In the case of the vegetation change from CN to C4, while the albedo values in the area where the vegetation was grassland do not change, albedo values in the area where the vegetation was the forest type significantly increase. Therefore, the value of the area mean albedo in the IND area in the C4 experiment is almost the same as that of the CN, and those in the ICP and MTC areas are greater than that of the CN.

In the green-less experiment (GR), the albedo values in the grassland area significantly increase, while those in the forest area significantly decrease, compared with that of the CN. Therefore, the mean value in the IND area increases, that in the MTC decreases, and that in the ICP slightly increases. In grass type vegetation, the transmittances and reflectances of a dead leaf or stem are greater than those of a green leaf. Therefore, the grassland albedo value in the green-less experiment becomes greater than that of the control. In forest type vegetation, although the reflectance of a dead leaf or stem for visible radiation is greater than that of a green leaf, the reflectance of a dead leaf or stem for near infrared radiation is less than that of a green leaf. In addition, the transmittances of a dead leaf or stem are very small. Through these effects, the albedo value for forest type vegetation in the green-less experiment decreases as a whole, compared with that in the control.

In the actual vegetation, the roughness length of the forest type vegetation is generally larger than that of grass vegetation. The roughness length of bare soil is very small and less than that of grassland. By the change of vegetation from the actual vegetation to bare soil (BS experiment), the values of roughness length in the MTC and ICP significantly decrease. Although the value of the change is small, the roughness length in the IND area also significantly decreases. In the C4 experiment, while the roughness length values in the area where the vegetation was grassland do not change, those in the area where the vegetation was the forest type significantly decrease. Consequently, the changes in the roughness length in the C4 experiment are similar to those in the BS experiment. In the GR experiment, the values of the morphological parameters for the vegetation do not change. Therefore, the values of roughness length do not change in each experiment area of the GR experiment.

3.3.2.3 Impact on heat and water balances at the land surface

Impacts on the heat and water balances for each experiment area were examined. From the results of Student's t-test, the changes appeared in each area were generally statistically significant.

In Table 1, when the vegetation changes to bare soil, the net radiation values in the areas where the actual vegetation was forest decrease, due to the increased albedo values in these areas. On the other hand, the net radiation values in the areas where the actual vegetation was grassland increase, due to the decreased values of the albedo. The latent heat fluxes in the experiment areas generally increase. The reason for this is that while the latent heat fluxes associated with transpiration and interception decrease, the latent heat flux due to direct evaporation from the soil surface increases. In the BS experiment, the value of soil water content of the overall experiment area mean increases slightly. Therefore, the increase of the direct evaporation from the soil surface is related to the increase in the surface wind speed over the land in the experiment areas. The change in the pattern of the sensible heat fluxes in the experiment areas is opposite that of the latent heat fluxes. The soil surface temperatures in the experiment areas generally increase because of the increase in the radiation that reaches the soil surface, due to the removal of the canopy. Although the changes of the soil water content vary according to locality, the mean values in the IND and ICP areas increase, while that in MTC decreases. The precipitation changes are discussed in the next section.

In the C4 experiment, the decreases in net radiation in the areas where the actual vegetation was forest are more significant, compared with the case of the vegetation change to bare soil. The reason for this is that the increased albedo values in the areas where the actual vegetation was forest are large, when the vegetation changes to C_4 grass. On the other hand, the changes in the areas where the actual vegetation type was grassland are not significant. The latent heat fluxes in the forest areas generally decrease. It is considered that one reason for this is the decrease in the net radiation in these areas, another reason being the decrease in the roughness length. The latent heat fluxes in the IND area generally increase because of the increase in the transpiration from the leaves of vegetation. The main vegetation type assigned to the IND area as the actual vegetation is C_3 type grass. Therefore, it is considered that by vegetation change to C_4 grass, photosynthesis becomes more active in the IND area. This is due to the fact that C_4 photosynthesis is more suitable than C_3 photosynthesis in a hot and dry environment, such as the IND area. In actuality, the grasses that exist in the IND area include both the C_3 type and C_4 type. Therefore, the possibility exists that the change in the latent heat flux in the IND area simulated by this study is overestimated. The sensible heat fluxes in the experiment areas generally decrease. The decreases in the sensible heat fluxes in the ICP and MTC areas are due to decreases in the net radiation, and that in the IND area to the increase in the latent heat flux.

The influences on the canopy temperature and on the soil surface temperature are somewhat complicated. The canopy temperatures in the ICP and MTC areas generally increase, and that in the IND area decreases. The soil surface temperatures in the ICP and IND areas generally decrease, and that in the MTC increases. In the IND, the latent heat fluxes associated with transpiration from the canopy leaves and direct evaporation from the soil surface increase. Therefore, both the canopy temperature and the soil surface temperature decrease. In the ICP area, the latent heat fluxes by the transpiration and the evaporation of intercepted water decrease; consequently the canopy temperature increases. The decrease in soil surface temperature in the ICP area is due to the increase in the latent heat flux by the direct evaporation from the soil surface. In the MTC area, the increase in the canopy temperature results from the same causes as found in the ICP area. The net radiation for the total vegetation layer of the MTC decreases. The radiation absorbed by the soil surface, however, increases. As a result, the temperature of the soil surface increases.

The soil water content in the IND area generally decreases, due to the increase in the latent heat flux. In the MTC area, the soil water content decreases, due to the decrease in precipitation. In the ICP area, the precipitation and the latent heat flux both decrease. Consequently, the change in the soil water content in the ICP is small, when compared with the control run.

The changes in the forest areas in the C4 experiment are fundamentally the same as those found in the results of the deforestation experiments of Franchito & Rao (1992) and Defries et al. (2002).

In the GR experiment, the net radiation values in the area where the vegetation type was grassland significantly decrease, as a result of an increase in the albedo. In the forest area of the ICP, although the albedo value decreases, the net radiation value does not change, due to a decrease in the downward short-wave radiation. Although no figures are shown, the decrease in the downward short-wave radiation in that area is due to the increase in low-level clouds. In the MTC, precipitation significantly increases and the downward short-wave radiation decreases. Therefore, although the value of the albedo decreases, the net radiation decreases.

The latent heat fluxes in the experiment areas generally decrease, due to the decrease in the transpiration from the leaves of vegetation. In the grassland areas, the direct evaporation from the soil surface increases. Therefore, the magnitude of the decrease in the latent heat flux in the grassland is less than that in the forest areas. The sensible heat fluxes in the experiment areas increase, especially in the forest areas, resulting from the decrease in transpiration. The temperatures of both the canopy and soil surface in the experiment areas generally increase especially in the forest areas, as a result of the decrease in the latent heat flux.

The values of the soil water content in the southern part of the IND, and in the MTC, increase due to the increase of precipitation and the decrease in the latent heat flux. The soil water content in the southwestern part of the ICP decreases because of the decrease in precipitation. In the northwestern part of the IND, the soil water content increases as a result of the increase in precipitation. In the other experiment areas, the change in soil water content is not clear as the result of decreases in both the precipitation and the latent heat flux. Consequently, in each experiment area mean, the soil water values in the MTC and IND areas significantly increase, and the change of that in the ICP is small.

3.3.2.4 Impact on the atmospheric circulation

The changes in the land surface vegetation lead to changes in the atmospheric circulation. The JJA mean atmospheric circulations simulated by the impact experiments were compared with those of the control integration (impact - control).

As for the differences between BS and CN (BS - CN), the strengthening of the Asian summer monsoon winds over the experiment areas is a direct effect of the vegetation change in these areas. The strengthening of the winds is due to the decrease in the roughness length by the vegetation change in the experiment areas. The strengthening of the monsoon winds induces the strengthening of the convergence at the lower atmospheric level over the southern part of China, and the weakening of those over the western coast of India, the western coast of the Indochina Peninsula, and over the islands of the maritime continent. At the 250-hPa level, the areas where the divergence is strengthened spread from China to the Middle East, and are related to the strengthening of the low level convergence. Ascending anomalies exist at the 500-hPa level over the low-level areas of stronger convergence. These changes in the atmospheric circulation induce changes in precipitation. The values of precipitation significantly increase over the areas from southeastern India to the area around the Philippine Islands. On the other hand, precipitation significantly decreases over the western coast of India, the western coast of the Indochina Peninsula, and the islands of the maritime continent.

In the case of C4 - CN, although the strengthening of the Asian summer monsoon winds in the C4 experiment is less than that in the BS case, the anomaly patterns of the atmospheric circulation are fundamentally the same as those in BS – CN. The change in the pattern of precipitation in the C4 case (Fig. 2) is also fundamentally the same as that in the BS case. The magnitudes of the decreases in the precipitation on the western coast of the Indochina Peninsula, and the islands of the maritime continent are larger than those in the BS case. The reasons for these phenomena are considered as follows. The albedo values in these areas significantly increase in the C4 case. The net radiation, the latent heat flux, and the sensible heat flux all decrease. These factors all lead to the local convective activity being suppressed.

In the case of GR - CN, the pattern of change in the atmospheric circulation differs from those of the other experiments. In the GR experiment, the Asian summer monsoon winds become somewhat weaker than those in the control. At the lower atmospheric level, the convergence over the southern part of India and over the islands of the maritime continent strengthens. At the upper atmospheric level, the divergence over these areas also strengthens. In Fig. 3, the precipitation increases over the areas of stronger convergence at the lower atmospheric level, and decreases over the surrounding areas. The precipitation anomaly pattern in the GR experiment is the opposite of those found in the BS and C4 experiments. The reason for this precipitation anomaly is considered as follows. In the GR experiment, the roughness lengths in the experiment areas do not change. Therefore, the effects due to changes in the roughness length on the wind field, such as in the BS or C4 experiments, do not occur. On the other hand, the temperatures of both the canopy and the soil surface increase, and the sensible heat flux increases as a result of the decrease in the latent heat flux by the transpiration from the leaves of vegetation, especially in the forest areas. The islands of the maritime continent are surrounded by the ocean, and have an abundant supply of water vapor. Under these conditions, the low-level convergence strengthens over the maritime continent islands, and convective precipitation over these areas increases.

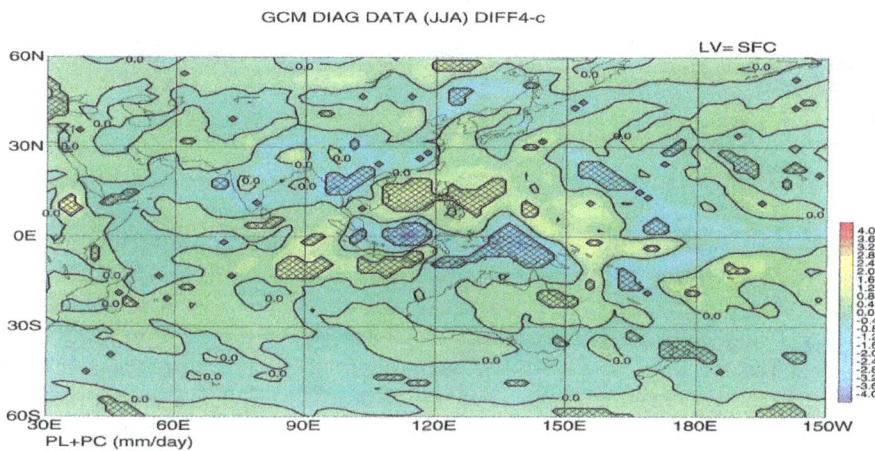

GCM DIAG DATA (JJA) DIFF4-c

Fig. 2. Comparison of the JJA mean values of precipitation of the impact experiments with those of the control integration (impact - control). The differences between C4 and CN are indicated. The results for the experiment areas and the surrounding areas are indicated. The colors toward red indicate relatively large values. The areas where the Student's *t*-test values indicate statistically significant differences (at the 95 % level) are hatched.

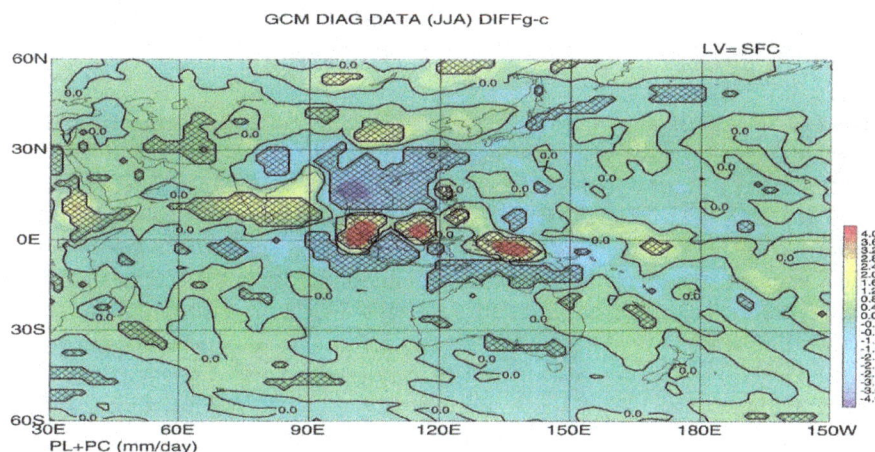

GCM DIAG DATA (JJA) DIFFg-c

Fig. 3. The same as in Fig. 2, except for the differences between GR and CN.

3.3.3 Results for DJF

3.3.3.1 Verification of the results of the model control integration

The comparison of the results of the model control integration with the analysis data for the DJF mean was performed. In general, as was found in the JJA case, the contrast of the pressure values of the model between the ocean and continent is clearer than those in the analysis data. For the surface wind vectors, the differences between the model results and the analysis data reflect the differences in the values of sea surface pressure. As for the geopotential heights and wind vector patterns of the model at 500-hPa, the distribution patterns are reasonable. The precipitation distribution pattern of the model roughly agrees with the CMAP data, and the consistency found in the DJF results is better than that in JJA.

3.3.3.2 Changes in surface albedo and roughness length

Table 2 indicates the comparison of the experiment area means of the calculated physical values for each simulation. The general features of the change in vegetation from the actual vegetation (CN) to BS for DJF, are almost the same as those found in JJA. The difference pattern in the DJF when changing the vegetation from CN to C4, is also similar to that in JJA. In the CN simulation of this study, the greenness values of the grassland and seasonal forest were reduced to almost zero during DJF in the Northern Hemisphere. Therefore, in the green-less experiment (GR), the albedo values in the IND and ICP areas were almost the same as those in the CN. These results may be somewhat extreme, compared with the actual situation. In the MTC area, originally covered mainly by evergreen tropical rain forest, the GR albedo values significantly decrease for the same reason as in the JJA case. The roughness length difference patterns between each experiment and the control for DJF are almost the same as those found in the JJA case.

3.3.3.3 Impact on the heat and water balances at the land surface

Impacts on the heat and water balances for each experiment area were examined (Table 2). From the results of Student's t-test, the changes appeared in each area were generally statistically significant.

In the BS experiment, the difference pattern of the net radiation is fundamentally the same as that found in the JJA case. As for the latent heat flux, the difference found in the ICP area is small compared with the JJA case, because of the relatively small DJF latent heat flux in this area. However, the overall patterns in the results are the same as those in the JJA case. The pattern of change in the sensible heat fluxes in the experiment areas is generally opposite that in the latent heat fluxes. The temperature increase in the ICP area is relatively small, due to the advection of cold air from the northern inland region. This DJF temperature difference pattern in the ICP area differs from that found in the JJA case. The pattern of change in the soil water content fundamentally corresponds with that of precipitation. The changes in the soil water content vary according to locality, and in the IND area the effects of the precipitation change in JJA also remain. The mean values increase in all experiment areas.

	IND (88)	Diff.	ICP (107)	Diff.	MTC (66)	Diff.
(ALB)						
CN	0.266	-	0.154	-	0.152	-
BS	0.166	-0.100	0.160	+0.006	0.156	+0.004
C4	0.279	+0.013	0.275	+0.121	0.244	+0.092
GR	0.267	+0.001	0.155	+0.001	0.124	-0.028
(Z0)						
CN	0.29	-	1.66	-	3.66	-
BS	0.0007	-0.29	0.0007	-1.66	0.0007	-3.66
C4	0.08	-0.21	0.08	-1.58	0.10	-3.56
GR	0.29	0.	1.66	0.	3.66	0.
(RNET)						
CN	4.19	-	6.08	-	10.11	-
BS	4.81	+0.62	5.51	-0.57	9.53	-0.58
C4	3.96	-0.23	4.97	-1.11	9.06	-1.05
GR	4.26	+0.07	6.11	+0.03	9.75	-0.36
(E)						
CN	3.62	-	4.43	-	8.02	-
BS	4.24	+0.62	4.44	+0.01	8.41	+0.39
C4	3.47	-0.15	4.04	-0.39	7.71	-0.31
GR	3.82	+0.20	4.42	-0.01	5.04	-2.98
(H)						
CN	0.779	-	1.88	-	2.04	-
BS	0.777	-0.002	1.32	-0.56	0.99	-1.05
C4	0.71	-0.07	1.21	-0.67	1.28	-0.76
GR	0.67	-0.11	1.92	+0.04	4.69	+2.65
(TC)						
CN	14.16	-	9.49	-	24.61	-
BS	-	-	-	-	-	-
C4	13.73	-0.43	7.99	-1.50	24.81	+0.20
GR	14.03	-0.13	9.39	-0.10	25.44	+0.83
(TG)						
CN	14.51	-	9.27	-	24.30	-
BS	16.52	+2.01	9.40	+0.13	25.83	+1.53
C4	14.23	-0.28	8.17	-1.10	24.70	+0.40
GR	14.35	-0.16	9.13	-0.14	24.56	+0.26
(WA)						
CN	72.59	-	94.99	-	99.78	-
BS	77.04	+4.45	95.41	+0.42	100.37	+0.59
C4	69.35	-3.24	95.00	+0.01	99.12	-0.66
GR	77.17	+4.58	94.83	-0.16	104.46	+4.68
(P)						
CN	0.94	-	1.98	-	7.81	-
BS	0.98	+0.04	2.11	+0.13	6.99	-0.82
C4	0.93	-0.01	1.72	-0.26	6.43	-1.38
GR	0.91	-0.03	1.88	-0.10	9.68	+1.87

Table 2. Comparison of the calculated physical values at the land surface in the experiment areas. Values are DJF means. The labels are the same as in Table 1.

In the C4 experiment, the same as in JJA, the decreases in the net radiation in the areas where the actual vegetation was forest are significant, due to the increased albedo values. The changes in the areas where the actual vegetation was grassland are not significant. The latent heat fluxes in the forest areas generally decrease as a result of the decreased net radiation and decreased roughness lengths. The change in the latent heat fluxes in the IND area is not significant, since the physiological activity of vegetation is weak during DJF. The decreases in the sensible heat fluxes in the ICP and MTC areas are due to the decreases in the net radiation. The change in the IND area is not significant. In DJF, the patterns of changes in the canopy temperature and the soil surface temperature are almost the same. The decreases of temperatures in the ICP and IND areas are due to the decreased net radiation. The cold air advection from the northern inland region also influences the temperature in the ICP area. In the MTC, the latent heat fluxes by the transpiration and the evaporation of intercepted water decrease, consequently the canopy temperature increases. Although the net radiation over the total vegetation layer in the MTC area decreases, the radiation absorbed by the soil surface increases. Therefore the soil surface temperature increases. The IND area mean soil water content decreases, due to the decrease in the northwestern part of the IND. It is considered that the effects of the increased latent heat flux in JJA in this area continue into DJF. The tendencies of the change in the DJF soil water content in the ICP and MTC areas are generally the same as those in JJA.

In the GR experiment, the greenness values were all set to zero, while the greenness values of the grassland and seasonal forest in the IND and ICP areas decrease in DJF during the control run. Therefore, the changes in the net radiation fluxes, the latent and sensible heat fluxes in the IND and ICP areas are generally not significant. In the MTC area, impacts on the heat and water balances are generally the same as in JJA. The temperature changes in the canopy and soil surface of the IND and ICP area are also not significant. In each experiment area mean, the values of soil water in the MTC and IND areas significantly increase, and the change of that in the ICP is small. The tendency of the change in soil water content in each experiment area in DJF is generally the same as those in the JJA case.

3.3.3.4 Impact on the low-latitude atmospheric circulation

In the case of BS - CN, the direct effect of the vegetation change is revealed as the strengthening of the northeasterly wind over the ICP area, the westerly wind over the islands of the MTC area, and the easterly wind over the northern equatorial Pacific. This strengthening of the winds is due to the decrease in the roughness lengths by the vegetation change in the experiment areas (see Table 2). The strengthening of the winds induces the strengthening of the low-level atmospheric convergence over the central equatorial Pacific and over the SPCZ. Over the ICP and MTC, although not significant, divergence anomalies are found. At the 250-hPa level, the areas where the divergence strengthens spread from the central equatorial Pacific to over Japan. These anomalies are related to the strengthening of the low-level convergence. Ascending anomalies exist at the 500-hPa level over the low-level areas of stronger convergence, and descending anomalies are found over the islands of the MTC. These changes in the atmospheric circulation induce changes in the precipitation. The precipitation significantly increases over the western part of the Indochina Peninsula, the central equatorial Pacific, and the SPCZ. On the other hand, the precipitation significantly decreases over the eastern part of the Indochina Peninsula, and the islands of the MTC.

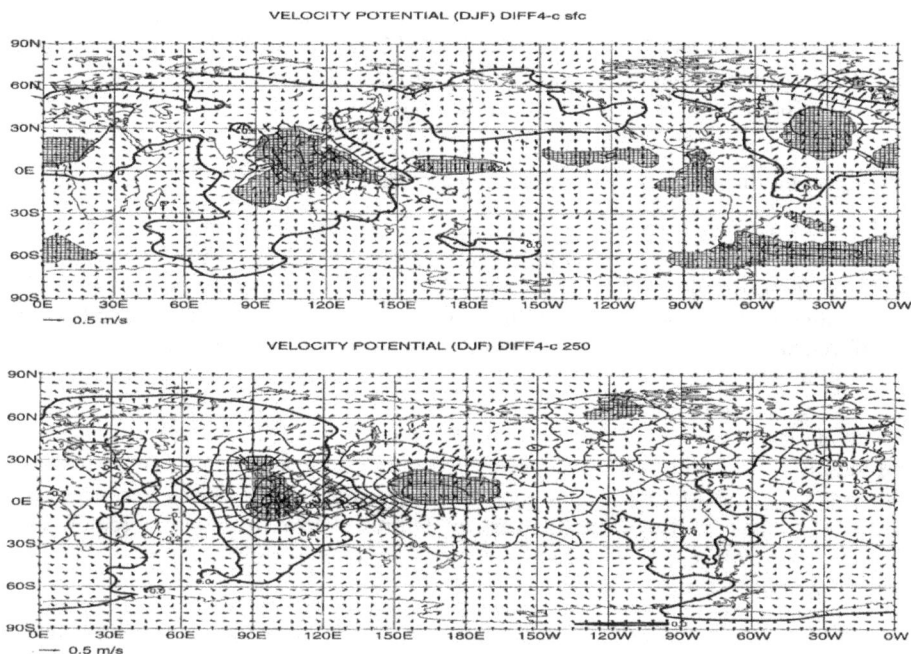

VELOCITY POTENTIAL (DJF) DIFF4-c sfc

VELOCITY POTENTIAL (DJF) DIFF4-c 250

Fig. 4. Comparison of the DJF mean results of the impact experiments with those of the control integration (impact - control). The differences in the velocity potential (10^6 m^2 s^{-1}) and the divergence/convergence of wind vectors between C4 and CN are indicated. The results at the (top) surface and (bottom) 250-hPa level. The areas where the Student's t-test values indicate statistically significant differences (at the 95 % level) are shaded.

The wind anomaly pattern in the C4 experiment is fundamentally the same as in the BS case. As in the BS experiment, the roughness lengths in the C4 experiment areas also decrease when compared with the control run, especially in the ICP and MTC areas. Compared with the BS case, the low-level convergence anomalies and upper-level divergence anomalies over the central equatorial Pacific and SPCZ are weaker, while the low-level divergence anomalies and the upper-level convergence anomalies over the ICP and MTC areas are more intense (Fig. 4). The effects of the change in the roughness length on the wind field in the C4 experiment are weaker than those in the BS experiment, as a result of the magnitude of the roughness length decrease in the C4 experiment being less than that in the BS experiment. The pattern of the change in precipitation in the C4 case is also fundamentally the same as that in the BS case. The magnitudes of the decreases in the ICP and MTC precipitation are greater than those in the BS case. These results were also found in the JJA case.

In the case of GR - CN, the DJF pattern of the change in atmospheric circulation differs from those of the other experiments. Wind anomalies are mainly found over the islands of the MTC. The GR anomaly pattern of the atmospheric circulation is fundamentally the same as in JJA, but the low-level convergence and upper-level divergence anomalies that are considered direct effects of the vegetation change are limited to over the islands of the MTC. The reason for this is considered that the influence of vegetation change in the GR experiment on the DJF circulation is mainly confined to the islands of the MTC (see Table 2). GR precipitation increases over the islands of the MTC, and decreases over the surrounding areas. This precipitation anomaly over the MTC is opposite those found in the BS and C4 experiments. The reason for this precipitation anomaly is considered to be the same as that in JJA.

3.3.3.5 Impact on the mid-latitude atmospheric circulation

There have been several studies on the effects of vegetation changes on the atmospheric circulation. The studies of Chase et al. (1996), Chase et al. (2000), and Zhao et al. (2001) investigated the effect of land cover change on the global atmospheric circulation. These studies simulated the impacts of the difference between actual vegetation conditions and potential vegetation conditions on climate. These studies indicated that the land cover changes in the tropics induce changes in the extratropic atmospheric circulation, especially in the winter season. Gedney & Valdes (2000) showed that complete Amazonian deforestation could result in changes in the climate far afield from the region of deforestation. In particular, the model predicted statistically significant changes to winter rainfall over the North Atlantic, extending towards Western Europe. Werth & Avissar (2002) also detected a noticeable impact of the Amazon deforestation in several other regions of the world, several of which showed a reduction in rainy season precipitation that exhibited a high signal-to-noise ratio.

Among other vegetation change studies, Zhang et al. (1996a, 1996b) performed numerical simulations of the potential impact of tropical deforestation in South America, Africa, and Southeast Asia using a climate model coupled with a realistic land surface model. Zhang et al. (1996b) discussed the influence of tropical deforestation on the large-scale climate system. It was concluded that the modification of the model surface parameters to simulate tropical deforestation produced significant modifications in both the Hadley and Walker circulations. A mechanism for the propagation of disturbances arising from tropical deforestation to middle and high latitudes was proposed, based on the mechanisms of Rossby wave propagation.

These mechanisms are similar to those associated with extratropical influences of ENSO events. There have been numerous studies of the global teleconnections associated with the tropical sea surface temperatures (SST): for example, Horel & Wallace (1981), Trenberth & Hurrell (1994), Latif & Barnett (1994), Hurrell (1996), Zhang et al. (1997), Renshaw et al. (1998), Enfield & Mestas-Nunez (1999), and Kobayashi et al. (2000). In the tropical atmosphere, anomalous SSTs force anomalies in convection and large-scale overturning, with subsidence in the descending branch of the local Hadley circulation. The resulting strong upper tropospheric divergence in the tropics and convergence in the subtropics act as a Rossby wave source. The climatological stationary planetary waves and associated jet streams, especially in the Northern Hemisphere, can make the total Rossby wave sources somewhat insensitive to the position of the tropical heating that induces them, and thus can create preferred teleconnection response patterns, such as the Pacific-North American (PNA) pattern. Anomalous SSTs and tropical forcing have tended to be strongest in the northern winter, and teleconnections in the Southern Hemisphere are weaker and more variable and thus more inclined to be masked by the natural variability of the atmosphere (Trenberth et al., 1998).

In this study, impacts of the deforestation in the Asian tropical region on the mid-latitude atmospheric circulation were also examined. The C4 experiment in this study is the most realistic case of deforestation among the three impact experiments. Therefore the influences of the vegetation changes in the C4 experiment on the mid-latitude atmospheric circulation were examined. In the C4 experiment, which differs from previous studies, the vegetation changes were applied only in the Asian tropical region, while the vegetations in South America and Africa were maintained as the actual vegetation. There is, however, the possibility of an influence of the vegetation changes of only those in the Asian tropical region, on the mid-latitude atmospheric circulation.

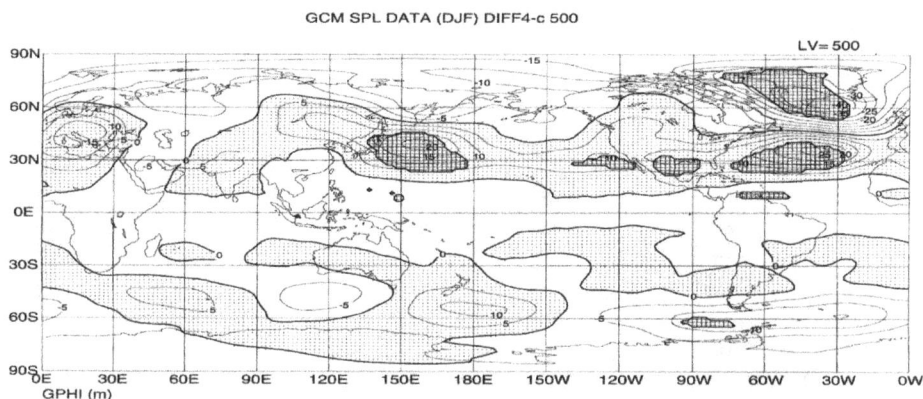

Fig. 5. Latitude-longitude distribution of the differences in the DJF 500-hPa geopotential heights (m) between C4 and CN (C4 - CN). The positive difference areas are shaded. The areas where the Student's t-test values indicate statistically significant differences (at the 95 % level) are densely shaded.

In Fig. 4 (C4 - CN), additional results are found of changes in the mid-latitude circulation that may be considered as resulting from the vegetation change in the Asian tropical region. There are areas of significant differences of the wind, not only over the ICP and MTC, but also around Japan and over the Atlantic Ocean. At the lower atmospheric level, with significant convergence anomalies found over the central equatorial Pacific and divergence anomalies over the ICP and MTC areas, there are significant divergence anomalies over the Atlantic Ocean that coincide with significant wind differences over the same area. At the upper atmospheric level, although not statistically significant, convergence anomalies also exist over the Atlantic Ocean. The same anomaly patterns exist more clearly in BS - CN, but are not found in GR - CN. Therefore, it is considered that these atmospheric circulation anomalies are due to modifications of the Hadley and Walker circulations, and are induced by the vegetation changes (morphological, physiological, and physical changes from forest to grassland or bare soil) in the Asian tropical region. In particular, the divergence/convergence anomaly pattern that appears at the upper atmospheric level in the C4 experiment (lower panel in Fig. 4) is very similar to that of an ENSO event (see Fig. 3 of Trenberth et al., 1998).

Figure 5 indicates the DJF differences of the 500-hPa geopotential heights between C4 and CN (C4 - CN). Over the Northern Hemisphere, areas of positive difference exist from Japan to Europe, and areas of negative difference are found over the Aleutian Islands to Greenland. Statistically significant areas of positive differences exist over the western part of the northern mid-latitudes of the Pacific Ocean and the northern mid-latitudes of the Atlantic Ocean. Over Greenland, an area of statistically significant negative differences can be seen. These height anomalies at the 500-hPa level are similar to those found during an ENSO event. Therefore, the possibility exists that the deforestation in the Asian tropical region induces teleconnections similar to those associated with ENSO events.

3.3.4 Discussion

Among the three experiments of this study, the assumption in the C4 experiment is the most similar assumption to those of the other deforestation experiments, although the C_4 photosynthesis process was assumed in the study. In particular, Henderson-Sellers et al. (1993) examined the impact of vegetation change in Southeast Asia, and found that during the wet season (July), the surface temperatures significantly decreased over the Indochina Peninsula and the island of Borneo, and the evaporation decreased over the islands of Borneo and New Guinea, and the Indochina Peninsula. Concerning precipitation, however, there was no great change in the basic pattern of rainfall, and few of the changes were statistically significant. During the dry season (January), significant decreased evaporation and net radiation were indicated over land points. The changes in the surface temperatures and precipitation, however, were not statistically significant. In their experiment, the impact of vegetation change in Southeast Asia on the atmospheric circulation and moisture convergence were also small, and the changes were not identified. Zhang et al. (1996a) also discussed the seasonal variation of impacts of deforestation over Southeast Asia. It was concluded that the evapotranspiration and the net radiation indicated statistically significant decreases, but the precipitation changes were not statistically significant. These results were almost the same as those of Henderson-Sellers et al. (1993). The results of this study differed somewhat from those of the above mentioned studies. The results of the C4 experiment in this study indicated statistically significant differences.

Sud & Smith (1985) examined the influence of local land surface processes on the Indian Monsoon. One of the numerical experiments included the case of no evapotranspiration from the land surface. In the results, there was very little change in the rainfall due to the enhanced moisture convergence, produced as a consequence of the increased sensible heating over land largely compensating for the lack of evapotranspiration. For the month of July, the moisture supply for precipitation over India was advected from the nearby Indian Ocean. Without evapotranspiration, the increased PBL heating by the sensible heat flux promotes this process by producing a thermal low. Polcher (1995) studied the relationship between land surface process changes and variations in the frequency of convective events, and indicated that the highest sensitivity was found for the sensible heat flux and its increase leads to deeper convective events. Although the design and the results of the GR experiment in this study somewhat differ from those of previous studies, the mechanism of the precipitation increase over the islands of the maritime continent in the GR experiment was consistent with those of Sud & Smith (1985), and Polcher (1995).

In the C4 experiment, vegetation changes were assumed only in the Asian tropical region. However, it was possible that the vegetation changes influenced the mid-latitude atmospheric circulation. In particular, the divergence/convergence anomaly pattern that appeared at the upper atmospheric level in the C4 experiment was very similar to that of an ENSO event. In the differences of the DJF mean 500-hPa geopotential heights between the C4 and CN, the height anomalies at the 500-hPa level were similar to those of an ENSO event. The possibility exists that the deforestation of the Asian tropical region could induce similar teleconnections as those associated with an ENSO event.

3.4 Changes in carbon cycle balances under vegetation transition due to deforestation in the Asian tropical region

3.4.1 Experiment design

Spin-up integration was carried out to estimate the initial values of the soil water content, the ice content in the soil, the soil temperature, and the carbon storage of the vegetation and that in the soil. This integration used the actual global vegetation and climatic sea surface temperature (SST) values. The SST and sea ice values were taken from the GISST2.2 dataset (Rayner et al., 1996). The initial values of the carbon dioxide concentration in the atmosphere were set to about 360 ppmv.

The fluxes of anthropogenic emission of carbon dioxide were taken into account during the integration. The Global, Regional, and National Fossil Fuel CO_2 Emissions database produced by the Carbon Dioxide Information Center (CDIAC) (http://cdiac.esd.ornl.gov/) was used. In this database, the 1950 to present CO_2 emission estimates were derived primarily from energy statistics published by the United Nations (2008), using the methods of Marland & Rotty (1984). From the database, the regional fluxes of anthropogenic emissions in 1995 were extracted (6.4 PgC year [-1] for the total global land area), and regarded as standard values of anthropogenic emissions. Furthermore, the regional amounts of annual increase of emissions were estimated using the regional average rate of increase during the period from 1991 to 2004. These estimations were performed separately for nine regions: Africa, South America, China (east Asia), Eastern Europe, Southeast Asia, the Middle East, North America, Oceania (including Japan), and Western Europe. The regional flux values of anthropogenic emissions for each model year during the integrations were estimated from these data by adding the annual increments to the standard values described above.

The monthly carbon dioxide fluxes between the sea surface and the atmosphere were derived from model-calculated data. The average flux values at each ocean grid point from 1990 to 1999 were calculated from the model average values of carbon dioxide fluxes between the sea surface and the atmosphere produced by the OCMIP-Phase2 experiments (http://c4mip.lsce.ipsl.fr/protocol.html), and regarded as the standard monthly flux values (Coupled Carbon Cycle Climate Model Intercomparison Project (C4mip) (Friedlingstein et al., 2006)). The mean annual changes were estimated using data of the same period. From these data, the monthly carbon dioxide fluxes between the sea surface and the atmosphere at each ocean grid point for each model year during the integrations were estimated by adding the annual changes to the standard fluxes.

Prior to the vegetation change impact experiment, a control time integration (CN) was performed using the initial values obtained from the spin-up calculation. The control integration was run for 100 model years; vegetation change impact experiments were then performed. To investigate the impact of vegetation changes in the Asian tropical region, two experiment areas were defined: ICP and MTC (Fig. 1). Under the control conditions, the ICP area was covered mainly by tropical seasonal forest and the MTC area was covered mainly by tropical rain forest vegetation. Strictly speaking, the ICP area is not presently homogeneously covered by tropical seasonal forest. However, wet and dry seasons, induced by the East Asian monsoon system, clearly exist in the ICP area. Therefore, in contrast to the tropical rain forest vegetation classified for the MTC area, tropical seasonal forest type vegetation was assigned to the ICP area as the typical vegetation in the control simulation.

Two type deforestation experiments were carried out. In the C4 experiment, the forest type vegetations in the experiment areas (ICP and MTC) were changed to grass vegetation. In the BS experiment, the forest type vegetations in those areas were changed to bare-soil. In the grassland (C4) experiment, it was assumed that the forest type vegetation was changed into C_4 grass vegetation, which is the most potent natural grass in the tropical region (Sage et al., 1999). Each experiment was run for 100 model years, starting with the initial values that were used in the control integration. In these experiments, all types of forest vegetation in the experiment areas were changed into non-forest type vegetations at the rate of 1.6 % year $^{-1}$, starting from the control vegetation distribution (Fig. 1). The 114 forest grids (57 forest grids in each area) were finally changed into the non-forest type vegetations. Forest reduction continued until the 58th model year, after which the vegetation distribution was not changed. These processes more realistically simulated the temporal progress of forest reduction by deforestation, and the temporal changes of both the energy balance and the carbon cycle balance under the vegetation transition with the progress of deforestation were examined. The FAO (2007) reported that the annual rate of net forest area change in Southeast Asia was - 1.3 % year $^{-1}$ from 2000 to 2005. The deforestation rate in the experiments (- 1.6 % year $^{-1}$) was somewhat greater than the value reported by the FAO. However, it is assumed that the state of deforestation in the Asian tropical region was, on the whole, reproduced in the experiments.

In the impact experiments, the carbon storage values for plants at two chosen forest grids were set to almost zero at the beginning of each model integration year. The natural growth process for C_4 grass was reproduced in the C4 experiment, whereas the plant growth process was deterred in the remaining period of model integration in the BS experiment. In the C4 experiment, the physiological parameters associated with photosynthesis for C_4 plants were employed (see Mabuchi et al., 1997). The results of the impact time integrations were then compared with those of the control integration.

During the control and the impact integrations, the same conditions of carbon dioxide fluxes produced by anthropogenic emission and those between the sea surface and the atmosphere were applied. In these simulations, the effects of global warming by increased atmospheric carbon dioxide concentration were not considered, because the purpose of this investigation is to detect the pure impact of deforestation on the energy and carbon cycle balances in the Asian tropical region, and to enable clear analysis of the impact mechanisms.

3.4.2 Results

3.4.2.1 Verification of model results in the control simulation

This section presents the model characteristics with a focus on the global carbon cycle reproduced by the model. The energy and carbon balances at the present time in the experiment areas are also verified.

For the first 10-year period (representing the present era), the vegetation carbon storage (VC) is about 670 PgC, the soil carbon storage (SC) is 1,846 PgC, the gross primary production (GPP) is 157 PgC year $^{-1}$, the net primary production (NPP) is 90 PgC year $^{-1}$, and the net ecosystem production (NEP) is 3 PgC year $^{-1}$ for the total land area (148.89 × 10 ^{6} km ^{2}). In these model results, SC and NPP are relatively larger than those of other estimations although VC is almost the same (e. g., Field et al., 1998; Ito, 2002; Arora et al., 2009). The SC values estimated by previous plot scale field investigations, on average, are 1,730 PgC (Ito, 2002). Other representative values are 1,567 PgC (IGBP-DIS, 2000), and 1,500 PgC (IPCC, 2001). However, the estimated SC values range widely. Furthermore, these estimated values are generally those found in the layer of soil near the surface. Therefore, it is assumed that the SC values estimated by the present model are within the range of actual values.

The features of the global atmospheric carbon dioxide concentration simulated by the model are briefly described. Discussion of the verification of the carbon dioxide concentration simulated by the control run can also be found in Mabuchi & Kida (2006). The values of the atmospheric carbon dioxide concentration calculated by the model were verified using in situ data observed at NOAA/GMD stations (Conway et al., 2007; Thoning et al., 2007). Figure 6 compares the model results with the data of nine observations. The nine stations for verification were selected as typical points in each latitude from the Northern to the Southern Hemispheres. The mean value of model grid points (10 × 10 grids) at the low vertical level surrounding each observatory was compared with the measured values at each observatory. Figure 6 compares the time series of the monthly mean values of the results from the second to the eleventh model year with those of the observed data from 1997 to 2006. The typical patterns of seasonal change of the observed carbon dioxide concentration in each zone are as follows. In the high latitudes in the Northern Hemisphere (Alert and Ocean Station M), the amplitude is about 15 ppmv. The maximum appears in April or May, and the minimum appears in July or August. The seasonal cycle amplitudes gradually decrease in the low latitudes (Terceira Island, Mauna Loa, and Mahe Island), and those in the Southern Hemisphere are small (Easter Island, Cape Grim, Palmer Station, and Halley Bay). The seasonal cycle patterns in the Southern Hemisphere are opposite those in the Northern Hemisphere. In each zone in the Northern Hemisphere, including the equatorial zone and the low latitude zone in the Southern Hemisphere, the model successfully reproduces the features of the seasonal cycle patterns and the increasing trend found in the observed data described above. In the middle and high latitude zones of the Southern Hemisphere, the amplitudes of the seasonal cycles calculated by the model are about 5 ppmv, which is somewhat greater than those in the observed data. It is necessary to verify the large-scale circulation of carbon dioxide, especially in the upper level in the Southern Hemisphere. Currently, these are pending questions for the model.

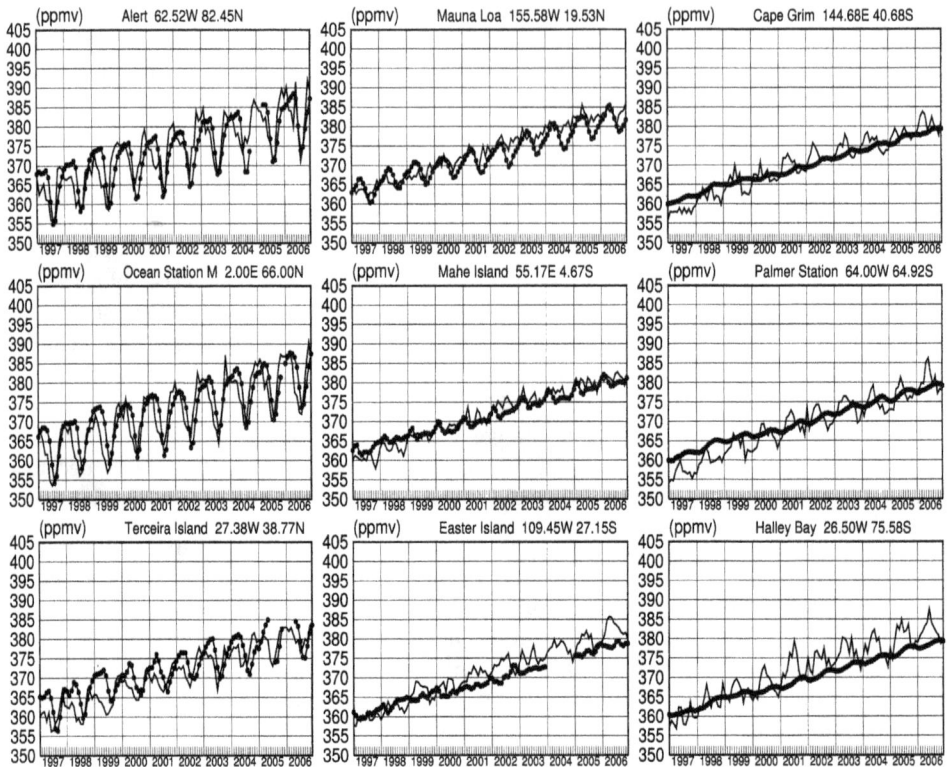

Fig. 6. Comparison of the atmospheric carbon dioxide concentration (ppmv) calculated by the model control run with the data of nine in situ observations. The dotted lines indicate the observed data, and the solid lines indicate the model results.

Table 3 indicates the values of essential elements of energy balance in the experiment areas calculated by the model control simulation. Mean values in the first 10-year period are listed. Table 4 indicates the observed values of those at three observation points in the Asian tropical region: Mae Klong (Thailand) (MKL), Sakaerat (Thailand) (SKR), and Bukit Soeharto (East Kalimantan/Indonesia) (BKS) (Gamo, personal communication). The land surface albedo (ALB) values simulated by the control simulation for the ICP and the MTC areas are consistent with those observed values. The values of the downward short-wave radiation at the land surface (DSW) are also consistent with the observed values, though the observed values at BKS are smaller than those at the other two points. The net radiation (RNE) values simulated by the model are close to the observed values, though the simulated values are about 1 MJ m^{-2} day^{-1} smaller than those observed. While the values of the sensible heat flux (H) by the model are close to the observed data, the values of the latent heat flux (E) by the model are larger than the observed data. Judging from the energy balance, it is assumed that the values of E in the observation are too small.

	ICP	MTC
ALB	13.8	14.0
DSW	17.2	16.7
RNE	9.6	10.1
H	2.4	1.5
E	7.1	8.3

Table 3. Essential elements of the energy balance in the experiment areas calculated by the model control simulation. Mean values in the first 10-year period are listed. The labels denote the land surface albedo (ALB) (%), the downward short-wave radiation at the land surface (DSW), the net radiation (RNE), the sensible heat flux (H), and the latent heat flux (E), all given in units of MJ m $^{-2}$ day $^{-1}$.

| | MKL | | SKR | | BKS | | |
	2003	2004	2001	2002	2003	2001	2002
ALB	13.7	14.3	14.3*	14.1	14.8*	14.9	13.9
DSW	17.4	17.5	16.5*	16.5	17.9*	13.8	14.7
RNE	11.9	11.5	10.7*	10.6	11.2	10.6	12.2*
H	1.5	1.7	-	3.9	3.5	2.1	2.3
E	3.8	3.3	-	4.9	6.2	2.4	2.9

Table 4. Observed values of essential elements of the energy balance at three observation points in the Asian tropical region. The labels are the same as those in Table 3. The values with an asterisk (*) indicate those for continuous period of data missing existed during the year.

Saigusa et al. (2008) gives the carbon fluxes observed from 2002 to 2005 in a tropical mixed deciduous forest (Mak Klong in Thailand), a tropical dry evergreen forest (Sakaerat in Thailand), and a tropical rain forest (Pasoh in Malaysia). Their results indicate clear seasonal changes in the forest points in Thailand, with a maximum GPP of about 10 gC m $^{-2}$ day $^{-1}$ and a minimum of about 6 gC m $^{-2}$ day $^{-1}$. Although the GPP values produced by the model were smaller than the observed data, the seasonal change patterns correspond to those observed (figure not shown). For the values of observed RES, the maximum is about 11 gC m $^{-2}$ day $^{-1}$ and the minimum is about 5 gC m $^{-2}$ day $^{-1}$. Also, the RES values predicted by the model were smaller than those observed; however, the seasonal change patterns were the same. The peak NEP observed is about ± 1.7 gC m $^{-2}$ day $^{-1}$. The order of model peak values and the seasonal change patterns were almost the same as those observed. Detailed descriptions are omitted, but the values and seasonal patterns of GPP, RES, and NEP produced by the model for the Maritime Continent indicated good agreement with those observed at the tropical rain forest in Malaysia.

3.4.2.2 Impact of deforestation on carbon balance in the Asian tropical region

This section discusses the results of the impact experiment, considering the control results including those discussed in the previous section. The results of C4 experiment are mainly discussed, contrasting with the results of BS experiment that is extreme case.

Figure 7 indicates the temporal distributions of the annual means of GPP and NEP for the 100-year period of the control simulation, and the differences between the control and the impact experiment. For GPP, the values of the control gradually increase in both areas.

Compared with the values in the first 10-year period, the values from the third period to the last are significantly large in the ICP area, as are the values from the second period to the last in the MTC area. These results are due to the increased carbon dioxide uptake by the land surface vegetation, resulting from the effect of carbon dioxide fertilization on C_3 type vegetation (forests).

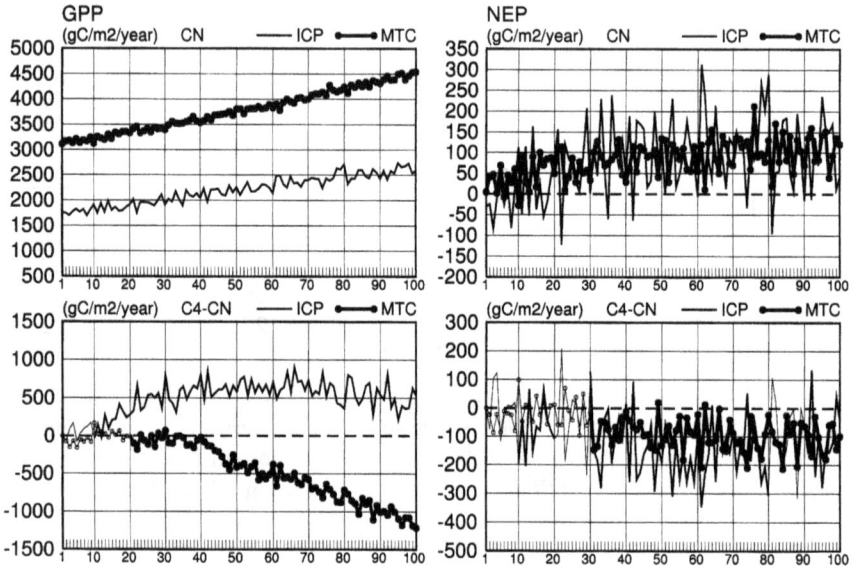

Fig. 7. Temporal distributions of the annual mean values of GPP (left) and NEP (right) (gC m^{-2} year^{-1}) for the 100-year simulations. The horizontal axes indicate the model years. The solid lines are for the Indochina peninsula area (ICP), and the lines with circle are for the Maritime Continent area (MTC). Top panel is for the control, and bottom panel is for the differences between the control and the C4 experiment (C4 – CN). In the bottom panel, thick lines for each area indicate statistically significant differences in each 10-year period.

In Fig. 7, the ICP area GPPs of the C4 experiment gradually increase compared with those of the control until the fifth 10-year period. However, the increasing tendencies gradually weaken as the vegetation changes to C_4 grass, and the positive differences decrease in the later period. The positive differences occur because the annual mean GPPs for C_4 grass vegetation are fundamentally greater than those for tropical deciduous forest vegetation. The increasing tendencies attenuate because the influence of carbon dioxide fertilization on C_4 vegetation is less than that on C_3 vegetation, due to differences in photosynthesis mechanisms. The basic limitations for C_4 photosynthesis consist of the limitations on the maximum carboxylation velocity and on light. The sensitivity of C_4 vegetation to changes in atmospheric carbon dioxide concentration is generally less than that of C_3 vegetation (Jones, 1992; Mabuchi et al., 1997). MTC area GPPs of the C4 experiment gradually decrease, compared with those of the control, and the differences are almost all negative. The annual mean GPPs for tropical rain forest vegetation and those for C_4 vegetation are potentially almost the same. Therefore, the unilateral negative differences are due to the direct effect of the difference between the influence of carbon dioxide fertilization on C_4 vegetation and that on C_3 vegetation. The temporal changes in differences of NPP between the control and the impact experiments accord with those of GPP in both areas, although vegetation respiration affects the NPP values (figure not shown).

In Fig. 7, the positive NEP values in the control indicate carbon absorption from the atmosphere, and the negative values indicate carbon release to the atmosphere. The control NEPs gradually increase in the early stage of integration and remain positive in both areas. Compared with the values in the first 10-year period, the values from the third to the last periods are significantly greater in the ICP area, as are those from the second to the last periods in the MTC area. The interannual changes in the ICP area exceed those in the MTC area because the vegetation activity for the tropical deciduous forest is more sensitive to varying environmental conditions than that for the tropical evergreen rain forest in the MTC area. The increasing tendencies are attenuated due to the increase of the respiration from soil (RRS) because of the increased carbon stock in the soil layer, resulting from the effect of carbon dioxide fertilization on C_3 vegetation.

The NEPs in the C4 experiment are less than those of the control in both areas, especially from the fourth 10-year period. The interannual fluctuations are intense in the ICP area, and the differences in the ninth 10-year period are not statistically significant. While in the MTC area, the negative differences are more systematical. One reason for this is that the GPPs and NPPs for the MTC area in the C4 experiment are systematically smaller than those in the control. Another reason is that the surface conditions in the MTC area become warmer and relatively more arid with vegetation change from forest to grassland. These results indicate that the change from forest vegetation to C_4 grass vegetation induces the reduction of carbon absorption by the land surface and, as a result, the increase in the atmospheric carbon dioxide concentration. These impacts are more distinct in the MTC area.

Although figures for the BS experiment are not shown, GPP and NPP values gradually decreased with the progress of deforestation in both areas due to the disappearance of vegetation. The NEPs were smaller than those of the control in both areas. The negative differences gradually increased until deforestation was complete, and then gradually decreased in the subsequent period, due to the decrease of RRS with the reduction of carbon stock in the soil layer. These results were naturally assumed phenomena. Hirano et al. (2007) gave the annual NEP as - 382 gC m ^{2} year $^{-1}$ for 2003 and - 313 gC m ^{2} year $^{-1}$ for 2004, observed in a tropical peat swamp forest disturbed by drainage (except for the peculiar 2002, an ENSO year). Although the results could not be directly compared due to the differences between the design of model simulation and the situations of observation, the peak values of the negative NEP in the BS experiment (- 250 to - 300 gC m ^{2} year $^{-1}$) were reasonable judging from those observed values.

3.4.3 Discussion

It is important that the continuous deforestation in the Asian tropical region certainly induces the elevations of the global atmospheric carbon dioxide concentrations. When forest vegetation is replaced by bare soil, the carbon dioxide uptake by land surface vegetation disappears, and only respiration from the soil occurs. As a result, NEPs in these areas become negative, until the carbon stored in the soil is completely released. The carbon balance under this condition is simple. However, when alternative vegetation (e. g., C_4 grass vegetation in this study) replaces forest vegetation, the carbon balance in the deforested area becomes more complicated. The results of this study indicated that continuous deforestation of the tropical forest could potentially induce a continuous decrease in carbon dioxide uptake by the land surface from the atmosphere. This will consequently produce an increased tendency of carbon dioxide concentrations in the atmosphere, even if the deforested area is not replaced by bare soil surface condition.

4. Simulation using the regional climate model

4.1 Introduction

The present state of environmental problems due to global warming resulting from increases of greenhouse gases has reached new levels. The international treaty known as the United Nations Framework Convention on Climate Change (UNFCCC) was adopted in 1992 to begin to consider what can be done to reduce global warming (URL: http://unfccc.int/). The Kyoto Protocol, adopted in 1997 at the third Conference of the Parties to the UNFCCC (COP 3), proposed a world-wide reduction of greenhouse gas emissions. Under these conditions, it became necessary to monitor the increases of greenhouse gases, especially carbon dioxide, and to conduct research to further understand the mechanisms of interactions between environmental changes and the carbon balance.

Estimations of the carbon dioxide budget are of great importance in taking the proper steps to deal with increased concentrations due to anthropogenic emissions, and in predictions of future concentration levels. To clarify the role of the terrestrial ecosystem, observational studies have been carried out using the flux tower network (Baldocchi et al., 2001; Falge et al., 2002). In the Asian region, the flux tower network (AsiaFlux) was established, and studies concerning the variations in the mechanisms of the net ecosystem carbon dioxide exchange have been performed (Yamamoto et al., 2005; Yu et al., 2006). Recently, Saigusa et al. (2008) discussed the characteristics of the seasonal and inter-annual changes in the net ecosystem production, the gross primary production, and the ecosystem respiration during the period from 2000 to 2005, employing more than ten flux observation sites in Asia, and demonstrated how ecosystems respond to the meteorological anomalies widely observed in East Asia.

While an integrated study based on observed data is effective, a numerical simulation study is another potent method to explain the features of observed results, and to better understand the mechanism and role of the heterogeneous terrestrial ecosystem. Several studies using models that can operate with short time-scales, and can resolve seasonal and diurnal variations in the carbon exchange between terrestrial ecosystems and the atmosphere, have been conducted (Bonan, 1995; Denning et al., 1996a, 1996b). These models are physical climate models with a terrestrial biosphere, and can simulate nonlinear biosphere/atmosphere interactions by on-line calculations. Bonan (1995) simulated the diurnal and annual cycles of biosphere-atmosphere carbon dioxide exchange, and investigated the geographic patterns of annual net primary production and the diurnal range and seasonality of the net carbon dioxide flux. Denning et al. (1996a, 1996b) also simulated the annual net primary productivity, the amplitude of the seasonal cycle of the net carbon dioxide flux, and the amplitude and phase of the diurnal and seasonal cycles of atmospheric carbon dioxide concentration.

Furthermore, many studies focused on vegetation physiology and carbon circulation associated with vegetation activity and climate have been conducted: for example, Foley et al. (1996), Bounoua et al. (1999), Cox et al. (2000), Friedlingstein et al. (2001), Tsvetsinskaya & Mearns (2001a, 2001b). Ito (2008) evaluated the regional scale carbon budget of East Asia at a high resolution, comparable with the scale of flux measurements, using a process-based terrestrial carbon cycle model driven by meteorological reanalysis data.

To treat the heterogeneity of the land surface as accurately as possible, it is desirable that climate models have a high resolution. The advantage of a regional climate model is that it can satisfy these conditions, using fewer computational resources. The use of regional climate models for climate studies was first employed by Dickinson et al. (1989), and Giorgi (1990). Since then, there have been many studies making use of regional climate models, among them Kidson & Thompson (1998), Noguer et al. (1998), Seth & Giorgi (1998), Giorgi et al. (1999), and Small et al. (1999). Mabuchi et al. (2000) clarified the relationships between climate and the carbon dioxide cycle over the Japanese Islands and surrounding area.

In this section, several results of numerical study using a regional climate model that includes a realistic biological land surface model are introduced. The purposes of the study are to clarify not only the carbon budget, but also the mechanism of the carbon cycle between the terrestrial ecosystem and the atmosphere, and to investigate climate factors impact on the carbon cycle in the East Asian terrestrial ecosystem (Mabuchi et al., 2009).

4.2 Regional climate model

The atmospheric model used in the experiment is a regional spectral model (Japan Spectral Model (JSM)) developed by the JMA (Segami et al., 1989). The domain of JSM originally covered only the Japanese Islands and the surrounding area (Mabuchi et al., 2000, 2001, 2002). In this study, the model domain of JSM is extended to cover the area of East Asia that includes not only the Japanese Islands, but also Mongolia, China, the Indian subcontinent, Indochina, and the Philippine Islands. The model employs sigma coordinates, with 23 layers in the vertical. It has a regular 151 × 111 square transform grid on a Lambert projection plane, the reference longitude being 105°E, which translates to a horizontal resolution of 60 km at the reference latitudes (15°N, 50°N). The model adopts the primitive equations as basic equations. The atmospheric prognostic variables are the virtual temperature, specific humidity, the zonal and meridional components of the wind, the carbon dioxide concentration of each atmospheric layer, and surface pressure. The model includes short-wave and long-wave radiation processes (Sugi et al., 1990; Lacis & Hansen, 1974). Precipitation is estimated by three processes, namely large scale condensation, moist convective adjustment, and evaporation of raindrops. Vertical diffusion is calculated by the turbulent closure model (level 2.0) proposed by Mellor & Yamada (1974). The time step interval of the integration is about two minutes.

BAIM2 was integrated into JSM to reproduce the energy and carbon dioxide exchange process between the land surface ecosystem and the atmosphere and to investigate the mechanism of the relationship between land surface vegetation activity and the climate. This regional climate model is termed JSM-BAIM2. In this model, BAIM2 is connected on-line to the atmospheric model. At each grid point of JSM, the vegetation type is specified, and the interactions between terrestrial ecosystems and the atmosphere are estimated by BAIM2.

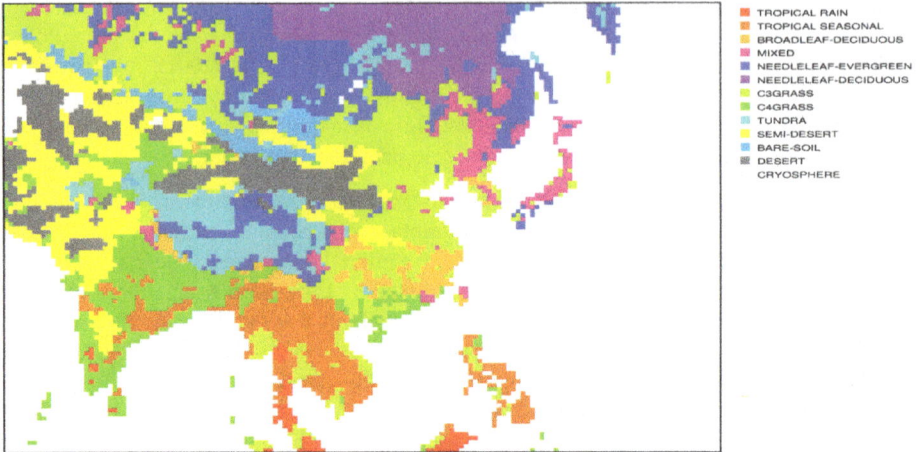

Fig. 8. Distribution of vegetation used in JSM-BAIM2. Vegetation types are indicated by the color legend on the right side of the figure.

The distribution of the types of vegetation used in JSM-BAIM2 is indicated in Fig. 8. In the simulation, 12 types of vegetation, excluding the cryosphere are used in the model domain. The vegetation map for JSM-BAIM2 was constructed from two vegetation data sets. For the Japanese Islands, the actual vegetation map compiled by the Environment Agency of Japan was used. For the Asian continent, reference was made to the Major World Ecosystem Complexes Ranked by Carbon in Live Vegetation data set (Olson et al., 1983). From the data sets, the actual vegetation of a given land surface grid point was classified as one of the 12 types. Several modifications were made to the vegetation distribution to conform to JSM-BAIM2. In this experiment, crop vegetation was regarded as C_3 grassland vegetation.

4.3 Climate factors impact on carbon cycle in the East Asian terrestrial ecosystem

4.3.1 Experiment design

Using JSM-BAIM2, a numerical simulation is performed under actual vegetation conditions (Fig. 8). The spectral boundary coupling (SBC) method (Kida et al., 1991) was used in the time integration of the regional climate model simultaneously with the time-dependent lateral boundary coupling (LBC) method (Tatsumi, 1986). The SBC is a nesting technique for a long-period integration of a regional climate model. With this coupling method, the regional model can simulate synoptic scale phenomena without phase deviation from the boundary data. Therefore, it can simulate the regional climate more accurately (Kida et al., 1991; Sasaki et al., 1995), so that can gain a more accurate description of the interactions between climate and terrestrial ecosystems. By using these coupling methods, JSM-BAIM2 was nested one-way in the global reanalysis data field. This simulation used the JRA-25 reanalysis data created by the JMA (Onogi et al., 2007). The grid size of the data set is 1.25°. Using 12-hour interval reanalysis data (0000 and 1200 UTC of each day), the JSM-BAIM2 grid point data were interpolated, and then used for initial conditions, lateral boundary conditions, and spectral boundary conditions of the meteorological data fields.

For the atmospheric carbon dioxide concentration field, ideal increase data was prescribed for the model boundary condition to clearly investigate the interaction mechanism between the regional climate and the carbon cycle in the model domain. The carbon dioxide concentrations have been measured at many in situ observatories around the world. The World Data Centre for Greenhouse Gases (WDCGG), one of the World Data Centres (WDC) under the Global Atmosphere Watch (GAW) program of the World Meteorological Organization (WMO), has been operating since 1990 under the control of JMA (URL: http://gaw.kishou.go.jp/wdcgg/). The boundary data increase rate for the simulation was set at 1.8 ppm year $^{-1}$ according to a recently observed rate of increase. The data had typical seasonal cycles of which the amplitude and mean values depended on the latitude of the model grid point. From these data, 12 hourly interval data sets were made and given for the boundary conditions of the atmospheric carbon dioxide concentration field.

The sea surface temperature (SST) and sea ice data for the model sea grid points were taken from the HadISST data set (Rayner et al., 2003). The monthly $1° × 1°$ SST and sea ice data during the period of the experiment were interpolated to obtain each value for the model sea area grid points. The carbon dioxide fluxes between the sea surface and the atmosphere were prescribed from estimated values of observed data. The data were the carbon dioxide exchange data at the sea surface observed in the Northwestern Pacific by the JMA (Japan Meteorological Agency, 1994).

Spin-up repetition time integrations were carried out using the one year period data from 1200-UTC 31 July 1998 to 1200-UTC 31 July 1999. Using the physical and carbon storage values obtained by the spin-up integrations as the initial conditions for the land area, the experiment time integration was started at 1200-UTC 31 July 1999 and continued until 1200-UTC 31 December 2005, for a total of six years and five months. The integration results for the six-year period from 1 January 2000 to 31 December 2005 were examined in the study.

4.3.2 Results

4.3.2.1 Verification of model results

Several results of verification of JSM including BAIM Version 1 (JSM-BAIM) were presented in Mabuchi et al. (2002). The performance of JSM-BAIM was confirmed regarding the reproducibility of the main atmospheric variables, and that of the seasonal and interannual variations of the principal elements (that is, precipitation, temperature, and radiation) which influence the heat, water, and carbon dioxide balances at the land surface through vegetation activity. It was found that JSM-BAIM had sufficient accuracy to allow for investigations of the interaction mechanisms between terrestrial ecosystems and climate; temporally at least on the level of the seasonal and interannual variations, and spatially at least on the level of the climatic classification of the Japanese Islands.

This section presents the verification results for JSM-BAIM2 with a focus on precipitation and vegetation phenology, the essential elements for the objectives of this study. The analysis regions for verification of the model are indicated in Fig. 9. The seven analysis regions are set in the model domain considering the vegetation type and the regional climate conditions. The control schemes of vegetation phenology used in the model are presented in Mabuchi et al. (2009).

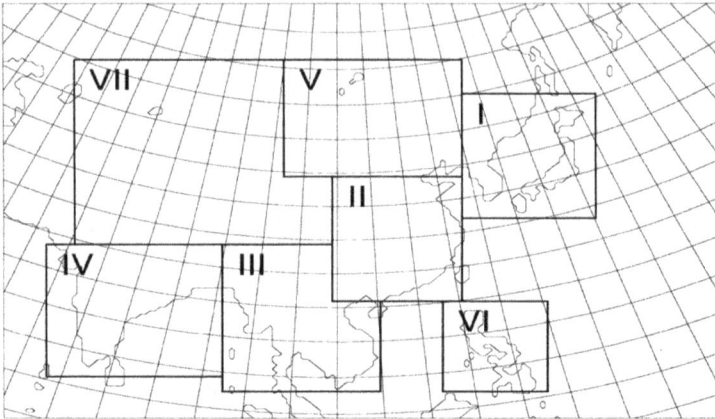

Fig. 9. Analysis regions of the model. Analysis regions I to VII are labeled as NE-Asia, China, Indochina, India, Mongolia, Philippines, and Inland.

The accuracy of precipitation is important for simulating the soil wetness and associated land surface processes. A more accurate estimation of precipitation (in strict sense cloudiness) also produces a better estimation of the downward short-wave radiation reaching the Earth's surface. These, in turn, induce a more precise reproduction of the physical and biological processes, including the carbon cycle at the land surface. The comparison of the model precipitation with the Climate Prediction Center (CPC) Merged Analysis of Precipitation (CMAP) data (Xie & Arkin, 1997) was performed. The values of the correlation coefficient for the variation of the anomalies for all months of the six years are listed in Table 5. These correlation coefficient values all exceed the 95 % significance level (absolute value is 0.232).

Region	Correlation coefficient	
	RAIN-CMAP	GLAI-NDVI
NE-Asia	0.54	0.36
China	0.50	0.31
Indochina	0.56	0.03
India	0.64	0.29
Mongolia	0.66	0.29
Philippines	0.57	0.16
Inland	0.77	0.32

Table 5. Values of the correlation coefficient between the anomaly of the model precipitation (RAIN) and those of CMAP and those between the anomaly of the model GLAI and those of NDVI in each analysis region. The values are for variations of monthly anomalies for all months of the six-year period. (Sample number is 72, significant absolute value at 95 % level is 0.232)

Reproducibility of the variations of vegetation phenology by the model was verified using the satellite Normalized Difference Vegetation Index (NDVI) data (SPOT-Vegetation NDVI) (Free VEGETATION Products). The verification was performed by comparing the model green leaf area index (GLAI) with the SPOT-Vegetation NDVI data. The base data for GLAI are the monthly mean values calculated from the output of daily grid point values, while that for NDVI is the maximum monthly value selected from 10-day composited grid point data (316 X 200 grid data; horizontal resolution 0.25°). The model GLAI and satellite NDVI are both indicators of vegetation activity. Therefore, the reproducibility of the temporal and spatial variations of vegetation phenology by the model can be verified by comparing the variations of model GLAI with those of satellite NDVI.

The values of correlation coefficient for the variation of anomalies for all months of the six years are listed in Table 5. The correlation coefficient values exceed a significance level of 95 % (absolute value is 0.232), except for the Indochina and the Philippines regions. For the yearly mean anomaly, the values of the correlation coefficient of the inter-annual variations in the regions of the NE-Asia and China are 0.87 and 0.96, respectively. These exceed a significance level of 95 %.

In the middle and high latitudes, the model reproduced the variation of vegetation phenology on the regional scale for the period when the vegetation activity was high and the change of vegetation phenology was large. During the cold season at high latitudes, the variation of GLAI predicted by the model and that of NDVI was not sufficiently consistent. One reason for this is that the variation of model GLAI is small in the cold season. Another reason is considered that the disturbance by snow cover affects the values of NDVI in this season. In low latitudes, the model performance was relatively poor. One reason for this is that the variation signal of GLAI is small because the main vegetation in this area is evergreen forest. Another reason to be considered is that disturbance by convective clouds can influence the values of NDVI at low latitudes.

Station	Latitude	Longitude	Elevation (m)
Ryori (Japan)	39 ° 02 ' N	141 ° 49 ' E	260
Yonagunijima (Japan)	24 ° 28 ' N	123 ° 01 ' E	30
Takayama (Japan)	36 ° 08 ' N	137 ° 25 ' E	1420
Tae-ahn Peninsula (Korea)	36 ° 43 ' N	126 ° 07 ' E	20
Ulaan Uul (Mongolia)	44 ° 27 ' N	111 ° 05 ' E	914
Mt. Waliguan (China)	36 ° 17 ' N	100 ° 54 ' E	3810

Table 6. Locations of the in situ observatories used for analysis of the atmospheric carbon dioxide concentration calculated by the model.

To validate the inter-annual variations of the atmospheric carbon dioxide concentration (CDC) reproduced by the model, the model results were compared with the data from the six in situ observatories located in the model calculation domain (Table 6). These observatories are stations that contribute to the WMO WDCGG. Of these observatories, Ryori and Yonagunijima are operated by the JMA (Watanabe et al., 2000; Japan Meteorological Agency, 2007). Takayama station is operated by the National Institute of Advanced Industrial Science and Technology (AIST) (Murayama et al., 2003). In addition, data observed at stations of the NOAA/GMD, located at the Tae-ahn Peninsula, Ulaan Uul, and Mt. Waliguan (Conway et al., 2007) were used. The mean value of the model grid points at the second vertical level (sigma coordinate 0.990) surrounding each observatory was compared with the measured values at each observatory.

Figure 10 presents the time series of the monthly mean CDC values for the six years from 2000 to 2005 from each observatory. The time series of the model values that correspond to each observatory is also plotted. The temporal variations of the observed data at each station indicate the seasonal cycles that exhibit higher values in the cold seasons and lower values in the warm seasons. The amplitudes of the seasonal cycles at each observatory indicate the inter-annual variation. On the other hand, the temporal variations of the model results indicate almost the same patterns as those observed at each observatory, although the amplitudes of the seasonal cycles differ in several years.

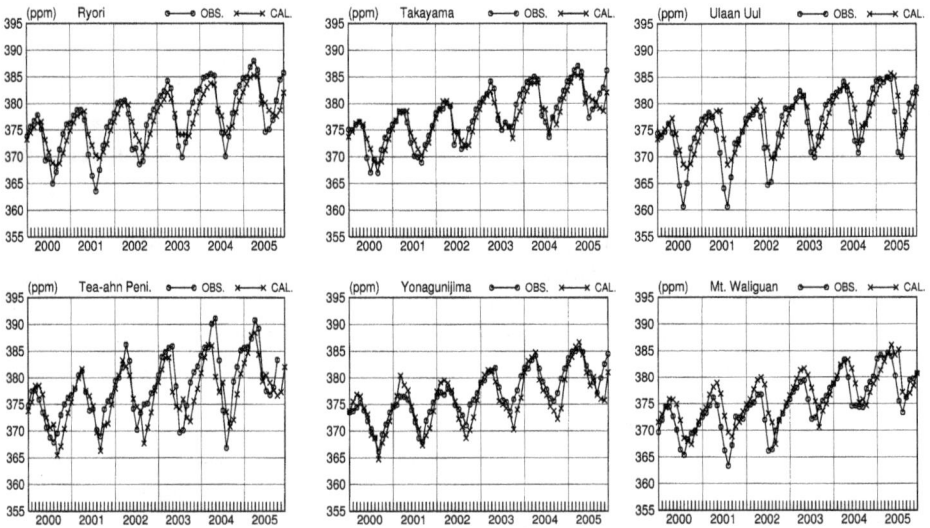

Fig. 10. Time series of monthly mean CDC values (ppm) for six years (2000 to 2005) observed at each observatory (OBS.) and those calculated by the model (CAL.). The model values are mean values of model grids at the second vertical level surrounding each observatory.

From the comparisons of the horizontal distributions of monthly anomalies of the atmospheric CDC with those of the net ecosystem production (NEP) simulated by the model, it is found that the variations of the atmospheric CDC at the near surface level are closely connected with the variations of NEP in the high vegetation activity season, including the effects of advection of the atmospheric carbon dioxide (see Figs. 9 to 14 in Mabuchi et al., 2009). The variations of NEP are closely related to those of GPP, which is the total amount of carbon that the vegetation absorbs by the photosynthesis process, and is directly affected by climate conditions. The differences between the values of NEP and those of GPP are influenced by the variations of ecosystem respiration. However, the variations of NEP are almost consistent with those of GPP in the high vegetation activity conditions (figure not shown). The relationship between climate and GPP is described in detail in the next section.

4.3.2.2 Regional characteristics of the relationship between climate and GPP

In this section, the regional features of the climate factors impact on GPP are described, using the model simulation results. The relationship between the variations of GLAI and GPP is also examined.

Table 7 lists the values of the correlation coefficient between the anomaly values of climate factors and those of GPP for each season (JFM, AMJ, JAS, and OND) for the seven analysis regions (see Fig. 9). In this study, the seasonal means are defined as: January, February, and March (JFM); April, May, and June (AMJ); July, August, and September (JAS); and October, November, and December (OND). The anomalies are deviations from the six-year means for each analysis regional mean. The downward short-wave radiation (DSW), the soil wetness (SW), and the surface temperature (TA) were chosen as the effective climate factors. The correlation coefficients between the variations of GLAI and those of GPP are also indicated. The coefficients in Table 7 indicate the correlation between the inter-annual variation of each factor and that of GPP for each season in the seven analysis regions. The features in each region are as follows.

In the NE-Asia region (Region I), the inter-annual variations of GPP in AMJ and OND are positively correlated with those of GLAI. This suggests that the variation of GLAI in the seasons of leaf expansion and leaf defoliation strongly affects the variation of GPP in the area where seasonal changes of the leaf area are great. In JAS, GLAI becomes mature, and the correlation between the variation of GLAI and that of GPP becomes unclear. After the vegetation leaves have expanded (especially in JAS and OND), variations of DSW and GPP become positively correlated, indicating that DSW in these seasons strongly affects the values of GPP. In AMJ and JAS, the correlation of the variations of SW and GPP also becomes high and positive. In JFM and OND when the temperatures are low, the correlation of the variations of TA and GPP becomes high and positive, indicating that the high anomalous temperatures in the cold season induce large values of GPP. In the China region (Region II), the inter-annual variations of GPP are highly correlated with those of GLAI, except in JAS. There is a large seasonal change of leaf area in this region, and the relationship between the variation of GLAI and that of GPP is similar to that in the NE-Asia region. The variations of DSW and those of GPP are positively correlated, except in AMJ. The values of DSW are also important for the values of GPP in this region. In AMJ and OND, the variations of SW and GPP are positively correlated. In AMJ in particular, the coefficient between SW and GPP exceeds those between DSW and GPP and between TA and GPP, indicating that the effect of SW on GPP is important for this region in the spring. The same relationship also exists in the NE-Asia region. The correlation between the variation of TA and that of GPP is positive and high in JFM. However, the correlation between TA and GPP in AMJ, JAS, and OND is negative, indicating that the positive anomalies of surface temperature in the warm seasons tend to induce negative anomalies of GPP in this region.

In the Indochina region (Region III), the seasonal variation of GLAI is relatively small. However, the correlation between the variations of GLAI and those of GPP tends to be positive. DSW is an important climate factor for GPP throughout the year in this region. In contrast, the correlation between SW and GPP is negative in all seasons, indicating that GPP tends to become a negative anomaly when the precipitation is a positive anomaly (that is when DSW is a negative anomaly). The correlations between TA and GPP are small, except in AMJ, because the seasonal variation of the surface temperature is small in this region.

	JFM	AMJ	JAS	OND
NE-Asia (Region I)				
DSW-GPP	-0.02	0.37	0.84	0.83
SW-GPP	-0.28	0.71	0.69	0.36
TA-GPP	0.94	0.37	0.00	0.87
GLAI-GPP	0.38	0.84	-0.37	0.91
China (Region II)				
DSW-GPP	0.52	-0.32	0.55	0.86
SW-GPP	-0.57	0.79	-0.07	0.70
TA-GPP	0.97	-0.75	-0.46	-0.31
GLAI-GPP	0.88	0.94	0.39	0.93
Indochina (Region III)				
DSW-GPP	0.59	0.85	0.94	0.99
SW-GPP	-0.24	-0.81	-0.28	-0.92
TA-GPP	0.39	0.82	-0.43	0.44
GLAI-GPP	0.77	0.03	0.68	0.46
India (Region IV)				
DSW-GPP	-0.17	0.54	0.87	0.50
SW-GPP	0.65	-0.25	-0.74	0.55
TA-GPP	-0.68	0.32	0.69	-0.83
GLAI-GPP	0.95	0.74	0.76	0.77
Mongolia (Region V)				
DSW-GPP	0.33	0.32	-0.65	-0.01
SW-GPP	0.33	0.73	0.86	0.96
TA-GPP	0.84	-0.07	-0.85	0.93
GLAI-GPP	0.56	0.71	0.88	0.82
Philippines (Region VI)				
DSW-GPP	0.97	0.85	0.95	0.98
SW-GPP	-0.87	-0.22	-0.21	-0.67
TA-GPP	0.06	0.33	-0.27	-0.23
GLAI-GPP	-0.31	0.49	0.01	0.67
Inland (Region VII)				
DSW-GPP	-0.38	-0.07	0.02	0.64
SW-GPP	0.66	0.69	0.87	0.59
TA-GPP	0.54	-0.43	0.45	0.59
GLAI-GPP	0.80	0.89	0.92	0.88

Table 7. Values of the correlation coefficient between the anomaly values of climatic factors and those of the model GPP for each season (JFM, AMJ, JAS, and OND) for the seven analysis regions. The anomaly values are deviations from the six-year mean values for each analysis region mean. The labels are for downward short-wave radiation (DSW), soil wetness (SW), and surface temperature (TA). The correlation coefficients between the variation of green leaf area index (GLAI) and that of GPP are also indicated. (Sample number is six, significant absolute value at 95 % level is 0.811, that at 90 % level is 0.729)

The high correlation of TA in AMJ is related to the high correlation of DSW in this season (when DSW deviates positively, TA also generally deviates positively). In the India region (Region IV), the correlation between GLAI and GPP is relatively high and positive in all seasons. In this area, the amplitude of the seasonal cycle of GLAI is not large, but GLAI tends to change with variations in climate factors, and the GLAI changes induce changes in GPP. Actually, the component of irregular variation is greater in the seasonal cycles of GLAI in this region than in the other regions (figure not shown). In JAS, the correlation between DSW and GPP is positive and high. This corresponds to the negative correlation between SW and GPP and the positive correlation between TA and GPP in this season. In JFM and OND, while the correlations between SW and GPP are positive, those between TA and GPP are negative, indicating that the unusually high temperatures and dry climate conditions in the cold season tend to reduce GPP in this region. As a whole, the downward short-wave radiation in the warm season and the soil water in the cold season are important climate factors for increasing GPP in this region.

In the Mongolia region (Region V), the correlations between GLAI and GPP in AMJ, JAS and OND are relatively high and positive. In this region, the correlations between DSW and GPP are small in all seasons, and the correlation in JAS is negative. On the other hand, the correlations between SW and GPP are positive and high, especially from AMJ to OND. In JAS, while the correlation between SW and GPP is positive, the correlations between DSW and GPP and between TA and GPP are negative. In JFM and OND, the correlations between TA and GPP are high and positive. These facts suggest that soil water is a more important climate factor than the downward short-wave radiation for increasing GPP in this region, and that anomalously high temperatures in the cold season are also important for the gain of GPP. It is clear that unusually high temperatures and dry climate conditions reduce GPP in this region, especially in the summer season. The Inland region (Region VII) has characteristics similar to those of the Mongolia region. Soil water in the summer season is an especially important climate factor for GPP in this region.

In the Philippines region (Region VI), the correlations between GLAI and GPP are small, because the changes in GLAI are small in this region and the change in the signal of GLAI is not clear (figure not shown). The correlations between TA and GPP are also small, because the change in surface temperature is small. On the other hand, the correlation between DSW and GPP is positive and high in all seasons, and those between SW and GPP are all negative, indicating that when the level of downward short-wave radiation is high (i.e., when the amount of precipitation is small), GPP becomes large in this region. The main climate factor affecting GPP in this region is the downward short-wave radiation.

4.3.3 Discussion

The value of GPP estimated by the model for the six years from 2000 to 2005 is 11.33 GtC year $^{-1}$, that of NPP is 6.65 GtC year $^{-1}$, and that of NEP is 0.25 GtC year $^{-1}$ for the values over the entire model domain, except the boundary areas (15 grids areas in the border of the model domain). Ito (2008) estimated the regional carbon budget in East Asia for the same period (2000 to 2005) using a process-based terrestrial carbon cycle model. According to the results, the values of GPP, NPP, and NEP are 1861, 996, and 58 TgC year $^{-1}$, respectively. For the area mean, these values are 1187, 635, and 37 gC m $^{-2}$ year $^{-1}$, respectively. The experiment model domain in Ito (2008) almost coincides with the region of NE-Asia (Region

I) in this study. In the results of this study, the values of GPP, NPP, and NEP are 989, 642, and 33 gC m $^{-2}$ year $^{-1}$, respectively, for the area mean of the NE-Asia region. These values are comparable with the results of Ito (2008).

Saigusa et al. (2008) measured NEP over forest stands in the sub-arctic, temperate, and tropical regions of East Asia over the period from 2000 to 2005. Their paper discussed three cases of how the forest ecosystems responded to meteorological anomalies during the study period. In the first case, they concluded that the negative anomaly of solar radiation in summer decreased GPP at forest sites in central Japan. At one site, decreased water stress suppressed the reduction of GPP due to unusually low summer solar radiation. Although the results in this study are general characteristics of the relationship between climate and GPP, the relationships between the variation of DSW and that of GPP and the variation of SW and that of GPP in summer in the NE-Asia region are consistent with the results of Saigusa et al. (2008).

In the second case described in Saigusa et al. (2008), they indicated that, for temperate forests, the unusually warm winter and early spring lead to increased NEP in the early spring. Also, in deciduous forests, the high air temperature resulted in early leaf expansion that enhanced NEP at the beginning of the growing period. In the results of this study for the NE-Asia region, the relatively high temperature during the cold season induced large values of GPP, and the inter-annual variations of GPP in spring were positively correlated with those of GLAI. These results are also consistent with the results of Saigusa et al. (2008).

In the third case discussed in Saigusa et al. (2008), it was indicated that the decreased precipitation during the season from January to March significantly decreased GPP and NEP in the tropical seasonal forest, due to a long period of dry conditions and severe drought stress. In the results of this study for JFM, a similar correlation between SW and GPP was found in the India region. However, the relationship between SW and GPP in JFM was unclear for the Indochina region, and the correlation between GLAI and GPP was positive and maximum. The spatial and temporal distributions of precipitation and soil water in the tropical area are very complicated. Therefore, there is the possibility that a gap in the scale between the model resolution and in situ observations induced this discrepancy.

5. Concluding remarks

The present studies obtained several results concerning the mechanism of relationship between the terrestrial ecosystem and the atmosphere. It is necessary to investigate as many cases as possible to resolve the problem of the gap in the scale between the model resolution and the in situ observations. Moreover, further study is needed with verifications using observed data as much as possible. Research work should continue on not only the steady state condition but also the mechanism of the variation process, which would be useful for improving future predictions. The interactions between the land surface ecosystem and climate are nonlinear, and the relationships between climate change and land surface processes are complicated due to the heterogeneity of the terrestrial ecosystem. Nonetheless, it is necessary to investigate the terrestrial ecosystem responses to climate changes and the climate responses to terrestrial ecosystem changes to better understand the universal mechanism of environmental changes on the Earth. Studies using the physical climate model, including the biological scheme, are useful for understanding the physical and biological mechanisms in regional and global climate systems. Through the accurate reproduction of actual phenomena and with accurate interpretation, the mechanisms that produce such phenomena can be clarified to improve estimates for future situations.

6. Acknowledgements

The author wishes to express special thanks to Dr. Thomas. J. Conway of the NOAA/ESRL for permission to use the atmospheric carbon dioxide concentration data acquired at NOAA/GMD stations. Special thanks are extended to Dr. Shohei Murayama and Dr. Minoru Gamo of the National Institute of Advanced Industrial Science and Technology for permission to use the atmospheric carbon dioxide concentration data acquired at Takayama station (Dr. Murayama) and the data observed at three observation points in the Asian tropical region (Mae Klong, Sakaerat, and Bukit Soeharto) (Dr. Gamo). The author is grateful to the collaborators for their helpful suggestions, discussions, and support for a series of the researches. The CMAP Precipitation data used in this research was provided by the NOAA/OAR/ESRL PSD, Boulder, Colorado, USA (http://www.cdc.noaa.gov/).

7. References

Arora, V. K., Boer, G. J., Christian, J. R., Curry, C. L., Denman, K. L., Zahariev, K., Flato, G. M., Scinocca, J. F., Merryfield, W. J., & Lee, W. G. (2009). The effect of terrestrial photosynthesis down regulation on the twentieth-century carbon budget simulated with the CCCma Earth System Model. *J. Climate*, 22, 6066-6088.

Avissar, R., & Pielke, R. A. (1989). A parameterization of heterogeneous land surfaces for atmospheric numerical models and its impact on regional meteorology. *Mon. Wea. Rev.*, 117, 2113-2136.

Baldocchi, D., Falge, E., Gu, L., Olson, R., Hollinger, D., Running, S., Anthoni, P., Bernhofer, C., Davis, K., Evans, R., Fuentes, J., Goldstein, A., Katul, G., Law, B., Lee, X., Malhi, Y., Meyers, T., Munger, W., Oechel, W., Paw K. T., Pilegaard, U, K., Schmid, H. P., Valentini, R., Verma, S., Vesala, T., Wilson, K., & Wofsy, S. (2001). FLUXNET: A new tool to study the temporal and spatial variability of ecosystem-scale carbon dioxide, water vapor, and energy flux densities. *Bull. Amer. Meteor. Soc.*, 82, 2415-2434.

Barnett, T. P., Dümenil, L., Schlese, U., Roeckner, E., & Latif, M. (1989). The effect of Eurasian snow cover on regional and global climate variations. *J. Atmos. Sci.*, 46, 661-685.

Betts, R. A., Cox, P. M., Lee, S. E., & Woodward, F. L. (1997). Contrasting physiological and structural vegetation feedbacks in climate change simulations. *Nature*, 387, 796-799.

Bonan, G. B., Pollard, D., & Thompson, S. L. (1992). Effects of boreal forest vegetation on global climate. *Nature*, 359, 716-718.

Bonan, G. B. (1995). Land-atmosphere CO_2 exchange simulated by a land surface process model coupled to an atmospheric general circulation model. *J. Geophys. Res.*, 100, 2817-2831.

Bounoua, L., Collatz, G. J., Sellers, P. J., Randall, D. A., Dazlich, D. A., Los, S.O., Berry, J. A., Fung, I., Tucker, C. J., Field, C. B., & Jensen, T. G. (1999). Interactions between vegetation and climate: Radiative and physiological effects of doubled atmospheric CO_2. *J. Climate*, 12, 309-324.

Cao, M., Prince, S. D., Tao, B., Small, J., & Li, K. (2005). Regional pattern and interannual variations in global terrestrial carbon uptake in response to changes in climate and atmospheric CO_2. *Tellus*, 57B, 210-217.

Chase, T. N., Pielke, R. A., Kittel, T. G. F., Nemani, R. R., & Running, S. W. (1996). Sensitivity of a general circulation model to global changes in leaf area index. *J. Geophys. Res.*, 101, 7393-7408.

Chase, T. N., Pielke, R. A., Kittel, T. G. F., Nemani, R. R., & Running, S. W. (2000). Simulated impacts of historical land cover changes on global climate in northern winter. *Climate Dyn.*, 16, 93-105.

Clark, D. B., Xue, Y., Harding, R. J., & Valdes, P. J. (2001). Modeling the impact of land surface degradation on the climate of tropical North Africa. *J. Climate*, 14, 1809-1822.

Conway, T. J., Lang, P. M., & Masarie, K. A. (2007). Atmospheric carbon dioxide dry air mole fractions from the NOAA ESRL Carbon Cycle Cooperative Global Air Sampling Network, 1968-2006, Version: 2007-09-19, Path: ftp://ftp.cmdl.noaa.gov/ccg/co2/flask/event/.

Cox, P. M., Betts, R. A., Jones, C. D., Spall, S. A., & Totterdell, I. J. (2000). Acceleration of global warming due to carbon-cycle feedbacks in a coupled climate model. *Nature*, 408, 184-187.

Defries, R. S., Bounoua, L., & Collatz, G. J. (2002). Human modification of the landscape and surface climate in the next fifty years. *Global Change Biology*, 8, 438-458.

Denning, A. S., Collatz, G. J., Zhang, C., Randall, D. A., Berry, J. A., Sellers, P. J., Colello, G. D., & Dazlich, D. A. (1996a). Simulations of terrestrial carbon metabolism and atmospheric CO_2 in a general circulation model. Part 1: Surface carbon fluxes. *Tellus*, 48B, 521-542.

Denning, A. S., Randall, D. A., Collatz, G. J., & Sellers, P. J. (1996b). Simulations of terrestrial carbon metabolism and atmospheric CO_2 in a general circulation model. Part 2: Simulated CO_2 concentrations. *Tellus*, 48B, 543-567.

Dickinson, R. E., Henderson-Sellers, A., Kennedy, P. J., & Wilson, M. F. (1986). Biosphere-atmosphere transfer scheme (BATS) for the NCAR Community Climate Model, *NCAR Tech. Note NCAR/TN-275+STR*, 69 pp., Boulder.

Dickinson, R. E., & Henderson-Sellers, A. (1988). Modeling tropical deforestation: A study of GCM land-surface parameterizations. *Quart. J. Roy. Meteor. Soc.*, 114, 439-462.

Dickinson, R. E., Errico, R. M., Giorgi, F., & Bates, G. T. (1989). A regional climate model for the western United States. *Clim. Change*, 15, 383-422.

Enfield, D. B., & Mestas-Nunez, A. M. (1999). Multiscale variabilities in global sea surface temperatures and their relationships with tropospheric climate patterns. *J. Climate*, 12, 2719-2733.

Falge, E., Baldocchi, D., Tenhunen, J., Aubinet, M., Bakwin, P., Berbigier, P., Bernhofer, C., Burba, G., Clement, R., Davis, K. J., Elbers, J. A., Goldstein, A. H., Grelle, A., Granier, A., Guðmundsson, J., Hollinger, D., Kowalski, A. S., Katul, G., Law, B. E., Malhi, Y., Meyers, T., Monson, R. K., Munger, J. W., Oechel, W., Paw, K. T., Pilegaard, U. K., Rannik, Ü., Rebmann, C., Suyker, A., Valentini, R., Wilson, K., & Wofsy, S. (2002). Seasonality of ecosystem respiration and gross primary production as derived from FLUXNET measurements. *Agric. For. Meteorol.*, 113, 53-74.

FAO (2007). *State of the world's forests 2007*, FAO (Food and Agriculture Organization of the United Nations), Rome, 144 pp.

Field, C. B., Behrenfeld, M. J., Randerson, J. T., & Falkowski, P. (1998). Primary production of the biosphere: integrating terrestrial and oceanic components. *Science*, 281, 237-240.

Foley, J. A., Prentice, I. C., Ramankutty, N., Levis, S., Pollard, D., Sitch, S., & Haxeltine, A. (1996). An integrated biosphere model of land surface processes, terrestrial carbon balance, and vegetation dynamics. *Global Biogeochem. Cycles*, 10, 603-628.

Franchito, S. H., & Rao, V. B. (1992). Climatic change due to land surface alterations. *Climatic Change*, 22, 1-34.

Friedlingstein, P., Bopp, L., Ciais, P., Dufresne, J.-L., Fairhead, L., LeTreut, H., Monfray, P., & Orr, J. (2001). Positive feedback between future climate change and the carbon cycle. *Geophys. Res. Lett.*, 28, 1543-1546.

Friedlingstein, P., Cox, P., Betts, R., Bopp, L., Von Bloh, W., Brovkin, V., Cadule, P., Doney, S., Eby, M., Fung, I., Bala, G., John, J., Jones, C., Joos, F., Kato, T., Kawamiya, M., Knorr, W., Lindsay, K., Matthews, H. D., Raddatz, T., Rayner, P., Reick, C., Roeckner, E., Schnitzler, K.-G., Schnur, R., Strassmann, K., Weaver, A. J., Yoshikawa, C., & Zeng, N. (2006). Climate carbon cycle feedback analysis: Results from the C4MIP model intercomparison. *J. Climate*, 19, 3337-3353

Garcia-Quijano, J. F., & Barros, A. P. (2005). Incorporating canopy physiology into a hydrological model: photosynthesis, dynamic respiration, and stomatal sensitivity. *Ecol. Model.*, 185, 29-49.

Gedney, N., & Valdes, P. J. (2000). The effect of Amazonian deforestation on the northern hemisphere circulation and climate. *Geophys. Res. Lett.*, 27, 3053-3056.

Giorgi, F. (1990). On the simulation of regional climate using a limited area model nested in a general circulation model. *J. Climate*, 3, 941-963.

Giorgi, F., Huang, Y., Nishizawa, K., & Fu, C. (1999). A seasonal cycle simulation over eastern Asia and its sensitivity to radiative transfer and surface processes. *J. Geophys. Res.*, 104, 6403-6424.

Hahmann, A. N., & Dickinson, R. E. (1997). RCCM2-BATS model over tropical south America: Applications to tropical deforestation. *J. Climate*, 10, 1944-1964.

Hales, K., Neelin, J. D., & Zeng, N. (2004). Sensitivity of tropical land climate to leaf area index: Role of surface conductance versus albedo. *J. Climate*, 17, 1459-1473.

Hasler, N., Werth, D., & Avissar, R. (2009). Effects of tropical deforestation on global hydroclimate: A multimodel ensemble analysis. *J. Climate*, 22, 1124-1141.

Henderson-Sellers, A., Dickinson, R. E., Durbidge, T. B., Kennedy, P. J., McGuffie, K., & Pitman, A. J. (1993). Tropical deforestation: Modeling local- to regional-scale climate change. *J. Geophys. Res.*, 98, 7289-7315.

Hirano, T., Segah, H., Harada, T., Limin, S., June, T., Hirata, R., & Osaki, M. (2007). Carbon dioxide balance of a tropical peat swamp forest in Kalimantan, Indonesia. *Global Change Biology*, 13, 412-425.

Horel, J. D., & Wallace, J. M. (1981). Planetary-scale atmospheric phenomena associated with the Southern Oscillation. *Mon. Wea. Rev.*, 109, 813-829.

Hurell, J. W. (1996). Influence of variations in extratropical wintertime teleconnections on Northern Hemisphere temperature. *Geophys. Res. Lett.*, 23, 665-668.

IGBP-DIS (International Geosphere-Biosphere Program, Data and Information Services) (2000). *Global Soil Data Products CD-ROM*. Oak Ridge National Laboratory, Oak Ridge, TN.

IPCC (Intergovernmental Panel on Climate Change) (2001). *Climate Change 2001: the Scientific Basis*. Cambridge University Press, Cambridge, United Kingdom. 881 pp.

Ito, A. (2002). Soil organic carbon storage as a function of the terrestrial ecosystem with respect to the global carbon cycle. *Japanese J. Ecol.*, 52, 189-227 (in Japanese).

Ito, A., & Oikawa, T. (2002). A simulation model of the carbon cycle in land ecosystems (Sim-CYCLE): a description based on dry-matter production theory and plot-scale validation. *Ecol. Model.*, 151, 143-176.

Ito, A. (2003). A global-scale simulation of the CO_2 exchange between the atmosphere and the terrestrial biosphere with a mechanistic model including stable carbon isotopes, 1953-1999. *Tellus*, 55B, 596-612.

Ito, A. (2008). The regional carbon budget of East Asia simulated with a terrestrial ecosystem model and validated using AsiaFlux data. *Agric. For. Meteorol.*, 148, 738-747.

Jang, C. J., Nishigami, Y., & Yanagisawa, Y. (1996). Assessment of global forest change between 1986 and 1993 using satellite-derived terrestrial net primary productivity. *Environmental Conservation*, 23, 315-321.

Japan Meteorological Agency (1994). The observation of carbon dioxide in air over the sea and surface sea water in the western North Pacific Ocean and estimation of air/sea flux. *J. Meteor. Res.*, 46, 63-69. (in Japanese).

Japan Meteorological Agency (2007). *Annual Report on Atmospheric and Marine Environment Monitoring No.7, Observation Results for 2005*. (http://www.data.kishou.go.jp/obs-env/cdrom/report2005/html/7_0.htm).

Jones, H. G. (1992). *Plants and microclimate*. Cambridge University Press, Cambridge, UK, 428 pp.

Kanae, S., Oki, T., & Musiake, K. (2001). Impact of deforestation on regional precipitation over the Indochina Peninsula. *J. Hydrometeor.*, 2, 51-70.

Kida, H., Koide, T., Sasaki, H., & Chiba, M. (1991). A new approach for coupling a limited area model to a GCM for regional climate simulations. *J. Meteor. Soc. Japan*, 69, 723-728.

Kidson, J. W., & Thompson, C. S. (1998). Comparison of statistical and model-based downscaling techniques for estimating local climate variations. *J. Climate*, 11, 735-753.

Kobayashi, C., Takano, K., Kusunoki, S., Sugi, M., & Kitoh, A. (2000). Seasonal predictability in winter over eastern Asia using the JMA global model. *Quart. J. Roy. Meteor. Soc.*, 126, 2111-2123.

Kuo, H. L. (1974). Further studies of the influence of cumulus convection on large scale flow. *J. Atmos. Sci.*, 31, 1232-1240.

Lacis, A. A., & Hansen, J. E. (1974). A parameterization for the absorption of solar radiation in the earth's atmosphere. *J. Atmos. Sci.*, 31, 118-133.

Latif, M., & Barnett, T. P. (1994). Causes of decadal climate variability over the Northern Pacific and North America. *Science*, 266, 634-637.

Lean, J., & Rowntree, P. R. (1997). Understanding the sensitivity of a GCM simulation of Amazonian deforestation to the specification of vegetation and soil characteristics. *J. Climate*, 10, 1216-1235.

Lofgren, B. M. (1995). Surface albedo-climate feedback simulated using two-way coupling. *J. Climate*, 8, 2543- 2562.

Mabuchi, K., Sato, Y., Kida, H., Saigusa, N., & Oikawa, T. (1997). A Biosphere – Atmosphere Interaction Model (BAIM) and its primary verifications using grassland data. *Papers in Meteor. Geophys.*, 47, 115-140.

Mabuchi, K. (1997). A Biosphere-Atmosphere Interaction Model (BAIM) for physical climate models. *Japanese Journal of Ecology*, 47, 327-331. (in Japanese)

Mabuchi, K. (1999). The Biosphere-Atmosphere Interaction Model (BAIM) for physical climate models. *Meteorological Research Note: Present and Future of Land Surface Process Research*, K. Mabuchi (ed.), 195, pp. 19-29. (in Japanese)

Mabuchi, K., Sato, Y., & Kida, H. (2000). Numerical study of the relationships between climate and the carbon dioxide cycle on a regional scale. *J. Meteor. Soc. Japan*, 78, 25-46.

Mabuchi, K., Sato, Y., & Kida, H. (2001). Numerical simulation study using a climate model including a sophisticated land surface model. *Present and Future of Modeling Global Environmental Change: Toward Integrated Modeling*, T. Matsuno & H. Kida (eds.), TERRAPUB, Tokyo, pp. 449-456.

Mabuchi, K., Sato, Y., & Kida, H. (2002). Verification of the climatic features of a regional climate model with BAIM. *J. Meteor. Soc. Japan*, 80, 621-644.

Mabuchi, K., Sato, Y., & Kida, H. (2005a). Climatic impact of vegetation change in the Asian tropical region Part I: Case of the Northern Hemisphere summer. *J. Climate*, 18, 410-428.

Mabuchi, K., Sato, Y., & Kida, H. (2005b). Climatic impact of vegetation change in the Asian tropical region Part II: Case of the Northern Hemisphere winter and impact on the extratropical circulation. *J. Climate*, 18, 429-446.

Mabuchi, K., & Kida, H. (2006). On-line climate model simulation of the global carbon cycle and verification using the in situ observation data. In: Voinov, A., Jakeman, A., Rizzoli, A. (eds). Proceedings of the iEMSs Third Biennial Meeting: "Summit on Environmental Modelling and Software". International Environmental Modelling and Software Society, Burlington, USA, July 2006. CD ROM.

Mabuchi, K., Takahashi, K. & Nasahara, K. N. (2009). Numerical investigation of climate factors impact on carbon cycle in the East Asian terrestrial ecosystem. *J. Meteor. Soc. Japan*, 87, 219-244.

Mabuchi, K. (2011). A numerical investigation of changes in energy and carbon cycle balances under vegetation transition due to deforestation in the Asian tropical region. *J. Meteor. Soc. Japan*, 89, 47-65.

Marland, G., & Rotty, R. M. (1984). Carbon dioxide emissions from fossil fuels: A procedure for estimation and results for 1950-82. *Tellus*, 36(B), 232-261.

Matala, J., Ojansuu, R., Peltola, H., Sievänen, R., & Kellomäki, S. (2005). Introducing effects of temperature and CO_2 elevation on tree growth into a statistical growth and yield model. *Ecol. Model.*, 181, 173-190.

Matthews, H. D., Weaver, A. J., & Meissner, K. J. (2005). Terrestrial carbon cycle dynamics under recent and future climate change. *J. Climate*, 18, 1609-1628.

Mellor, G. L., & Yamada, T. (1974). A hierarchy of turbulent closure models for planetary boundary layers. *J. Atmos. Sci.*, 31, 1791-1806.

Murayama, S., Saigusa, N., Chan, D., Yamamoto, S., Kondo, H., & Eguchi, Y. (2003). Temporal variations of atmospheric CO_2 concentration in a temperate deciduous forest in central Japan. *Tellus* 55B, 232-243.

Nobre, P., Malagutti, M., Urbano, D. F., De Almeida, R. A. F., & Giarolla, E. (2009). Amazon deforestation and climate change in a coupled model simulation. *J. Climate*, 22, 5686-5697.

Noguer, M., Jones, R. G., & Murphy, J. (1998). Sources of systematic errors in the climatology of a nested regional climate model (RCM) over Europe. *Clim. Dyn.*, 14, 691-712.

Olson, J. S., Watts, J. A., & Allison, L. J. (1983). *Carbon in Live Vegetation of Major World Ecosystems, ORNL-5862, Environmental Sciences Division Publication No. 1997*, Oak Ridge National Laboratory, Oak Ridge, Tennessee.

Onogi, K., Tsutsui, J., Koide, H., Sakamoto, M., Kobayashi, S., Hatsusika, H., Matsumoto, T., Yamazaki, N., Kamahori, H., Takahashi, K., Kadokura, S., Wada, K., Kato, K., Oyama, R., Ose, T., Mannoji, N., & Taira, R. (2007). The JRA-25 Reanalysis, *J. Meteor. Soc. Japan*, 85, 369-432.

Polcher, J. (1995). Sensitivity of tropical convection to land surface processes. *J. Atmos. Sci.*, 52, 3143-3161.

Prentice, I. C. (2001). The carbon cycle and atmospheric carbon dioxide. In: *Climate Change 2001: The scientific basis* (eds. J. T. Houghton, Y. Ding, D. J. Griggs, M. Noguer, P. J. van der Linden, X. Dai, K. Maskell, & C. A. Johnson). Cambridge University Press, Cambridge, United Kingdom, 183-237.

Rayner, N. A., Horton, E. B., Parker, D. E., Folland, C. K., & Hackett, R. B. (1996). *Version 2.2 of the Global Sea-Ice and Sea Surface Temperature data set, 1903-1994. CRTN 74*, Hadley Centre for Climate Prediction and Research, Meteorological Office, London Road, Bracknell, Berkshire, RG12 2SY.

Rayner, N. A., Parker, D. E., Horton, E. B., Folland, C. K., Alexander, L. V., Rowell, D. P., Kent, E. C., & Kaplan, A. (2003). Global analyses of sea surface temperature, sea ice and night marine air temperature since the late nineteenth century, *J. Geophys. Res.*, 108(D14), 4407, doi:10.1029/2002JD002670.

Renshaw, A. C., Rowell, D. P., & Folland, C. K. (1998). Wintertime low-frequency weather variability in the North Pacific-American sector 1949-93. *J. Climate*, 11, 1073-1093.

Sage, R. F., Wedin, D. A., & Li, M. (1999). The biogeography of C_4 photosynthesis: patterns and controlling factors, 313-373. In C_4 *plant Biology* (eds. R. F. Sage and R. K. Monson). Academic Press, San Diego, USA, 596 pp.

Saigusa, N., Yamamoto, S., Hirata, R., Ohtani, Y., Ide, R., Asanuma, J., Gamo, M., Hirano, T., Kondo, H., Kosugi, Y., Li, S.-G., Nakai, Y., Takagi, K., Tani, M., & Wang, H. (2008). Temporal and spatial variations in the seasonal patterns of CO_2 flux in boreal, temperate, and tropical forests in East Asia. *Agric. For. Meteorol.*, 148, 700-713.

Sasaki, H., Kida, H., Koide, T., & Chiba, M. (1995). The performance of long-term integrations of a limited area model with the spectral boundary coupling method. *J. Meteor. Soc. Japan*, 73, 165-181.

Sato, N., Sellers, P. J., Randall, D. A., Schneider, E. K., Shukla, J., Kinter, J. L., Hou, Y.-H., & Albertazzi, E. (1989). Effects of implementing the simple biosphere model in a general circulation model. *J. Atmos. Sci.*, 46, 2757-2782.

Schwalm, C. R., & Ek, A. R. (2004). A process-based model of forest ecosystems driven by meteorology. *Ecol. Model.*, 179, 317-348.

Segami, A., Kurihara, K., Nakamura, H., Ueno, M., Takano, I., & Tatsumi, Y. (1989). Operational mesoscale weather prediction with Japan Spectral Model. *J. Meteor. Soc. Japan*, 67, 907-924.

Sellers, P. J., Mintz, Y., Sud, Y. C., & Dalcher, A. (1986). A simple biosphere model (SiB) for use within general circulation models. *J. Atmos. Sci.*, 43, 505-531.

Sellers, P. J., Berry, J. A., Collatz, G. J., Field, C. B., & Hall, F. G. (1992). Canopy reflectance, photosynthesis, and transpiration. III. A reanalysis using improved leaf models and a new canopy integration scheme. *Remote Sens. Environ.*, 42, 187-216.

Sellers, P. J., Randall, D. A., Collatz, G. J., Berry, J. A., Field, C. B., Dazlich, D. A., Zhang, C., Collelo, G. D., & Bounoua, L. (1996). A revised land surface parameterization (SiB2) for atmospheric GCMs. Part I: Model formulation. *J. Climate*, 9, 676-705.

Sen, O. L., Wang, Y., & Wang, B. (2004). Impact of Indochina deforestation on the east Asian summer monsoon. *J. Climate*, 17, 1366-1380.

Seth, A., & Giorgi, F. (1998). The effects of domain choice on summer precipitation simulation and sensitivity in a regional climate model. *J. Climate*, 11, 2698-2712.

Shukla, J., Nobre, C., & Sellers, P. (1990). Amazon deforestation and climatic change. *Science*, 247, 1322-1325.

Small, E. E., Giorgi, F., & Sloan, L. C. (1999). Regional climate model simulation of precipitation in central Asia: Mean and interannual variability. *J. Geophys. Res.*, 104, 6563-6582.

Snyder, P. K., Foley, J. A., Hitchman, M. H., & Delire, C. (2004). Analyzing the effects of complete tropical forest removal on the regional climate using a detailed three-dimensional energy budget: An application to Africa. *J. Geophys. Res.*, 109, D21102, doi:10.1029/2003JD004462.

Sud, Y. C., & Smith, W. E. (1985). Influence of local land-surface processes on the Indian monsoon: A numerical study. *J. Climate Appl. Meteor.*, 24, 1015-1036.

Sugi, M., Kuma, K., Tada, K., Tamiya, K., Hasegawa, N., Iwasaki, T., Yamada, S., & Kitade, T. (1990). Description and performance of the JMA operational global spectral model (JMA-GSM89). *Geophys. Mag.*, 43, 105-130.

Tatsumi, Y. (1986). A spectral limited-area model with time-dependent lateral boundary conditions and its application to a multi-level primitive equation model. *J. Meteor. Soc. Japan*, 64, 637-663.

Thoning, K.W., Kitzis, D.R., & Crotwell, A. (2007). Atmospheric carbon dioxide dry air mole fractions from quasi-continuous measurements at Barrow, Alaska; Mauna Loa, Hawaii; American Samoa; and South Pole, 1973-2006, Version: 2007-10-01, Path: ftp://ftp.cmdl.noaa.gov/ccg/co2/in-situ/.

Trenberth, K. E., & Hurrell, J. W. (1994). Decadal atmosphere-ocean variations in the Pacific. *Climate Dyn.*, 9, 303-319.

Trenberth, K. E., Branstator, G. W., Karoly, D., Kumar, A., Lau, N.-C., & Ropelewski, C. (1998). Progress during TOGA in understanding and modeling global teleconnections associated with tropical sea surface temperatures. *J. Geophys. Res.*, 103, 14291-14324.

Tsvetsinskaya, E. A., & Mearns, L. O. (2001a). Investigating the effect of seasonal plant growth and development in three-dimensional atmospheric simulations. Part I: Simulation of surface fluxes over the growing season. *J. Climate*, 14, 692-709.

Tsvetsinskaya, E. A., & Mearns, L. O. (2001b). Investigating the effect of seasonal plant growth and development in three-dimensional atmospheric simulations. Part II: Atmospheric response to crop growth and development. *J. Climate*, 14, 711-729.

United Nations (2008). *2006 Energy Statistics Yearbook*. United Nations Department for Economic and Social Information and Policy Analysis, Statistics Division, New York. 616 pp.

Vernekar, A. D., Zhou, J., & Shukla, J. (1995). The effect of Eurasian snow cover on the Indian monsoon. *J. Climate*, 8, 248-266.

Watanabe, F., Uchino, O., Joo, Y., Aono, M., Higashijima, K., Hirano, Y., Tsuboi, K., & Suda, K. (2000). Interannual variation of growth rate of atmospheric carbon dioxide concentration observed at the JMA's three monitoring stations: Large increase in concentration of atmospheric carbon dioxide in 1998. *J. Meteor. Soc. Japan*, 78, 673-682.

Werth, D., & Avissar, R. (2002). The local and global effects of Amazon deforestation. *J. Geophys. Res.*, 107(D20), 8087, doi:10.1029/2001JD000717.

Werth, D., & Avissar, R. (2005a). The local and global effects of African deforestation. *Geophys. Res. Lett.*, 32, L12704, doi:10.1029/2005GL022969.

Werth, D., & Avissar, R. (2005b). The local and global effects of Southeast Asian deforestation. *Geophys. Res. Lett.*, 32, L20702, doi:10.1029/2005GL022970.

Xie, P. P., & Arkin, P. A. (1997). Global precipitation: A 17-year monthly analysis based on gauge observations, satellite estimates, and numerical model outputs. *Bull. Am. Met. Soc.*, 78, 2539-2558.

Xue, Y. (1997). Biosphere feedback on regional climate in tropical north Africa. *Quart. J. Roy. Meteor. Soc.*, 123, 1483-1515.

Yamamoto, S., Saigusa, N., Murayama, S., Gamo, M., Ohtani, Y., Kosugi, Y., & Tani, M. (2005). Synthetic analysis of the CO2 fluxes at various forests in East Asia. In: Omasa, K., I. Nouchi, & L. J. DeKok (Eds.), *Plant Responses to Air Pollution and Global Change*, Springer-Verlag Tokyo, pp. 215-225.

Yasunari, T., Kitoh, A., & Tokioka, T. (1991). Local and remote responses to excessive snow mass over Eurasia appearing in the northern spring and summer climate - a study with the MRI GCM. *J. Meteor. Soc. Japan*, 69, 473-487

Yu, G.-R., Wen, X.-F., Sun, X.-M., Tanner, B. D., Lee, X., & Chen, J.-Y. (2006). Overview of ChinaFLUX and evaluation of its eddy covariance measurement. *Agric. For. Meteorol.*, 137, 125-137.

Zhang, H., Henderson-Sellers, A., & McGuffie, K. (1996a). Impacts of tropical deforestation. Part I: Process analysis of local climatic change. *J. Climate*, 9, 1497-1517.

Zhang, H., McGuffie, K., & Henderson-Sellers, A. (1996b). Impacts of tropical deforestation. Part II: The role of large-scale dynamics. *J. Climate*, 9, 2498-2521.

Zhang, Y., Wallace, J. M., & Battisti, D. S. (1997). ENSO-like interdecadal variability: 1900-93. *J. Climate*, 10, 1004-1020.

Zhao, M., Pitman, A. J., & Chase, T. (2001). The impact of land cover change on the atmospheric circulation. *Climate Dyn.*, 17, 467-477.

Part 4

Aerosols

11

A Review of Modeling Approaches Accounting for Aerosol Impacts on Climate

Abhilash S. Panicker and Dong-In Lee
Department of Environmental Atmospheric Sciences,
Pukyong National University, Busan,
South Korea

1. Introduction

Atmospheric aerosols are tiny particles suspended in the air, in solid as well as in liquid particle forms. Aerosols are known to cause serious air pollution problems and adverse health effects. Apart from these effects, aerosols have now been revealed as a key component in changing climate scenarios. Along with green house gases, aerosols also play an important role in modulating both the global and regional climate balance. Aerosols are found to influence the climate directly and indirectly through radiative forcing (Panicker et al. 2008). Aerosols directly influence the climate by scattering and absorbing incoming solar radiation and indirectly through modifying the cloud microphysical processes (Ramanthan et al. 2001). The quantitative estimation of aerosol radiative forcing is more complex than the radiative forcing estimation of greenhouse gases, attributed to the large spatio-temporal variability of aerosol particulates. This variability is largely due to the much shorter atmospheric lifetime of aerosols compared with the life time of important greenhouse gases. The typical life time of aerosols varies from days to weeks, where as greenhouse gases sustain in the atmosphere for years. The optical properties and microphysical processes of cloud aerosol interactions are also different among various aerosol species and hence are scarcely understood. The quantitative estimation of their effects, therefore, has been highly uncertain (Takemura et al. 2002). Intensive field experiments, new surface and remote-sensing observations, and improved representation of aerosol processes in models have provided new insights into the controlling mechanisms, radiative effects, and the influence of aerosols on climate (Bollasina and Nigam, 2006). The influence of anthropogenic aerosols on the Earth's radiation budget however is still considered as the largest uncertainty in radiative forcing under climate change (IPCC, 2007). According to the latest International Panel for climate change report (IPCC, 2007), the global mean direct aerosol radiative forcing is -0.5±0.4 Wm^{-2} and the estimate for the global annual radiative forcing of the first indirect effect is -0.7 Wm^{-2} with an uncertainty range of -1.8 to -0.3 Wm^{-2}. This negative radiative forcing is supposed to partly offset the the warming caused by greenhouse gases. The aerosol induced surface cooling namely 'White house effect', counter acting to greenhouse effect is revealed to generate a 'global dimming 'in the past century. It is proposed that this reduction of solar radiation at the surface and associated cooling can alter the tropospheric temperature profile and can even change the cloud processes and hence precipitation patterns (Satheesh and Ramanathan, 2000). Absorbing aerosols such as black carbon are found to be responsible for a new phenomenon known as 'Atmospheric Brown clouds' (Ramanathan et al. 2005), which is considered as a threat for Himalayan glaciers due to heating in the lower troposphere, leading to melting of glaciers. Radiative effects of aerosols had generally been neglected in climate models while simulating different large scale phenomenon. However studies have pointed out that such omission of aerosol component could induce large bias in model outputs. Also recent studies proved that aerosols are influential climate components in modulating large scale phenomenon like Indian summer monsoon (eg: Lau et al. 2006; Menon et al. 2002). Several studies have been published in the aspect of aerosol radiative effects and their climate impacts. In this chapter we describe a compact review of aerosol radiative forcing studies reported using modeling approaches. Due to length constraint of the chapter, we present only most highlighted and recent findings in aerosol direct and indirect effect estimations carried out using atmospheric model simulations. Aerosol models vary from one dimension to three dimensions. This chapter focuses on the modeling aspects of aerosol optical properties and their direct and indirect radiative effects using one dimensional and three dimensional modeling studies. The chapter has been divided in to five sections including description of modeling of aerosol optical properties, aerosol direct radiative forcing estimates in local scale using one dimensional models, findings of aerosol direct radiative forcing in regional and global scales using three dimensional models, aerosol indirect forcing estimates in regional and global scale and the final section describes the important findings of role of aerosols in modulating the large scale phenomenon, the monsoon.

2. Modeling of aerosol optical properties

Aerosol models represent a simple, generalized description of typical atmospheric conditions (Shetle and Fenn., 1979). The definition of an aerosol model is a complex process because of the great variability in the physical, chemical, and optical properties of aerosols in both time and space. Two fundamental approaches toward defining an aerosol model exist: first, a direct measurement of all the necessary optical properties as a function of space and time and, second, a computation of the optical characteristics of aerosols after the microphysical properties have been collected and averaged by use of data derived from different sources (Levoni et al. 1997). The sparcity of adequate observational data sets makes the first approach much complex. Hence the second approach is used to model aerosol optical parameters. Several authors have developed models for atmospheric aerosols: beginning with the early works of Toon and Pollack (1976) and Fenn (1964). The aerosol component models has been popularized after the developmet of World Climate Research Program aerosol models (Deepak et al.1983) and the new version known as to as WCP-112 (WCP, 1986). In this section, we briefly decribe the fundamental aerosol optical models popularly used by aerosol community for modeling aerosol optical properties.

The essential aerosol optical parameters required for estimation of radiative forcing includes, Aerosol optical depth (AOD), Single Scattering Albedo (SSA), Asymmetry parameter (ASP) and Angstrom exponent (ANG). AOD represents the aerosol concentration in terms of attenuation of light. It is defined as the height integrated extinction coefficient and is a unit less quantity. A low value of AOD indicates an atmosphere with less aerosol loading and its typical value above 0.7 is considered as very high aerosol loading condition.

SSA indicates the aerosol radiative absorption nature,viz. absorbing type or scattering type. SSA is also a unit less quantity and its typical value above 0.9 indicates a majority scattering type aerosols (eg: sulphate, nitrate) and value below 0.8 represents significant presence of absorbing aerosols (eg: Black carbon, dust). ASP represents the direction of scattering by aerosols, (Panicker et al.2008; Pandithuarai et al.2004) ie. forward or back ward scattering. The value of ASP ranges between -1 to +1. The range -1 to 0 is indicates backward scattering by aerosols and 0 to 1 is assumed to be for forward scattering fraction of aerosols. Angstrom wavelength exponent indicates the size distribution of aerosol fractions. Higher value of ANG indicates an atmosphere with abundance of smaller particles (fine mode) and smaller values of ANG indicates the dominance of bigger particles (Coarse mode). The observations of these important aerosol optical parameters have been carried out using hand held/ automatic sunphotometers/radiometers. However when, the direct observations of these aerosol optical parameters are un available, chemical sampling data sets has been used to model the aerosol optical parameters. In this regard, one dimensional aerosol models has been utilized. In general, aerosol optical models utilize aerosol chemical data sets as inputs and provide aerosol optical parameters as output, which are necessary for estimating radiative forcing.

There are a few models, which utilizes aerosol chemistry data sets to derive aerosol optical properties. Out of these, Optical properties of aerosols and clouds (OPAC) is one of the most popular and widely used model. OPAC is a software package used to derive aerosol optical properties such as AOD, SSA, ASP, ANG and extinction coefficient at a given location. OPAC provides the optical properties of aerosol particles, water droplets and ice crystals, both in solar and terrestrial spectral range, and on the other hand include software to handle data (Hess et al., 1998). OPAC utilizes aerosol chemical composition data sets. It uses generally six different types of aerosol composition viz. Water soluble, Insoluble, soot, mineral aerosol, seasalt particles, sulfates chemistry data sets as its input (defaut or user defined). OPAC also has been characterized with ten different environmental conditions over continental and maritime regions. It provides outputs of aerosol optical parameters such as AOD, SSA,ASP, extinction coefficient, Angstrom expeont etc. at 8 different relative humidity conditions (0%, 50%, 70%, 80%, 90%, 95%, 98%, 99%).

Several investigators Several investigators utilize OPAC model to estimate aerosol optical properties from chemical composition data sets. A study by Panicker et al. (2010b) showed a promising comparison of aerosol optical properties derived potential comparison of OPAC derived AOD to Sun/Sky radiometer derived AOD. Another study by von Huene et al.(2011) also has reported fair comparison results between experimental aeorosol data sets and OPAC modeled aerosol components for different atmospheric environments. Several studies over the globe use Aerosol optical properties derived by OPAC as reference parameters for estimating aerosol raditive effects over the concerned environments. (eg: Satheesh et al. 1999; Ramachandran et al. 2006). A few literature report their radiative forcing computations using OPAC modeled aerosol optical poroperties (eg: Panicker et al. 2008; Sreekanth et al. 2007; Vinoj et al. 2004; Kim et al. 2006; Ramachandran et al. 2006). Satheesh et al. (1999) developed an aerosol optical model for natural and anthropogenic aerosols over Indian Ocean during Indian ocean experiment (INDOEX) based on OPAC model and was used to derive aerosol optical properties and hence the radiative forcing estimations during INDOEX. Smirnov et al. (2003) developed another model for accounting aerosol optical properties and size distributions exclusively over maritime regions using AERONET datasets.

Levoni et al. (1997) developed an optical aerosol model to derive aerosol optical properties from chemistry data sets. In this model, Users can select a default aerosol class or component or they can input a user-defined aerosol class or component by setting mixing ratios, size distribution, refractive indices of aerosol components, and mixing type. The user can choose a grid within the following parameter ranges: wavelength, 0.2–40 mm; scattering angle, 0°–180°; and RH, 0%–99%. It provides 15 different environmental conditions over continental and maritime areas. The model also has been provided with several aerosol scattering and absorbing species as its inputs (eg: Sulphate,Soot, Watersoluble, insoluble etc.) The main output parameters from this model includes, Extinction coefficient, scattering coefficient, SSA,ASP and phase function. This model found to have shown satisfactory comparison with well tested WCP-112 model (Levoni et al. 1997). The data sets described in this model has been used by several investigators for validation as well as for radative transfer calculations (eg: Guzzi et al. 2007; Nessler et al. 2005).

Global Aerosol Data Set (GADS) provides the aerosol optical parameters globally using aerosol chemical composition data sets. It is a completely revised version of the aerosol climatology by d'Almeida et al. (1991). GADS and OPAC use the same set of optical data of aerosol components. GADS provides aerosol optical parameters on a grid of 5 degrees longitude and latitude, with 7 differentiating height profiles. GADS consists of number distribution, mass per volume and optical properties of aerosols. In this package aerosol particles are described by 10 main aerosol components which are representative for the atmosphere and characterized through their size distribution and their refractive index depending on the wavelength. These aerosol particles are based on components resulting from aerosol emission, formation and removal processes within the atmosphere, so that they exist as mixture of different substances, both external and internal (Kopke et al. 1997). Typical components include water-soluble, water-insoluble, soot, sea-salt and mineral aerosols. The sea-salt particles are defined in two classes and the mineral particles in four. Aerosols are modeled as 10 components which are described with size distribution and spectral refractive index. From these chemical composition data sets, the optical properties are calculated with Mie-Theory at wavelengths between 0.25 µm and 40 µm and for 8 values of relative humidity similar as in OPAC, while over entire globe at 5^0 X 5^0 grid spacing. GADS has been cited several times in the literature (eg: Chin et al. 2003; Manoj et al.2010).

3. One dimensional modeling of direct aerosol radiative forcing

Direct Aerosol radiative forcing (DARF) indicates the reduction in surface reaching solar radiation due to its scattering and absorption by aerosols. Observed aerosol properties such as AOD, SSA and ASP and ANG have been the most important parameters required for estimating DARF. DARF is defined as the difference in solar fluxes (irradiance), with aerosol condition and fluxes with clean condition in the atmosphere. It requires, down welling and up welling irradiances to calculate radiative forcing and is calculated as,

$$\text{DARF}= [F{\downarrow} \_F{\uparrow}] \text{ aerosol} - [F{\downarrow}\_ F{\uparrow}] \text{ no aerosol}$$

For calculating DARF, it is necessary to have the explicit data sets of aerosols, gaseous components, albedo of the observing surface, verical profiles of meteorological parametrs and aerosols. Radiative transfer models are used to compute the aerosol direct radiative forcing at any level in the atmosphere with the appropriate input values for aerosol properties. The core of a radiative transfer model lies the radiative transfer equation that is numerically solved using a solver such as a discrete ordinate method or a Monte Carlo method. The radiative transfer equation is a monochromatic equation to calculate radiance in a single layer of the Earth's atmosphere. To calculate the radiance for a spectral region with a finite width, one has to integrate this over a band of wavelengths. The most exact way to do this is to loop through the wavelengths of interest, and for each wavelength, calculate the radiance at this wavelength. For this, one needs to calculate the contribution of each spectral line for all molecules in the atmospheric layer; this is called a *line-by-line* calculation.

In general, Radiative transfer models derive the radiative fluxes at specified spectral wavelengths. It derives fluxes at different levels with aerosol condition and no aerosol condition, difference of which, provides aerosol radiative forcing. A list of most important atmospheric radiative transfer models and their wavelength bands of simulation are depicted in table 1. Radiative transfer models in general uses a set of programs based on a collection of highly developed and reliable physical models, which have been developed by the atmospheric science community over the past few decades. The basic programs in radiative transfer codes consists of consists of (a) cloud model, containing the optical datasets of clouds,(b) Gas absorption model, describing the gaseous absorption in the atmosphere,(c) Extra terrestrial source spectra, consisting of model spectrums incorporating solar spectral wavelengths. (d)standard atmospheric models to characterize typical climate conditions (eg: Mc Clathey et al.(1972) contianing tropical, midlatitude winter and summer; sub artic winter and summer and artic winter and summer over the globe). The models also facilitate the option of using user defined atmospheric profiles of temperature and humidity for concerned environments. Also these models uses standard aerosol models for defining aerosol vertical profiles (eg: SBDART (Richiazzi et al. 1998) uses the MC Clathey et al. (1972) models, of vertical aerosol profiles for visibility 5 and 23 km).

Model name	Source	Spectral wavelength of simulation	Mode of operation
DISORT	Stamnes et al. (1998)	UV, Visible, IR	Offline
LBLRTM	Clought et al. (2005)	UV, Visible, IR	Offline
libRadtran	Mayer and Kylling (2005)	UV, Visible, IR	offline
MODTRAN	Berk et al. (1998)	UV, Visible, IR	offline
SBDART	Ricchiazzi et al. (1998)	UV, Visible, IR	offline
Fast Rt	Engelsen and Kylling(2005)	Exclusive for UV	online

Table 1. Popular atmospheric radiative transfer models and their wavelength bands of simulation.

Typically radiative forcing is estimated at the surface, Top of the Atmosphere (TOA) and in the atmosphere. DARF results in short wave region (0.3-3µm) report a negative forcing at the surface, attributed to radiation reduction at the surface by scattering and absorption of aerosols, depending on the aerosol species. The forcing at TOA also depends on aerosol species, which used to be positive for absorbing fractions and negative for scattering species. The difference between TOA and surface forcing, estimates the net gain of radiation by aerosols in the atmosphere, known as atmospheric aerosol forcing, which used to be a positive quantity in short wave domain. Several studies report aerosol shortwave radiative forcing estimates using ground based observations and also by modeling approaches. In this section, we brief the most important findings of one dimensional model based aerosol radiative forcing estimates over different regions in the globe.

Indian Ocean Experiment (INDOEX) (Jayaraman et al. 1998; Satheesh et al. 1999) was one of the initial attempts, which largely used the observed aerosol datasets in dimensional models to estimate regional aerosol radiative forcing. Tropospheric Aerosol Raditaive Forcing Observational experiment (TARFOX) over Atlantic Ocean (Hignett et. al., 1999) also contributed estimates of aerosol forcing using observation-model coupling analysis. Mean Clear-sky aerosol surface forcing reported was about -26W m^{-2} for Atlantic Ocean during TARFOX (Hignett et. al., 1999) and -29 W m^{-2} for tropical Indian Ocean during INDOEX (Satheesh and Ramanathan, 2000).

Conant (2000) showed an aerosol surface forcing of -7.6 ± 1.5 W m^{-2} during INDOEX period in the 400–700 nm spectral region (visible region of short wave band). He introduced two new promising methods for estimating Aerosol radiative forcing (ARF) and forcing efficiency. One a pure experimental method and another Hybrid method, utilizing Montercarlo radiative transfer model in conjunction with experimental data sets of observed fluxes. Both the methods found to show results with minimal bias. It is also shown that, a 0.1 change in aerosol optical depth produces a -4.0 ± 0.8 W m^{-2}; change in the 400–700 nm surface flux and 55% of this forcing is confined in the 400–540 nm region.

Satheesh et al. (1999) used Montecarlo radiative transfer model in conjunction with the INDOEX aerosol model and showed that anthropogenic aerosols reduces the incoming solar radiation up to -50 Wm^{-2} at the surface. Using OPAC in conjunction with SBDART, Vinoj et al. (2004) estimated radiative forcing over Bay of Bengal (BOB) and found that Regionally averaged aerosol (net) forcing over the Bay of Bengal was in the range -15 to -24 W m^{-2} at the surface, -2 to -4 W m^{-2} at the top of the atmosphere, leading to an atmospheric absorption of 13 to 20 W m^{-2}. Pandithurai et al. (2004) found a surface solar reduction of 33 Wm^{-2} over an urban Indian terrain, Pune. They used the observed aerosol optical properties from a Sun/Sky radiometer in SBDART to derve fluxes and hence forcing. The TOA forcing was found to be Zero, and the atmospheric absorption was found to be 33 Wm^{-2}. The surface aerosol forcing efficiency, which is the standardized aerosol radiative forcing for unit increase in AOD were computed using two methods and the forcing efficiency values found to be comparable ie. -88 and -84 Wm^{-2}, respectively, indicating a reduction of around 84 Wm^{-2} of short wave radiation at the surface for unit increase in AOD.

Aerosol forcing estimates across East Asia, which experiences large dust out breaks, have provided valuable information about the contribution of different chemical species to total forcing in the past decade. Kim et al. (2006) estimated DARF over a typical east Asian site

(Gosan, Korea) by using chemistry data sets in OPAC model in conjunction with a radiative transfer model (Fu–Liou RTM)(Fu and Liou, 1993). On estimating species segregated raditive forcing, they obtained the value of - 38.3Wm^{-2} at the surface attributable to mineral dust (45.7%). water-soluble components (sum of sulfate, nitrate, ammonium, and water-soluble organic carbon (WSOC)) induced a forcing of 26.8% and elemental carbon (EC) contributed up to 26.4% of the surface forcing. However, sea salt does not observed to play a major role in surface forcing. For the cases of Asian dust and smoke episodic events a diurnal averaged forcing of -36.2Wm^{-2} was observed at the surface over Gosan, contributed by mineral dust(-18.8Wm^{-2}), EC (-6.7Wm^{-2}), and water-soluble components (-10.7 Wm^{-2}). The results of this important study conclude that water-soluble and EC components as well as a mineral dust component are responsible for a large portion of the aerosol radiative forcing at the surface in the continental outflow region of East Asia. Another study by Li et al. (2009), estimated DARF using SBDRAT model over 25 stations in East Asia (China). ARF was determined at all the stations at the surface, inside the atmosphere, and at the top of atmosphere (TOA). Nationwide annual and diurnal mean ARF over China is found to be -15.7 ± 8.9 Wm^{-2} at the surface, 0.3 ± 1.6 Wm^{-2} at the TOA, and 16.0 ± 9.2Wm^{-2} in the atmosphere.

Extensive campaigns have been conducted to explore aerosol forcing over Middle East, where widespread sand/dust storm events are frequent. A typical study by Marcowicz. et al. (2008) found that relatively small value of the aerosol forcing efficiency (forcing per unit optical depth) at the surface (-53 W m^{-2}) persisting over United Arab Emirates (UAE)) during UAE2 campaign. They used MODTRAN based on DISORT method for the study and found a mean diurnal forcing of -20Wm^{-2} during campaign period. It is also found that aerosol forcing during the UAE2 campaign leads to a significant reduction of the incoming solar radiation at the surface (about 9%). The land–sea circulation found to have strong influence on the aerosol optical thickness and the aerosol diurnal variability. Larger aerosol radiative forcing is observed during the land breeze in comparison to the sea breeze. Markowiz et al. (2002) reported a radiative forcing estimate of -17.9 Wm^{-2} over mediteranian coast of Greece using MoDTRAn model, attributing the forcing to aerosols of anthropogenic origin. The study also revealed an increase of 11.3 W m^{-2} in the atmospheric solar absorption, and also an increase of 6.6 W m^{-2} in the reflected solar radiation at the top-of-the atmosphere. Thus giving observational proof for the large role of absorbing aerosols in the Mediterranean. This negative surface forcing and large positive atmospheric forcing values observed for the Mediterranean aerosols is nearly identical to the highly absorbing south Asian haze observed over the Arabian Sea. Another study over Mediterranean zone by Saha et al. (2008) discussed aerosol induced maintainance of summer heat waves over Europe. Using Global Atmospheric ModEl (GAME) (Dubuisson et al., 1996, 2004) they have reported the reduction of surface solar radiation by 26±3.9 W m^{-2} by aerosols during summer. This shortwave reduction at surface found to generate an atmospheric heating of 2.5 to 4.6 K day^{-1}. Hence it is proposed that, this increase in heating rate could be one of the reason for maintenance of heat-waves frequently occurring over Mediterranean coastal region during summer. Marc et al. (2009) constrained the Impact of sea-surface dust radiative forcing on the oceanic primary production (PP) using GAME model. It is found that dust are able to induce a significant decrease of PP due to the attenuation of light by about 15–25% for dust optical depth (DOD) larger than 0.6–0.7 (at 550 nm). However for DOD lower than 0.2–0.3, the influence of dust on PP found to be weak (5%).

Regional estimates of DARF have been popularized after the establishment of Aerosol robotic Nerwork (AERONET), operated by NASA across various regions of the globe (http://aeronet.gsfc.nasa.gov/). AERONET has been active since 1993 (Smirnov et al. 2003). On parallel to AERONET, Chiba university of Japan established another aerosol network across Asia, known as SKYNET (http://atmos.cr.chiba-u.ac.jp/). Gonzi et al. (2007) reported DARF across 25 AERONET sites in Europe, spreading across the geographical borders of different European countries. They used sdisort radiative transfer code in libRadtran environment to simulate DARF. It is found that the annual median mean of this TOA, AERONET clear sky aerosol radiative forcing effect is around -3Wm^{-2} but with higher values in summer than in winter. The atmospheric aerosol forcing effect during summer is larger and has a typical median value of +3Wm^{-2}, essentially associated with high pollutant trapping in winter. The experiments were conducted for different albedo conditions and found to have a high sensitivity in TOA forcing with different albedo conditions. A median forcing efficiency (forcing per unit optical depth) of up to -25Wm^{-2} and -35Wm^{-2} were also found over Europe, respectively at TOA and surface.

SBDART model based studies were carried out across different parts of the world to estimate the contribution of most important absorbing aerosol species, Black carbon on the total aerosol short wave radiative forcing. It is revealed that, even though BC contributes only 3-6% of total aerosol mass, BC forcing contributes up to 55-80% of total aerosol atmospheric forcing (Sreekanth et al. 2007, Gaddavi et al.2010, Panicker et al. 2010a), attributed to strong atmospheric absorption by BC. This strong atmospheric warming along with surface cooling by BC aerosols can inhibit cloud formation processes by retarding convection by inducing lower atmospheric inversions.

Several studies across the globe have reported DARF in the short wave region. However, Aerosol long wave forcing effects has generally been neglected and estimates are sparse. Recent studies suggest that Aerosol long wave forcing is an important component in compensating short wave losses at the surface. It is also found that this long wave aerosol forcing has been of comparable magnitude as that of green house forcing at local scale (Vogelman et al.2003). Panicker et al. (2008) simulated aerosol long wave radiative forcing using SBDART model over a highly urbanized Indian city, Pune (A SKYNET Site). It is found that on contrary to short wave forcing, aeorols enhances the long wave radiation at the surface and compensates the shortwave cooling by up to 25%. It is also found that long wave radiaitve forcing is sensitive to temperature and humidity profiles. Increase in water vapour reduces aerosol long wave forcing while, an enhanced aerosol long wave forcing increases atmospheric temperature. Vogelman et al.(2003)using experimental data in conjunction with a radiative transfer model estimated a long wave forcing of up to 10 Wm^{-2} over pacific ocean during ACE-2 campaign. Dufresne et al. (2002) reported the long wave scattering effects of mineral aerosols. Using SBDART they showed that neglecting scattering by mineral aerosols in long wave region may unset estimates of long wave forcing, up to 50% at TOA and up to 15 % at the surface. It is also computed that for unit aerosol optical thinckness TOA forcing can reach up to +8 Wm^{-2}. Sathessh and Lubin (2003) analyzed the effect of wind speed in long wave forcing. It is showed that increased wind speed generating increased seasalt production can enhance long wave forcing due to absorption of long wave by sea salt aerosols. It is also found that even at moderate winds (6–10 m s^{-1}), the short wave forcing reduces by ~45% due to the dominance of sea-salt aerosols. At high winds (>10 m s^{-1}), a major fraction of the long wave forcing is contributed by sea-salt (more than 70%). Hence it is proved that neglecting long wave forcing can make large bias in climate models.

4. Three dimensional modeling of direct aerosol radiative forcing

Direct Aerosol raditaive forcing has been modeled in regional as well as in global scales in order to find its reduction potential of green house warming. Global Modeling studies indicate that the aerosol radiative forcing effect is similar in magnitude, but opposite in sign, to the radiative forcing effect due to greenhouse gases (Charlson et al, 1992; Haywood and Boucher, 2000; Ramanathan et al, 2001; IPCC, 2007). This section briefly discusses the important three dimensional modeling results constraining important findings on global and regional scales.

Using a modeling approach, Haywood and Shine (1995) reported direct aerosol radiative forcing values of -2Wm^{-2} for an external mixture of soot and sulfate. Kiehl et al.(2000) estimated the direct and indirect radiative forcing due to sulfate aerosols using the National Center for Atmospheric Research (NCAR) Community Climate Model (CCM3). This model includes a sulfur chemistry model and predicts the mass of sulfate. The estimated global annual mean DARF forcing was found to be -0.56 W m^{-2}. Takemura et al.(2002) using a three-dimensional aerosol transport-radiation model coupled with a general circulation model reported the global annual mean values of the total direct radiative forcing of anthropogenic carbonaceous plus sulfate aerosols as -0.19 and -0.75 respectively under whole-sky and clear-sky conditions. They also described the global mean radaitve forcing induced by individual aerosol species segregated as sulphate, sea salt, soil dust and carbonaceous aerosols for clear as well as for all sky conditions. Chung et al. (2005) reported a global estimtate of direct aerosol radiatve forcing by integrating satellite and ground based observations with models of aerosol chemistry, transport, and radiative transfer models (GOCART and Montecarlo models). The typical value for the global top-of-atmosphere (TOA) aerosol radiative forcing effect obtained was in the range of -0.1 to -0.6Wm^{-2} for all sky conditions and around -1Wm^{-2} for clear-sky conditions. The global annual mean surface forcing values obtained were a found to be an order of magnitude greater than TOA values, showing -3.4 Wm^{-2} reductions at the surface and the corresponding global atmospheric aerosol forcing obtained was +3 Wm^{-2}. Verma et al. (2006) using a GCM simulation estimated the sulphate forcing. The model results indicate that the change in the sulfate aerosols number concentration is negatively correlated to the indirect radiative forcing. The model simulated annual mean direct radiative forcing ranged from -0.1 to -1.2Wm^{-2}. The global annual mean direct effect estimated by the model was -0.48Wm^{-2}. International panel for climate change (IPCC) from the forward global modeling approach concluded that the mean aerosol direct forcing value is -0.5±0.4 Wm^{-2} (IPCC, 2007). The independent contribution by different aerosol species in IPCC AR4 (2007) are, as follows, sulfate -0.4±0.2 Wm^{-2}; fossil fuel organic carbon -0.05± 0.05 Wm^{-2} ; fossil fuel black carbon 0.2± 0.15 Wm^{-2} ; biomass burning 0.03± 0.12 Wm^{-2}; Nitrate -1±0.1 Wm^{-2} ; mineral dust -1±0.2 Wm^{-2}.

The aerosol model, 'Spectral Radiation-Transport Model for Aerosol Species (SPRINTARS)' (Takemura et al. 2002) has been now made available as an online tool for getting simulated global aerosol raditive forcing in daily as well as in monthly basis. Figure 1 shows global distribution of clear sky aerosol direct raditive forcing in different months representative of different seasons viz. March (spring), June (summer), October (Autumn) and December (winter)during 2010 simulated by SPRINTARS. The model also has been used as a forecast model (Takemura et al. 2005) for aerosol raiative effects as well as air pollution transport studies. The details of model are available in, http://sprintars.riam.kyushu-u.ac.jp/indexe.html

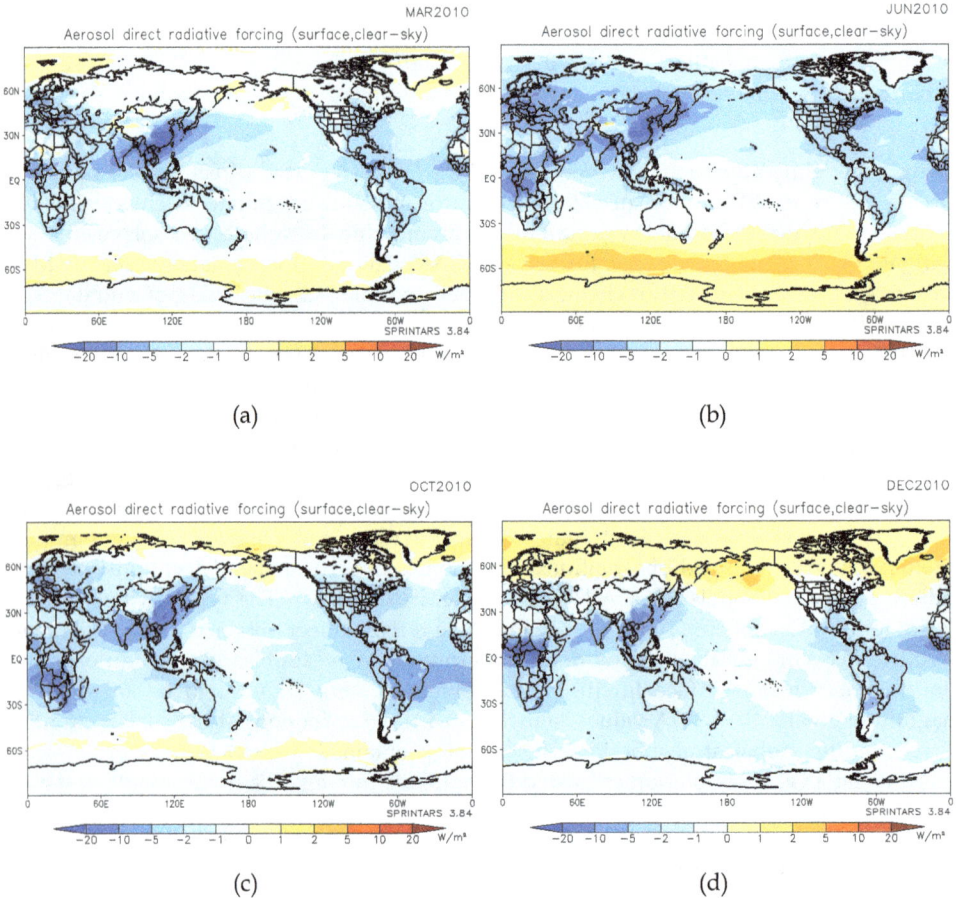

(a)   (b)

(c)   (d)

Fig. 1. Global aerosol clear sky direct radiative forcing distribution simulated by SPRINTARS in representative months during 2010 (a) March (Spring); (b) June (Summer); (c) October (Autumn); (d) December(winter).

Several modeling studies report the aerosol regional forcing over entire Europe. Based on a GCM (Global Circulation Model), Haywood and Ramaswamy (1998) estimated the radiative forcing for an external mixture of sulfate and black carbon (BC) aerosol to -5Wm^{-2} for Europe. Chung et al. (2005) reported the anthropogenic aerosol radiative forcing effect averaged over the year for clear-skies solely for Europe found to to lie between -2 and -

4Wm^{-2}. Based on a GCM in combination with a chemical aerosol model, Takemura et al (2002) specified the TOA aerosol radiative forcing effect to between -5Wm^{-2} and -10Wm^{-2} over the continent in summer months. Marmer et al. (2007) used a regional chemistry transport model REMOTE (Langmann, 2000) to analyze the seasonal variation of aerosol forcing over the entire European continent. Positive top-of-the-atmosphere forcing was simulated over eastern and southeastern Europe in spring and winter attributed to the contribution of black carbon. Its strength varies from +0.2 to + 1 W m^{-2}, depending on aerosol mixing assumptions. Sensitivity studies shows a mean European direct forcing of -0.3 W m^{-2} in winter and -2.5 W m^{-2} in summer, regionally ranging from -5 to + 4 W m^{-2}.

Using Spectral Radiation-Transport Model for Aerosol Species (SPRINTARS), Takemura et al. (2003) simulated the DARF over Asia pacific region. On comparison, the SPRINTARS model showed promising results with the observed aerosol optical data sets. It showed a forcing of over -10 W m^{-2} at the tropopause in the air mass during the large-scale dust storm, to which both anthropogenic aerosols and Asian dust contribute almost equivalently. This study emphasizes the enhanced presence of strong absorbing aerosols over East Asian region. It suggests that, the enhanced presence of black carbon and soil dust aerosols, which absorb solar and thermal radiation, make strong negative radiative forcing by the direct effect at the surface, which can exceed the positive forcing by anthropogenic greenhouse gases over the East Asian region.

Dust strom outbreaks over Asia are a significant phenomenon both in pollution and climate perspectives. Park and Jeong (2008) simulated direct radiative forcing over Asia during dust storm events in 2002 using Asian dust aerosol model (ADAM) and a column radiation model. It is found that, the direct radiative forcing contributed by the Asian dust aerosol is about 22% of the mean radiative forcing at the surface (-6.8Wm^{-2}), about 31% at the top of atmosphere (-2.9Wm^{-2}) and about 13% in the atmosphere (3.8Wm^{-2}), suggesting relatively inefficient contribution of the Asian dust aerosol on the direct radiative forcing contrary to as anticipated. The aerosol direct radiative forcing at the surface is mainly contributed by the mixed type aerosol (30%) and the Secondary inorganic aerosol (SIA) (25%) while at the top of atmosphere it is mainly contributed by the SIA aerosol (43%) and the Asian dust aerosol (31%) with positively (warming) contributed by BC and mixed type aerosols. The atmosphere is warmed mainly by the mixed type aerosol (55%) and the BC aerosol (26%). Chung et al. (2010) using a modeling study using assimilation of avaialbe observed data sets showed that the all-sky aerosol forcing over Asia (60– 138E and Eq. −45 N) is −1.3Wm^{-2} (TOA), +7.3Wm^{-2} (atmosphere) and −8.6Wm−2 (surface). Zhang et al. (2009) used a modeling approach to study the effect of carbonaceous aerosol in south Asia. They concluded that the carbonaceous aerosol induced radiative forcing can cause the surface temperature to increase in southern china and india, and decreased the total cloud cover and precipitation over this area. However, the opposite effects are caused for most of northern China and Bangladesh. Carbonaceous aerosol was proposed to induce summer precipitation to decrease in southern China but increase in northern China by radiative forcing.

Very few modeling studies reports exclusive aerosol radiative forcing estimates across global oceans. One of such study by Bates et al. (2006) investigated DARF over three regions downwind of major urban/population centers over North Indian Ocean (NIO), the Northwest Pacific Ocean (NWP), and the Northwest Atlantic Ocean (NWA) using chemistry transport models in conjunction with radiative transfer models. The resulting constrained clear-sky TOA anthropogenic aerosol forcng was −3.3±0.47, −14±2.6, −6.4±2.1Wm^{-2} for the NIO, NWP, and NWA, respectively. Malavelle et al.(2011) utilized RegCM3 model to assess optical properties and clear-sky direct radiative forcing (DRF) of mineral dust and carbonaceous aerosols over West Africa. The DARF calculations were found to be extremely sensitive to aerosol optical properties and underlying surface albedo. Over dark surfaces, the sum of shortwave (SW) and longwave (LW) top of the atmosphere (TOA) direct radiative forcing averaged found to −5.25 to −4.0 Wm-2 while over bright surfaces it was close to zero (−0.15 Wm-2). They concluded that, large differences between SW surface and SW TOA direct radiative forcing indicates that SW absorption had an important influence on the radiative budget. The SW radiative heating rate associated with the aerosol reached 1.2 K/d at local noon (diurnal mean of 0.40 K/d) at surface levels and it showed peak values in high altitudes.

Few studies report the influence of aerosol forcing in snow cover. A typical study by Kim et al. (2006) employed the Mesoscale Atmospheric Simulation (MAS) regional climate model (Soong and Kim, 1996) to investigate the impact of direct aerosol radiative forcing on surface insolation and snowmelt in the southern Sierra Nevada Mountains in United States. They found interesting results on interaction of aerosol forcing on snow melt by modulating surface temperature. With a prescribed aerosol optical thickness of 0.2, it is found that direct aerosol radiative forcing influences spring snowmelt primarily by reducing surface insolation and that these forcing on surface insolation and snowmelt vary strongly following terrain elevation. The direct aerosol radiative forcing on snowmelt is notable only in high altitudes and is primarily via the reduction in the surface insolation by aerosols. The effect of this forcing on low-level air temperature is as large as -0.3°C, but its impact on snowmelt is small because the sensible heat flux change is much smaller than the insolation change. The direct aerosol radiative forcing on snowmelt was found significant only when low-level temperature is near the freezing point, between -3° and 5°C. The elevation dependency of the direct aerosol radiative forcing on snowmelt is claimed to be related with this low-level temperature effect as the occurrence of the favored temperature range is most frequent in high elevation regions. Takemura et al. (2009) using simulations by aerosol model SPRINTARS analyzed the radiative forcing effects of soil dust aerosols at the Last Galcial Maxima (LGM) period and found that the direct radiative forcings of soil dust aerosols at the LGM was close to zero at the tropopause and −0.4Wm-2 at the surface. These radiative forcings are about twice as large as those in the present climate, attributed to higher dust flux during LGM period due to extended arid regions and wind speed. It is suggested that atmospheric dust might contributed to the cold climate during the glacial periods.

Inter comparison experiments of modeling results are necessary to find the causes in creating discrepancy in forcing estimates. Recently Rind et al. (2009) carried out an inter comparison of most popular three dimensional models used for aerosol direct forcing estimates and concluded that the difference among models in the direct radiative forcing of sulfate aerosols are primarily associated with the loading, different size distribution of sulfate aerosols and different relative humidity influences. Ruti et al. (2011) carried out a detailed inter comparison of models used for African Monsoon Multidisciplinary Analysis (AMMA) campaign, deploying the inter comparison of chemistry transport and chemistry climate models. Storelvmo et al. (2009) suggested that the discrepancy of 2 Wm⁻² spread in present day aerosol short wave forcing in coupled global atmospheric-ocean models used in IPCC AR4 (IPCC, 2007) is attributed to the different methods used to calculate cloud droplet number concentration (CDNC) from aerosol mass concentrations.

5. Aerosol indirect effect: Observational and modeling perspective

Aerosols influence the climate indirectly by altering the cloud microphysics, known as Aerosol indirect effect (AIE). There are different types of AIE have been proposed (Twomey, 1974; 1977; Albrecht, 1989; Kaufman and Fraser, 1997). The first aerosol indirect effect also known as the cloud albedo effect and the Twomey effect (Twomey,1974; 1977), is the reduction of solar radiation reaching the surface due to an increase in cloud albedo. If cloud's liquid water content (LWC) is kept constant and cloud condensation nuclei (CCN) concentration is increased via an increase in aerosol concentration, there will be a corresponding increase in cloud droplet concentration and a decrease in the effective radius of the droplets. This change results in an increase in the cloud's surface area and in turn, an increase in the cloud's albedo. The overall impact is a cooling effect as less solar radiation reaches the surface. The second indirect effect, also known as the cloud lifetime effect (Albrecht, 1989), occurs due to a reduction in precipitation efficiency as a result of increased CCN. As in the first indirect effect, given a constant LWC and an increase in CCN, a cloud will have a higher concentration of cloud droplets with a smaller effective radius. Due to this decrease in cloud droplet size, the development of precipitation will be retarded and may be suppressed altogether (Panicker et al. 2010a). Semi-direct effect (Hansen et al., 1997; Kaufman and Fraser, 1997) involves the absorption of solar radiation by aerosols such as black carbon and dust. In the presence of clouds, these aerosols absorb shortwave radiation at cloud top, causing a local warming. The absorption increases the temperature, thus lowering the relative humidity and producing evaporation, hence a reduction in cloud liquid water and hence results in the early dissipation of clouds.

Since after reporting the first aerosol indirect effect (Twomey , 1974; 1977), several attempts has been made to measure the indirect effect. One of the promising ground based observational estimation of AIE was reported by Feingold et al. (2003). They used remote sensing data sets of sub cloud Raman lidar aerosol extinction and cloud droplet effective radius to constrain AIE using the formula

$$AIE = -\frac{d\ln r_e}{d\ln \tau_a}$$

Where τ_a is aerosol proxy and r_e is the cloud effective radius for fixed LWC. They Obtained AIE values between 0.07 and 0.11 over the ARM site for liquid water paths between 100 and 130 gm⁻². Based on remote sensing measurements Pandithurai et al. (2009) estimated an indirect effect on thin non precipitating cirrus clouds over East China Sea region. The indirect effect estimates are made for both droplet effective radius different liquid water path ranges and they range 0.02–0.18. Generally indirect effect values found to be are positive. However negative AIE also has been reportd over some parts of the world at certain environmental conditions (eg: Yuan et al., 2008). McComiskey and Feingold (2009) proposed a method for estimating changes in radiatve forcing values from inferred AIE values by using a one dimension model (SBDART). Depending on anthropogenic aerosol perturbation, radiative forcing ranged from -3 to -10 W m^{-2} for each 0.05 increment in AIE, hence narrowing uncertainty in measures of AIE to an accuracy of 0.05.

A detailed review of AIE was presented by Lohman and Feichter (2005), in which they in depth discussed the Aerosol effects on different types of clouds and their magnitudes over land and oceans. Aerosol indirect effects are estimated from general circulation models (GCMs) by conducting a present-day simulation and a pre-industrial simulation in which the anthropogenic emissions are set to zero. The difference in the top-of-the atmosphere radiation budget of these multi-year simulations is then taken to be the anthropogenic indirect aerosol effect (Lohmann and Feichter 2005). The global radiative forcing due to AIE ranges between -0.5 to -1.9 Wm^{-2} by Twomey effect both at TOA and surface. However the second indirect values range between -0.3 to -1.4 Wm^{-2} at surface and TOA. Semi-direct effect values are found to range between -0.5 to +1 Wm^{-2} globally both at surface and TOA. Different simulations over land and ocenic areas (Menon et al. 2002; Quas et al. 2004., Takemura et al. 2005) confirmed a dominant AIE over land areas, attributed to enhanced anthropogenic aerosol loading. It is also found that the aerosol indirect effect is dominant over northern hemisphere (almost double) compared to southern hemisphere essentially due to large human inhabited land mass and associated anthropogenic aerosol emissions. From comparisons of different simulations, it is observed that, the magnitude of aerosol first indirect effect (Twomey effect) is larger than that of precipitation effect. There is a large discrepancy among modeling results of AIE. Lohmann and Feichter (2005) suggested that the main reason for this discrepancy between models could be associated with the dependence of the indirect aerosol effect on the background aerosol concentration. The latest estimates of IPCC AR4 reports global annual radiative forcing of the first indirect effect is -0.7 Wm$_{-2}$ with an uncertainty range of -1.8 to -0.3 Wm^{-2} (IPCC, 2007).

Dispersion effect, describing the spread of cloud droplet number distribution and aerosol number density as defined by Liu and Daum (2002) found to offset the cooling of the AIE or Twomey effect by 10-80%. Liu et al (2008) demonstrated that the first aerosol indirect effect (AIE) is the algebraic sum of the conventional Twomey effect and dispersion effect and that dispersion effect is proportional to Twomey effect in magnitude. However the factors that determine the dispersion effect is poorly understood. A better understanding of the dispersion effect may improve the aerosol cloud interaction in Global Climate Models (GCMS), which in turn reduces the uncertainty in AIE estimates.

Lohmann et al. (2007) discussed a detailed review of different approaches for constraining anthropogenic aerosol influence on indirect effect in global models. They in depth discussed the methods to derive aerosol parameterization from theoretical principles and also from observations. SPRINTARS model (Takemura et al. 2002) has been one of the active tools for estimating the global distribution of Aerosol Indirect forcing. Figure 2 shows the latest global aerosol surface indirect forcing during 2010 in various parts of the globe during representative months in spring, summer, autumn and winter.

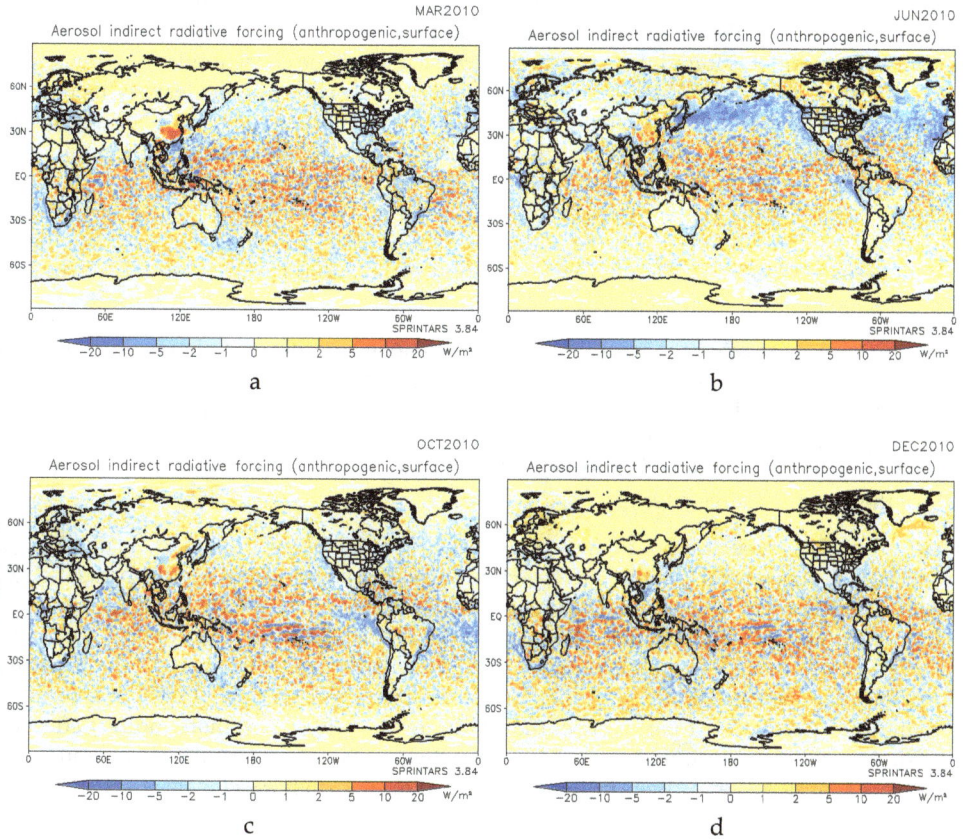

Fig. 2. Global aerosol clear sky indirect radiative forcing distribution simulated by SPRINTARS in representative months during 2010 (a) March (Spring); (b) June (Summer);(c) October (Autumn); (d) December(winter).

Myhre et al. (2007) used MODIS satellite data and two global models (Oslo CTM2 and CAMOslo) to investigate the relationship between AOD and cloud parameters. They found that there was an enhancement in cloud cover with increase in AOD. On analyzing the Angstrom exponent, it is concluded that hygroscopic growth is not likely to be a main contributor to the cloud cover enhancement; hence it could be attributed to the cloud life time effect. Strolvemo et al. (2006) compared the observed and modeled AIE values using MODIS satellite data and CAM-Oslo model over15 selected regions. They observed a large bias in Modeled and observationally estimated AIE values in both regional and global scales. Chylek et al. (2006) analyzed the AIE over Indian Ocean during pollution episodes. On contrary to the finding by Rosenfeld(1999), it is found that during pollution episodes, the radius of Ice crystals were shifted toward larger size than smaller size. ECHAM-4 modeling results of this study also reproduced an increase in ice crystal size. It is reasoned that this enhancement in ice crystal size could be associated with an"inverse aerosol effect', which need to be further investigated.

Recently Kurten et al. (2011) facilitated independent estimates for how the changes in the oxidant predicted by the TOMCAT chemical transport model (Chipper field, 2005) change the aerosol forcing. It is found that, a 10-fold increase in methane concentrations is predicted to significantly decrease hydroxyl radical (OH) concentrations, while moderately increases ozone (O3) concentration. These changes lead to a 70% increase in the atmospheric lifetime of methane, and an 18% decrease in global mean cloud droplet number concentrations (CDNC), and hence inducing an aerosol indirect effect. This CDNC change causes a radiative forcing that is comparable in magnitude to the longwave radiative forcing ("enhanced greenhouse effect") of the added methane. Hence it is suggested that, together, the indirect CH4-O3 and CH4-OH-aerosol forcing could more than double the warming effect of large methane increase.

Willox (2011) investigated the semi direct effect of smoke aerosols over clouds. By a model simulation experiment the negative radiative forcing associated with this semi-direct effect of smoke over clouds is estimated to be $-5.9\pm3.5 Wm^{-2}$. The average positive direct radiative forcing by smoke over an overcast scene is $9.2\pm6.6 Wm^{-2}$. Therefore it is suggested that the cooling associated with the semi-direct cloud thickening effect compensates for greater than 60% of the direct radiative effect.

Apart from the three important indirect effects reported above, Lohman et al. (2002) proposed a " glaciations indirect effect'. Here increases in ice nuclei in the present-day climate result in more frequent glaciation of supercooled clouds and increase the amount of precipitation via the ice phase. This reduces the cloud cover and the cloud optical depth of mid-level clouds in mid- and high latitudes of the Northern Hemisphere and results in more absorption of solar radiation within the Earth-atmosphere system. Therefore it is assumed that, this effect can at least partly offset the cloud lifetime effect. Two more indirect effects, viz. Riming indirect effect, and thermodynamic effects has been proposed (Lohmann and Feitcher. 2005). However further studies are required to confirm these climate effects.

6. Influence of aerosol radiative forcing on monsoon

Aerosols in the early 20th century have been considered only as particles inducing air pollution and detericious health effects. Later in mid 1960s it is revelaed that they influances the global climate by counter acting to global warming by enhancing global dimming. However recent studies suggest that aerosol contribution to the climate is more than what it was anticipated. Aerosols are found to influence the large scale phenomenon like monsoon by inducing atmospheric heating. Especially in this aspect, the influence of strongly absorbing species of aerosols such as black carbon in much debated. Studies using different general circulation models (GCMs) indicate that direct radiative forcing (DRF) of absorbing black carbon (BC) aerosols can influence the Indian summer monsoon.

Ramanathan et al. (2005) used the PCM ocean-Atmosphere coupled model to simulate the influence of Black carbon (BC) induced atmospheric brown cloud on monsoon. They found that an increase in the BC induced DRF over Indian Subcontinent and surrounding regions leads to a reduction of monsoon precipitation during June to September months while an enhancement to the pre-monsoon (March–April–May) precipitation. The study of Meehl et al. (2008) with a coupled model having more comprehensive treatment of aerosol-radiation interaction, supports the above findings. Collier and Zang. (2009) by a

GCM experiment showed that the, reduced shortwave aerosol heating and enhanced evaporation at the surface during April and May results in weakening of the near-surface cyclonic circulation and, consequently, has a negative feedback on precipitation during the active monsoon months of June and July. Menon et al. (2002) using a GCM simulation, showed that black carbon induced atmospheric warming changes the precipitation patterns over India and China. It is found that an atmosphere with enhanced BC, yield increased precipitation in southern China and over India and Myanmar where AOD was largest. There was a broad band of decreased precipitation to the south of the region with increased precipitation, with a lesser decrease to the north. Ji et al. (2011) using a coupled RCM-Chemistry aerosol model, showed that, in northeast India and Myanmar, aerosols lead the summer monsoon onset advancing 1–2 pentads, and delaying by 1–2 pentads in central and southeast India.

Lau et al. (2006) proposed one of the most debated mechanisms of aerosols influence on Indian monsoon. The hypothesis namely Elevated heat pump mechanism (EHP) was constrained with the NASA finite volume atmospheric general circulation Model simulation experiments over Indian region. They found that absorbing aerosols such as black carbon and dust, which accumulates over the foothills of Himalaya in Indo-Gangetic Plain (IGP) during pre-monsoon months (April-May) absorbs the solar radiation and heats the atmosphere. As the air warms, it rises over the Tibetan plateau; it draws in more moist air from Indian Ocean. Hence more moist air is drawn toward foothills of himalaya, producing anomalous rainfall. The increased condensation causes more upper troposheric heating, which draws in more low-level moisture from ocean and hence maintains a positive feedback namely heat pump. Hence according to this hypothesis, monsoon precipitation would be suppressed over central India due to aerosol-induced surface cooling. However, precipitation would come earlier and be enhanced over northern India and the southern slope of the Tibetan Plateau. Several modeling investigators supports this hypothesis adopts this idea in their research (e.g. Huang et al., 2007;;Meehl et al., (2008); Ramanathan and Carmichael, 2008; Randles and Ramaswamy, 2008). However debates have been enduring on this aspect. Bollasina et al. (2008) compared different studies of aerosol impact on the Asian Summer Monsoon and noticed that those using coupled ocean-atmosphere models yielded opposing results to those proposed by Lau et al. (2006). Kuhlmann and Quass. (2010) suggested three reasons for this discrepancy viz. simulation of Lau et al. (2006) was lacking interaction of atmospheric processes with the ocean, ignoring aerosol indirect effect and finally not considering vertical aerosol profiles in model. Kuhlmann and Quass. (2010) using CALIPSO satellite data of aerosol vertical profiles in conjunction with a radiative transfer model investigated the influence of aerosols in monsoon. They found that aerosol plumes reduce shortwave radiation throughout the Monsoon region in the seasonal average by between 20 and 30 Wm^{-2}. Peak shortwave heating in the lower troposphere reaches 0.2 K/day. In higher layers this shortwave heating is partly balanced by longwave cooling. Although high-albedo surfaces, such as deserts or the Tibetan Plateau, increase the shortwave heating by around 10%, the overall effect is strongest close to the aerosol sources. The simulated SW heating above the Tibetan Plateau does not exceed 0.05 K/day in the seasonal mean and is thus considerably weaker than in the surrounding regions. This result stands in contrast to the Elevated Heat Pump (EHP) hypothesis by Lau et al. (2006) who simulated SW heating of between 0.2 and 0.4 K/day above the Tibetan Plateau. Nigam and Bollasina (2010) confronted the observational feasibility of EHP hypothesis, and in response

Lau and Kim.(2011) defended their hypothesis by challenging the methodology of analysis prescribed in Nigam and Bollasina(2010).

Wang et al. (2009) explained the northward propagation of monsoon convective systems by coupling dynamic effects induced by aerosol heating. They used a community climate model (CCM3) to investigate the influence of absorbing aerosols on monsoon. The experiment was conducted for three different simulations for monsoon circulations viz. including only scattering aerosols, only absorbing aerosols, and both aerosols in the model. They found that, among different types of anthropogenic aerosols, scattering aerosols found to have only a very limited impact on monsoon circulation and precipitation. On the other hand, absorbing aerosols, with or without the co-existence of scattering aerosols, have a strong influence on the monsoon circulation and in the development of convective precipitation. They proposed that the influence of absorbing aerosols is reflected in a perturbation to the sub-cloud layer moist energy structure, initiated with a heating by absorbing aerosols of the planetary boundary layer, mostly over the land areas north of the Arabian Sea and also along the south slope of the Tibetan Plateau. This is then enhanced by the import of airmass with high water vapor concentration. The corresponding anomalous sub-cloud layer airflow that brings relatively humid air primarily originated from the Arabian Sea and hence the convective precipitation experiences a clear northward shift. Manoj et al. (2010) using a raditive transfer model (SBDART) in conjunction with MODIS/TOMS satellite data analyzed the influence of absorbing aerosols in modulating Indian summer monsoon break active cycles. It is shown those monsoon years, in which intense loading of absorbing aerosols were present over central India was able to heat the atmosphere over central India, creating enough pressure gradient between northern Indian Ocean and central India and could sustain the active condition after breaks. Whereas the monsoon years where less absorbing aerosols were present over central India, couldn't succeed in sustaining active spells after breaks.

Apart from these direct effects, aerosols also found to influence the Indian monsoon through indirect effect. Ravi kiran et al. (2009), Bhawar and Devara(2010), based on MODIS data qualitatively suggested that, there is a strong influence for aeorol- cloud interaction in influencing Indian summer monsoon. Patra et al. (2005) using MODIS satellite data in conjunction with a chemistry transport model (CTM) showed that prevailing monsoon dynamics in good and bad monsoon bringing air mass containg absorbing/ non absorbing aerosols can induce indirect effect and hence can influence precipitation. Bollasina et al. (2008) proposed that, Anomalous heating of the land surface by aerosol induced reduction in cloudiness (the "semidirect" effect) can increase in downward surface shortwave radiation. Stronger heating of the land surface in the month of May generates greater ocean atmosphere contrast and thus provides more monsoon rainfall in June. Panicker et al. (2010b) using MODIS data estimated AIE and found that there was a majority positive AIE (Twomey effect) in fixed Liquid and ice path bins during bad monsoons and a majority negative AIE (Anti-Twomey) effect in fixed cloud ice and water path bins during good monsoons, influencing the precipitation patterns. However more studies and more remote sensing observations are necessary to establish these results. An exclusive five year campaign Cloud aerosol interaction and precipitation enhancement experiment (CAIPEEX) based on aircraft observations of aerosol cloud parameters has been in operation over different Indian regions since 2009 to unravel the complex aerosol cloud interaction process on Indian monsoon.

7. Conclusions

Aerosols, which are tiny particles, present in the atmosphere, influence the climate through radiative forcing. Aerosol radiative forcing and associated surface cooling is found to compensate the green house gas warming in the past century. This chapter describes a brief review of aerosol radiative forcing estimates using modeling approaches. Aerosol models vary from one dimension to three dimensions. The one dimensional modeling studies on aerosol direct effects have been accelerated in the past two decades after the establishement of global aerosol networks across the world. Direct Aerosol radiative forcing has been estimated using ground based instruments in conjunction with radiative transfer models across maritime and continental environments in different parts of the globe. Several modeling studies reports the aerosol forcing estimates in regional as well as in global scales, both for combined fraction of aerosols and independently for different species. The latest estimate of aerosol radiative forcing by International panel for climate change (IPCC, AR4), reports a global mean aerosol direct forcing value of -0.5±0.4 Wm^{-2}. The independent contribution reported by different aerosol species in IPCC AR4 are, as follows, sulfate - 0.4±0.2 Wm^{-2}; fossil fuel organic carbon -0.05± 0.05 Wm^{-2} ; fossil fuel black carbon 0.2± 0.15 Wm^{-2} ; biomass burning 0.03± 0.12 Wm^{-2}; Nitrate -1±0.1 Wm^{-2} ; mineral dust -1±0.2 Wm^{-2}. Global aerosol indirect effect estimates have still been a challenge, as the uncertainty levels are high. Latest estimate of the global annual radiative forcing of the first indirect effect by IPCC AR4 is -0.7 Wm^{-2} with an uncertainty range of -1.8 to -0.3 Wm^{-2}. Out of the main identified indirect effects, first and second indirect effects exert a negative forcing at surface and TOA. However, semi direct effect exerts a positive forcing at TOA and negative forcing at surface. Global modeling studies also have given a new insight in revealing the role of aerosols in modulating large scale phenomenon, such as monsoon. Different modeling studies confirmed that aerosol direct forcing is a key factor modulating onset, break phases of Indian summer monsoon. Aerosols are proposed to increase or decrease monsoon rainfall depending up on its species of origin and its amount of presence during the season. Several mechanisms have been proposed regarding the influence of direct aerosol radiative forcing on Indian summer monsoon. However still the uncertainty in aerosol forcing estimates, especially indirect effect and its possible influence on different large scale phenomenon remains un answered. Hence assimilation of real time data sets of aerosol optical properties and aerosol vertical profiles should be included in global models to overcome this uncertainty.

8. Acknowledgements

This work is supported by National Research Foundation of Korea (NRF) through a grant provided by the Korean Ministry of Education, Science & Technology (MEST) in 2011 (No. K20607010000). The SPRINATRS archive obtained from http://sprintars.riam.kyushu-u.ac.jp/indexe.html is acknowledged with thanks.

9. References

Albrecht, B. A. (1989), Aerosols Cloud Microphysics, and Fractional Cloudiness, *Science*, 245, 1227– 1230.

Bates T.S et al. (2006), Aerosol direct radiative effects over the northwest Atlantic, northwest Pacific, and North Indian Oceans: estimates based on in-situ chemical and optical measurements and chemical transport modeling, Atmos. Chem. Phys., 6, 1657–1732

Berk, A., Bernstein, L. S., Anderson, G. P., Acharya, P. K., Robertson, D. C.,Chetwynd, J. H., Adler-Golden, S. M. (1998). "MODTRAN cloud and multiple scattering upgrades with application to AVIRIS". *Remote Sensing of Environment* (Elsevier) 65 (3,): 367–375.

Bhawar, R. L., Devara, P. C. S. (2010), Study of successive contrasting monsoons (2001–2002) in terms of aerosol variability over a tropical station Pune, India, Atmos. Chem. Phys., 10, 29–37.

Bollasina, M., Nigam, S., and Lau, K. M (2008), Absorbing Aerosols and Summer Monsoon Evolution over South Asia: An Observational Portrayal, *J. Climate*, 21, 3221–3239.

Charlson, R. J., S. E. Schwartz, J. M. Hales, R. D. Cess, J.A. Coakley, J. E. Hansen, and D. J. Holmann(1992), Climate forcing by anthropogenic aerosols, *Science*, 255, 423-430.

Chin, M., Ginoux, P., Lucchesi, R. , Huebert, B., Weber, R., Anderson, T., Masonis, S., Blomquist, B., Bandy, A., Thornton, D. (2003), A global aerosol model forecast for the ACE-Asia field experiment, *J. Geophys. Res.*, 108(D23), 8654, doi:10.1029/2003JD003642.

Chipperfield, M.: New version of the TOMCAT/SLIMCAT off-line chemical transport model: intercomparison of stratospheric tracer experiments (2006), *Q. J. Roy. Meteorol. Soc.* 132, 1179– 1203.

Chung CE, Ramanathan V, Kim D, Podgorny IA (2005), Global anthropogenic aerosol direct forcing derived from satellite and ground-based observations. *J Geophys Res* 110: D24207 (DOI: 10.10292005JD00006356).

Chung, C. E., Ramanathan, V., Carmichael, G., Kulkarni, S., Tang, Y., Adhikary, B., Leung, L. R., Qian, Y. (2010),Anthropogenic aerosol radiative forcing in Asia derived from regional models with atmospheric and aerosol data assimilation, *Atmos. Chem. Phys.*, 10, 6007–6024.

Chylek, P., Dubey, M. K., Lesins, G. (2006), Greenland warming of 1920–1930 and 1995–2005, *Geophys. Res. Lett.*, 33, L11707, doi:10.1029/2006GL026510.

Clough, S. A., Shephard, M. W., Mlawer, E. J., Delamere, J. S., Iacono, M. J., Cady-Pereira, K., Boukabara, S., Brown, P. D. (2005). "Atmospheric radiative transfer modeling: a summary of the AER codes", *J. Quant. Spectrosc. Radiat. Transfer* 91: 233–244.

Collier, J. C. and Zhang, G. J. (2009), Aerosol direct forcing of the summer Indian monsoon as Simulated by the NCAR CAM3, *Clim. Dynamics.*, 32, 313–332.

Conant, W. C. (2000), An observational approach for determining aerosol surface radiative forcing: Results from the first phase of INDOEX, *J. Geophys. Res.*, *105*, (D12), 15347-15360.

d'Almedia GA, Koepke P, Shettle EP (1991) Atmospheric aerosols: global climatology and radiative characteristics. Deepak Publishing, Hampton.

Deepak, A and. Gerbers, H. E ,eds., "Report of the experts' meeting on aerosols and their climatic effects, (1983)" WCP-55 World Climate Research Program, Geneva.

Dubuisson, P., Buriez, J.C., Fouquart, Y. (1996), High spectral resolution solar radiative transfer in absorbing and scattering media: application to the satellite simulation, *Journal of Quantitative Spectroscopy & Radiative Transfer* 55, 103–126.

Dubuisson, P., Dessailly, D., Vesperini,M., Frouin, R. (2004),Water vapor retrieval over ocean using near-infrared radiometry, *Journal of Geophysical Research* 109, D19106. doi:10.1029/2004JD004516.

Dufresne, J. L., Gautier, C., Ricchiazzi, P. (2002), Long wave scattering effects of mineral aerosols, *J. Atmos Sci.*, 59, 1959-1966.

Engelsen, O. and Kylling, A. (2005), Fast simulation tool for ultraviolet radiation at the Earth's surface, *Optical Engineering*, 44(4), 041012.

Fenn, R. W. (1964), Aerosol-Verteilungen und atmospharisches Streulicht, *Beitr. Phys. Atmos.* 37, 69-104.

Fu, Q., Liou, K-N. (1993). Parameterization of the radiative properties of cirrus clouds, *Journal of the Atmospheric   Sciences* 50, 2008-2025.

Gadhavi, H., Jayaraman, A.,(2010). Absorbing aerosols: contribution of biomass burning and implications for radiative forcing, Ann. Geophys., 28, 103-111

Gonzi, S., Dubovik, O., Baumgartner, D., Putz, E., (2007), Clear-sky aerosol radiative forcing effects based on multi-site AERONET observations over Europe, *Meteorology and Atmospheric Physics* 96, 277-291. doi:10.1007/s00703-006- 0212-9.

Guzzi, R., Ballista, G., Nicolantonio,W.D., Carboni,E., (2001), Aerosol maps from GOME data, *Atmospheric Environment* 35 5079-5091.

Hansen, J., Sato, M., Ruedy, R. (1997), Radiative forcing and climate response. *J. Geophys. Res.* 102, 6831-6864.

Haywood, J. M., Ramaswamy, V. (1998), Global sensitivity studies of the direct radiative forcing due to anthropogenic sulfate and black carbon aerosols. *J. Geophys. Res.*, 103, 6043-6058.

Haywood, J., and O. Boucher (2000), Estimates of the direct and indirect radiative forcing due to tropospheric aerosols: A review, Rev. Geophys., 38, 513- 543.

Hess, M., Koepke, P. and Schult, I (1998), Optical properties of aerosols and clouds: The software package OPAC, *Bull. Amer. Meteorol. Soc.*, 79, 831-844.

Hignett, P., J. P. Taylor, P. N. Francis, and M. D. Glew (1999), Comparison of observed and modeled direct aerosol forcing during TARFOX, *J. Geophys. Res.*, 104, 2279- 2287.

Huang, J., Minnis, P., Yi, Y., Tang, Q.,Wang, X., Hu, Y., Liu, Z., Ayers, K., Trepte, C., Winker, D.(2007), Summer dust aerosols detected from CALIPSO over the Tibetan Plateau, *Geophys. Res. Lett.*, 34, L18805, doi:10.1029/2007GL029938.

Huene, W,V,H., Yoon, J., Vountas, M., Istomina, L. G., Rohen, G. , Dinter, T. , Kokhanovsky, A. A., Burrows J. P.(2011), *Atmos. Meas. Tech.*, 4, 151-171.

IPCC: Climate change 2007 - The scientific basis, Contribution of working group I to the Fourth Assessment Report of the Intergovernmental Panel on Climate Change, Cambridge University Press, Cambridge, 2007.

Jayaraman, A., Lubin, D. Ramachandran, S. Ramanathan, V. Woodbridge, E. Collins, W. D. Zalpuri, K. S. (1998), Direct observations of aerosol radiative forcing over the tropical Indian Ocean during the January-February 1996 pre-INDOEX cruise, *J. Geophys. Res.*, 103 (D12), 13,827 -13,836.

Ji, A., Kang, S., Zhang, D., Zhu, C., Wu, J., Xu, Y. (2011), Simulation of the anthropogenic aerosols over South Asia and their effects on Indian summer monsoon, *Climate Dynamics* 36,1633-1647

Kaufman, Y. J., Fraser,R. S. (1997), The effect of smoke particles in clouds and climate forcing, *Science,* 277, 1636-1639.

Kiehl,J.T., Schneider, T.L., Rasch, P.J.,Barth, M.C .(2000), Radiative forcing due to sulfate aerosols from simulations with the National Center for Atmospheric Research Community Climate Model, Version 3, *J. Geophys. Research*, 105,1441-1457.

Kim,J., Yoon S.C., Kim, S.W., Brechtel, F., Jefferson,A., Dutton,E.G., Bower, K.N., Cliff,S., Schaue, J.S. (2006), Chemical apportionment of shortwave direct aerosol radiative forcing at the Gosan super-site, Korea during ACE-Asia, *Atmospheric Environment* 40 (2006) 6718–6729.

Koepke P, Hess M, Schult I, Shettle EP (1997) Global aerosol dataset. Max-Plank-Institut fu¨r Meteorologie Report 243, 44.

Kuhlmann, J., Quaas,J (2010), How can aerosols affect the Asian summer monsoon? Assessment during three consecutive pre-monsoon seasons from CALIPSO satellite data, *Atmos. Chem. Phys. Discuss.*, 10, 4887–4926.

Kurten, T. et al. (2011), Large methane releases lead to strong aerosol forcing and reduced cloudiness, *Atmos. Chem. Phys. Discuss.*, 11, 9057–9081, 2011.

Langmann, B. (2000), Numerical modelling of regional scale transport and photochemistry directly together with meteorological processes, *Atmos. Environ.*, 34, 3585–3598.

Lau, K. M., Kim, M. K., Kim, K. M. (2006), Asian monsoon anomalies induced by aerosol directteffects. *Climate dynamics* 26, 855-864, doi. 10.1007/ s00382-006-0114-z.

Lau, K. M., Kim, K. M. (2011), Comment on "'Elevated heat pump' hypothesis for the aerosol-monsoon hydroclimate link: 'Grounded' in observations?" by S. Nigam and M. Bollasina, *J. Geophys. Res.*, 116, D07203, doi:10.1029/2010JD014800.

Levoni, C., Cervino, M., Guzzi, R., Torricella,F (1997), Atmospheric aerosol optical properties: a database of radiative characteristics for different components and classes , *Applied optics*,36, 8031-8041.

Li, Z., Lee,K.H., Wang,Y., Xin,Y and HaoW ,M., (2010), First observation-based estimates of cloud-free aerosol radiative forcing across China, *J. Geophys. Res.*, 115, D00K18, doi:10.1029/2009JD013306.

Liu, Y., Daum, P. H. (2002), Indirect warming effect from dispersion forcing, Nature, 419, 580–581

Liu,Y., Daum,P.H., Guo, H., Peng, Y., Dispersion bias, dispersion effect, and the aerosol-cloud conundrum, Environ. Res. Lett. 3 (2008) 045021, doi: 10.1088/1748-9326/3/4/045021

Lohman,U and Feichter, J. (2005), Global indirect aerosol effects: A review. *Atmos. Chem. Phys.*, 5, 715–737.

Lohmann , U., Quass, J., Kinne, S., Feitcher, J. (2007), Different approaches for constraining GCMs of the anthropogenic indirect aerosol effect. BAMS 88: 243-249

Lohmann, U (2002), A glaciation indirect aerosol effect caused by soot aerosols, *Geophys. Res. Lett.*, 29, doi:10.1029/2001GL014 357.

Malavelle, F., Pont, V., Mallet, M., Solmon, F., Johnson, B., Leon, J.-F., and Liousse, C, (2011), Simulation of aerosol radiative effects over West Africa during DABEX and AMMA SOP-0, *J. Geophys. Res.*, 116, D08205, doi:10.1029/2010JD014829.

Manoj, M. G., Devara, P. C. S., Safai, P. D., Goswami, B. N. (2010), Absorbing aerosols facilitate transition of Indian monsoon breaks to active spells, *Climate Dynamics* DOI 10.1007/s00382-010-0971-3.

Marc, M., Chami, M. Gentili, B., Sempe´re´, R., Dubuisson. P. (2009), Impact of sea-surface dust radiative forcing on the oceanic primary production: A 1D modeling approach applied to the West African coastal waters, *Geophys. Res. Lett.*, 36, L15828, doi:10.1029/2009GL039053.

Markowicz K.M., Flatau, P.J., Ramana M.V., Crutzen P.J., Ramanathan, V (2002), Absorbing mediterranean aerosolslead to large reduction in the solar radiation at the surface. *Geophys Res Lett* 29(20) (DOI: 10.1029/ 2002GL015767)

Markowicz, K. M., Flatau, P. J., Remiszewska, J., Witek, M., Reid, E. A., Reid, J. S. , Bucholtz, A., Holben, B. (2008), Observations and modeling of the surface aerosol radiative forcing during the UAE2 experiment, *J. Atmos. Sci.*, 65, 2877–2891, doi:10.1175/2007JAS2555.1.

Marmer, E., Langmann, B., Hungersho"fer, K., Trautmann, T., (2007), Aerosol modeling over Europe: 2. Interannual variability of aerosol shortwave direct radiative forcing, *J. Geophys. Res.*, 112, D23S16, doi:10.1029/2006JD008040.

Mayer, B., Kylling, A. (2005). "Technical note: The libRadtran software package for radiative transfer calculations - description and examples of use". *Atmospheric Chemistry and Physics* 5: 1855–1877.

McClatchey, R. A., Fenn, R.W., Selby, J. E. A. et al. (1972), Optical properties of the atmosphere, 3rd ed., *Environ. Res. Pap.411*, 108 pp., Airforce Cambridge Res. Lab., Hanscom AFB, Mass.

McComiskey, A., Feingold, G. (2008), Quantifying error in the radiative forcing of the first aerosol indirect effect, *Geophys. Res. Lett.*, 35, L02810, doi: 10.1029/2007GL032667.

Meehl, G. A., Arblaster, J. M., Collins, W. D. (2008), Effects of Black Carbon Aerosols on the Indian Monsoon, J. Climate, 21, 2869–2882.

Menon, S., Hansen, J., Nazarenko, L., Luo, Y. (2002), Climate effects of black carbon aerosols in China and India. *Science*, 297, 2250–2253.

Myhre, G., Stordal, F., Johnsrud, M., Kaufman, Y. J., Rosenfeld, D., Storelvmo, T., Kristjansson, J. E., Berntsen, T. K., Myhre, A., and Isaksen, I. S. A.(2007), Aerosol-cloud interaction inferred from MODIS satellite data and global aerosol models, Atmos. Chem. Phys., 7, 3081–3101.

Nessler,R., Weingartner, E., Baltensperger., (2005) Effect of humidity on aerosol light absorption and its implications for extinction and the single scattering albedo illustrated for a site in the lower free troposphere, Journal of Aerosol Science, 36, 958-972.

Nigam, S., Bollasina, M. (2010), "Elevated heat pump" hypothesis for the aerosol -monsoon hydroclimate link: "Grounded" in observations?, J. Geophys. Res., 115, D16201, doi:10.1029/2009JD013800.

Pandithurai, G., Pinker, R. T., Takamura, T and Devara, P. C. S (2004), Aerosol radiative forcing over a tropical urban site in India, Geophys. Res. Lett., 31, L12107, doi:10.1029/2004GL019702

Pandithurai, G.,Takamura, T., Yamaguchi, J., Miyagi, K., Takano, T., Ishizaka, Y., Dipu, S., Shimizu, A. (2009), Aerosol effects on cloud droplet size as monitored from surface-based remote sensing over East China Sea region, *Geophy.Res. Lett.*, 36, *L10802* , doi: 10.1029/2009GL038451.

Panicker, A. S., Pandithurai, G., Dipu, S. (2010a), Aerosol indirect effect during Successive contrasting monsoon years over Indian sub continent: using MODIS data, *Atmospheric Environment*, 44, 1937-1943, doi:10.1016/j.atmosenv.2010.02.015.

Panicker, A. S., Pandithurai, G.. Safai, P. D., Kewat S. (2008), Observations of enhanced aerosol longwave radiative forcing over an urban environment, Geophys. Res. Lett., 35, L04817, doi:10.1029/ 2007GL032879.\

Panicker, A. S., Pandithurai,G., Safai, P. D., Dipu, S., Dong-In Lee. , (2010b). On the contribution of black carbon to composite aerosol radiative forcing over an urban environment, *Atmospheric Environment*, doi:10.1016/j.atmosenv.2010.04.047, 44, 3066- 3070.

Park,S.U., Jeong, J.I .(2008), Direct radiative forcing due to aerosols in Asia during March 2002 , Science of the total Environment 4 0 7, 3 9 4 – 4 0 4.

Patra, P. K., Behera, S.K., Herman, J. R., Maksyutov, S., Akimoto, H., Yamagata, T. (2005), The Indian summer monsoon rainfall: Interplay of coupled dynamics, radiation and cloud microphysics, *Atmos. Chem. Phys.*, 5, 2181-2188.

Quaas, J., Boucher,O., and Br´eon, F.-M. (2004), Aerosol indirect effects in POLDER satellite data and in the LMDZ GCM, J. Geophys. Res., 109, D08205, doi: 10.1029/2003JD004317.

Ramachandran, S., Rengarajan, R., Jayaraman, A., Sarin, M. M., Das, S. K. (2006), Aerosol radiative forcing during clear, hazy and foggy conditions over a continental polluted location in north India, *J .Geophys. Res.*, *111*, D20214, doi:10.1029/2006JD007142.

Ramanathan, V. and Carmichael, G. (2008), Global and regional climate changes due to black carbon, Nat. Geosci., 1, 221–227.

Ramanathan, V., Chung, C., Kim, D., Betge, T., Buja, L. Kiehl, J.T., Washington, W.M., Fu, Q., Sikka, D.R., Wild, M, (2005), Atmospheric Brown clouds: Impacts on south Asian climate and hydrological cycle, Proc. Natl.Acad.Sci. U.S.A., 102, 5326-5333, doi:10.1073/pnas.0500656102.

Ramanathan, V., Crutzen,, P.J., Kiehl, J.T., Rosenfeld, D (2001), Aerosols, climate, and the hydrological cycle, Science, 294, 2119– 2124.

Randles, C. A. and Ramaswamy, V.(2008), Absorbing aerosols over Asia: A Geophysical Fluid Dynamics Laboratory general circulation model sensitivity study of model response to aerosol optical depth and aerosol absorption, J. Geophys. Res., 113, D21203, doi:10.1029/2008JD010140.

Ravi Kiran, V., Rajeevan, M. Vijaya Bhaskara Rao S. and Prabhakara Rao, N. (2009),Analysis of variations of cloud and aerosol properties associated with active and break spells of Indian summer monsoon using MODIS data, *Geophys. Res. Lett.*, *36*, L09706, doi: 10.1029/2008GL037135.

Ricchiazzi, P., Yang, S., Gautier, C., and Sowle, D. (1998), SBDART: A research and teaching software tool for plane parallelradiative transfer in the Earth's atmosphere, *Bull. Amer. Meteorol. Soc.*, *79*, 2101-2114.

Rind, D., Chin, M., Feingold, G., Streets, D., Kahn, R.A., Schwartz, S.E., Yu, H.,(2009) Modeling the effects of aerosols on climate. Atmospheric Aerosol Properties and Impacts on Climate. A Report by the U.S. Climate Change Science Program and the Subcommittee on Global Change Research. National Aeronautics and Space Administration, Washington, D.C., USA.

Rosenfeld, D. (1999), TRMM observed first direct evidence of smoke from forest fires inhibiting rainfall, Geophys. Res. Lett., 26, 3105– 3108

Ruti, P. M., Williams, J. E., Hourdin, F., Guichard, F., Boone, A., Van Velthoven, P., Favot, F., Musat, I., Rummukainen, M., Domínguez, M., Gaertner, M. Á., Lafore, J. P., Losada, T., Rodriguez de Fonseca, M. B., Polcher, J., Giorgi, F., Xue, Y., Bouarar, I., Law, K., Josse, B., Barret, B., Yang, X., Mari, C. and Traore, A. K. (2011), The West African climate system: a review of the AMMA model inter-comparison initiatives. Atmospheric Science Letters, 12: 116–122. doi: 10.1002/asl.305

Saha, A., Mallet, M., Roger, J.C., Dubuisson, P., Piazzola, J., Despiau, S. (2008) One year measurements of aerosol optical properties over an urban coastal site: Effect on local direct radiative forcing Atmospheric Research 90, 195-202.

Satheesh, S. K. Ramanathan, V. Jones, X.L., Lobert, J.M., Podgorny, I.A., Prospero, J.M., Holben B.N., Loeb, N.G. (1999), A Model for Natural and Anthropogenic Aerosols over the Tropical Indian Ocean Derived from INDOEX data, J. Geophys. Res., 104, D22, doi: 10.1029/1999JD900478, 27,421-27,440.

Satheesh, S. K., and V. Ramanathan (2000), Large differences in the tropical aerosol forcing at the top of the atmosphere and Earth's surface, *Nature, 405,* 60-63.

Satheesh, S. K., Lubin, D. (2003), short - wave versus long -wave radiative forcing by Indian Ocean aerosols: Role of sea-surface winds, *Geophy. Res. Lett., 30(13),* 1695, doi: 10.1029/2003GL017499, 2003.

Shettle, E. P. , Fenn, R. W. (1079), "Models for the aerosol lower atmosphere and the effects of humidity variations on their optical properties," Rep. Tr-79-0214 U.S. Air Force Geophysics Laboratory, Hanscom Air Force Base, Mass.

Singh, S., Soni, K., Bano, T. , Tanwar, R. S., Nath, S., Arya, B. C.(2010),Clear-sky direct aerosol radiative forcing variations over mega-city Delhi, Ann. Geophys., 28, 1157–1166.

Smirnov, A., Holben, B. N., Dubovik, O., Frouin, R., Eck, T. F. and Slutsker, I. (2003), Maritime component in aerosol optical models derived from Aerosol Robotic Network data, J. Geophys. Res., 108(D1), 4033, doi:10.1029/2002JD002701.

Soong, S., and Kim, J.M. (1996), Simulation of a heavy winter time precipitation event in california, *Climate change,* 32, 55-77

Sreekanth, V., Niranjan, K., Madhavan, B. L. (2007), Radiative forcing of black carbon over eastern India, *Geophys. Res. Lett., 34, L17818,* doi: 10.1029/2007GL030377.

Stamnes, Knut. Tsay, S. C., Wiscombe, W., Jayaweera, K (1988). "Numerically stable algorithm for discrete-ordinate method radiative transfer in multiple scattering and emitting layered media". *Appl. Opt.* 27 (12): 2502–2509.

Storelvmo, T., Kristj´ansson, J. E., Myhre, G., Johnsrud, M., Stordal, F.(2006), Combined observational and modeling based study of the aerosol indirect effect, Atmos. Chem. Phys., 6, 3583–3601.

Storelvmo, T., Lohmann, U., Bennartz R. (2009), What governs the spread in shortwave forcings in the transient IPCC AR4 models?, Geophys. Res. Lett., 36, L01806, doi:10.1029/2008GL036069.

Takemura, T., Egashira, M., Matsuzawa, K., Ichijo, H., O'ishi, R., Abe-Ouchi, A. (2009), A simulation of the global distribution and radiative forcing of soil dust aerosols at the Last Glacial Maximum. *Atmospheric Chemistry and Physics,* 9, 3061-3073.

Takemura, T., Nakajima, T., Higurashi, A., Ohta, S., Sugimoto, N. (2003), Aerosol distributions and radiative forcing over the Asian-Pacific region simulated by Spectral Radiation-Transport Model for Aerosol Species (SPRINTARS). *Journal of Geophysical Research,* 108(D23), 8659, doi:10.1029/2002JD003210.

Takemura, T., Nozawa,T., Emori,S., Nakajima, T.Y., Nakajima, T. (2005), Simulation of climate response to aerosol direct and indirect effects with aerosol transport-radiation model. *Journal of Geophysical Research,* 110, D02202, doi:10.1029/2004JD005029.

Takemura, T., Nakajima, T., Dubovik, O., Holben, B.N, Kinne, S. (2002): Single-scattering albedo and radiative forcing of various aerosol species with a global three-dimensional model. *Journal of Climate*, 15, 333-352, doi:10.1175/1520-0442.

Toon, O. B., Pollack,J. B. (1976), A global average model of atmospheric aerosol for radiative transfer calculations, J. Appl. Meteorol. 15, 225–246.

Twomey, S. (1974), Pollution and the planetary albedo, Atmos. Environ., 8, 1251– 1256.

Twomey, S. (1977), The influence of pollution on the shortwave albedo of cluds, *J. Atmos. Sci., 34.*, 1149-1152.

Verma, S., Boucher, O., Upadhyaya, H.C., Sharma, O.P. (2006), Sulfate aerosols forcing: An estimate using a three-dimensional interactive chemistry scheme, *Atmospheric Environment* 40, 7953–7962

Vinoj, V., Babu ,S.S., Satheesh, S. K., Krishna Moorthy, K., and Kaufman, Y.J. (2004), Radiative forcing by aerosols over the Bay of Bengal region derived from ship-borne, island-based and satellite (Moderate-Resolution Imaging spectroradiometer) observations, *J. Geophys. Res.*, 109 (D5): Art. No. D05203, doi:10.1029/2003JD004329.

Vogelmann, A, M., Flatau, P. J., Szczodrak, M., Markowicz, K. M., Minnett, P. J. (2003), Observations of large aerosol infrared forcing at the surface, *Geophy. Res. Lett*, 30(12), 1655, doi: 10.1029/2002GL016829.

Wang, C., Kim, D., A. M. L. Ekman, M. C. Barth, and P. J. Rasch (2009), Impact of anthropogenic aerosols on Indian summer monsoon, *Geophys. Res. Lett.*, 36, L21704, doi:10.1029/ 2009GL040114.

Wolx, E.M. (2011), Direct and semi-direct radiative forcing of smoke aerosols over clouds, *Atmos. Chem. Phys. Discuss.*, 11, 20947–20972.

World Meteorological Organization (1986), "A preliminary cloudless standard atmosphere for radiation computation," WCP-112 (World Climate Research Program, CAS, Radiation Commission of IAMAP, Boulder, Colo.)

Yuan, T., Li, Z., Zhang, R., Fan, J. (2008), Increase of cloud droplet size with aerosol optical depth: An observation and modelling study, *J. Geophys. Res., 113*, D04201, doi: 10.1029/2007JD008632.

Zhang, H., Wang Z., Guo, P., Wang, Z., , (2009), A Modeling Study of the Effects of Direct Radiative Forcing Due to Carbonaceous Aerosol on the Climate in East Asia, *Advances in atmospheric sciences*, 26, 57–66.

12

Correcting Transport Errors During Advection of Aerosol and Cloud Moment Sequences in Eulerian Models

Robert McGraw

Atmospheric Sciences Division, Environmental Sciences Department
Brookhaven National Laboratory, Upton, NY
USA

1. Introduction

The method of moments (MOM) provides a highly efficient approach to tracking particle populations, be they aerosols or cloud droplets. In the case of aerosols the method is a statistically based alternative to bin-sectional and modal approaches [Wright et al., 2000]. In the case of clouds, where moments are often the desired product of a simulation for comparisons with radar and satellite observations, the MOM can replace the bin-sectional method. Recent studies have begun looking at the inclusion of higher-order moments, beyond droplet mixing ratio, for improving the representation of cloud microphysics in models [Van Weverberg et al., 2011; Milbrandt and McTaggart-Cowan, 2010].

Early applications of the MOM suffered from inability to close the moment evolution equations, except in the case of very special growth laws. This problem has been largely eliminated with introduction of the quadrature method of moments (QMOM), which allows one to obtain closure under very general conditions and to compute physical and optical properties of a particle population directly from its moments [McGraw, 1997]. Buoyed by this success, the attempt was made early on to incorporate the QMOM into a regional-scale chemical transport model (CTM) - the idea being to evolve and track the moments of several particle populations and transport these in the manner of chemical species during the advection step. Errors were soon encountered and attributed to the corruption of moment sequences during advective transport, which in this case was implemented using the Bott scheme, but any nonlinear transport scheme designed to reduce numerical diffusion would have the same effect. Wright examined invalid moment set generation by two representative advection schemes for ensembles of 10^4 test cases covering a range of initial moment sets and flow conditions. These tests revealed invalid moment set frequencies exceeding 0.7% for both schemes [Wright, 2007].

The paper of Wright, in addition to presenting a clear description of the problem, also analyzed its first solution: "vector transport", or VT, previously implemented in the chemical transport model [Wright et al., 2000]. In VT a sequence of moments is normalized to a selected lead tracer, typically number or volume, and only that one tracer is transported for each population. The remaining vector components (remaining moments) are transported with the same mixing coefficients as the lead tracer, thereby preserving moment ratios within the sequence. The main advantages of the VT schemes are that they preserve valid moment sequences, and the number of transported quantities is reduced, but the accuracy is not as good as found when more than one moment is transported e.g. as in Wright's number-volume VT scheme [Wright, 2007]. The main disadvantage, other than loss of accuracy, is that the VT schemes require modification of the transport algorithm to explicitly call out the cell-to-cell mixing coefficients at each time step. These modifications have proven tedious and go against the concept that aerosol and cloud modules should be interchangeable with any transport scheme. More recently a non-negative least squares (NNLS) method for preservation of moment sequences was developed and tested for transport of aerosol mixtures [McGraw, 2007]. The NNLS scheme makes use of all of the moments and the mixing coefficients (optimized in the least squares sense) are determined by the requirement that the final (post advection step) moment sequence be a non-negative linear combination of the moment sequence vectors in same cell and neighboring cells prior to the advection step – the idea here being based on the fact any linear combination of valid moment sequences with non-negative coefficients is itself a valid moment sequence. The approach proved to be much less diffusive than the VT schemes in tests of source apportionment for aerosol mixtures [McGraw, 2007], but has the disadvantage that valid moment sequences from the previous time step need to be carried forward, as these comprise the basis set for NNLS optimization of the updated moments. The new approaches developed here work, instead, on individual grid cells without requiring stored information from previous time-steps or neighboring cells. Cell-to-cell mixing coefficients are not required as the new methods test moment sequences and correct failed ones in a minimally disruptive way that preserves as many of the transport algorithm generated moments as possible.

2. Moment inequalities

We are interested in tracking, in an atmospheric model, the moments of a generally unknown distribution function, $f(r)$. The required methods are illustrated for a particle size distribution expressed in terms of particle radius but other coordinates, such as particle volume or mass, could just as easily be used. The radial moments are:

$$\mu_k \equiv \int_0^\infty r^k f(r) dr \tag{2.1}$$

for $k = 0, 1, 2, \cdots$. In order to have physical validity it is clear that both the particle distribution function and its domain need to be positive: $f(r) \geq 0$; $r \geq 0$. Identification of the necessary and sufficient conditions for a valid moment sequence - i.e., one consistent with a distribution function of this type – is attributed to Stieljes and these are usually expressed as inequalities involving the Hankel-Hadamard determinants constructed from the moments [Shohat and Tamarkin, 1943]:

$$\Delta_n = \begin{vmatrix} \mu_0 & \mu_1 & \cdots & \mu_n \\ \mu_1 & \mu_2 & \cdots & \mu_{n+1} \\ \cdots & \cdots & \cdots & \cdots \\ \mu_n & \mu_{n+1} & \cdots & \mu_{2n} \end{vmatrix} \geq 0 \tag{2.2a}$$

$$\Delta_n^1 = \begin{vmatrix} \mu_1 & \mu_2 & \cdots & \mu_{n+1} \\ \mu_2 & \mu_3 & \cdots & \mu_{n+2} \\ \cdots & \cdots & \cdots & \cdots \\ \mu_{n+1} & \mu_{n+2} & \cdots & \mu_{2n+1} \end{vmatrix} \geq 0 \qquad (2.2b)$$

Without loss of generality we will work with normalized distribution moments ($\mu_0 = 1$). The un-normalized moments are easily restored by multiplying each moment of the normalized sequence by μ_0. With normalized moments, the requirement that determinant $\Delta_1 \geq 0$, for example, is equivalent to the requirement that the variance be positive: i.e., $\mu_2 - \mu_1^2 \geq 0$. More generally, a moment sequence is valid if and only if all inequalities 2.2a and 2.2b are satisfied.

Two new approaches for testing valid moment sequences, and correcting invalid ones, are now presented. The first of these will be referred to as Positive Alpha Sequence Enforcement (PASE), where the "alpha sequence" consists of certain mathematical quantities introduced by Gordon [1968] that are related to the determinant sequence defined above. The PASE algorithm is introduced in Sec. 3. The second approach uses difference tables and has the advantage that it can pinpoint specific moment errors for sequences of six or more moments. This second approach is essentially a filtering method that smoothes moment sequences if and when an invalid sequence is found. The filter algorithm is introduced in Sec. 4.

3. Correcting invalid moment sequences by positive alpha sequence enforcement (PASE)

It is convenient to replace the determinants of Sec. 2 by another set of non-negative quantities, the alpha sequence, investigated by Gordon and generated by him using the product-difference (PD) algorithm [Gordon, 1968]. Inspection of the alpha sequence will (1) indicate immediately whether or not a given moment sequence, e.g. one obtained after the advection step of a model simulation, is valid and (2) provide a recipe for correction if the tested sequence proves to be invalid. PASE has the further advantage that it includes most of the steps en route to obtaining quadrature points for the weight function $f(r)$ directly from its moments. These points can then be used to approximate the physical and optical properties of $f(r)$ while providing the moment closure needed to track the evolution of $f(r)$ directly from its lower-order moments. Indeed this is the basis of the QMOM [McGraw, 1997]. In those cases where an invalid moment sequence is found, the (corrected) quadrature points easily provide corrected moments.

Consider the following ordering of the determinants defined in Sec. 2:

$$e = \{\Delta_0, \Delta_0^1, \Delta_1, \Delta_1^1, \Delta_2, \Delta_2^1, \cdots\} \qquad (3.1)$$

with $\Delta_0 = \mu_0 = 1$. The kth member of this sequence introduces the (k-1)th moment, as evidenced by inspection of Eqs. 2.2. Because its elements have an abundance of useful properties, it is more convenient to work with the alpha sequence:

$$\alpha = \{\alpha_1, \alpha_2, \alpha_3, \cdots\}, \qquad (3.2)$$

This sequence can be written in terms of moments and e-sequence elements as follows:

$$\alpha_1 = 1$$

$$\alpha_2 = \Delta_0^1 = \mu_1$$

$$\alpha_3 = \frac{\Delta_1}{\Delta_0^1} = \frac{\mu_2}{\mu_1} - \mu_1$$

$$\alpha_4 = \frac{\Delta_1^1}{\Delta_1 \Delta_0^1} = \frac{e_4 e_1}{e_3 e_2} = \frac{(\mu_2)^2 - \mu_1 \mu_3}{(\mu_1)^3 - \mu_1 \mu_2} \tag{3.3}$$

$$\vdots$$

$$\alpha_n = \frac{e_n e_{n-3}}{e_{n-1} e_{n-2}} = \frac{\Delta_{(n/2)-1}^1 / \Delta_{(n/2)-1}}{\Delta_{(n/2)-2}^1 / \Delta_{(n/2)-2}} \quad (n \text{ even; } n \geq 4)$$

$$= \frac{\Delta_{(n-1)/2} \, \Delta_{(n-5)/2}^1}{\Delta_{(n-3)/2}^1 \, \Delta_{(n-3)/2}} \quad (n \text{ odd; } n \geq 5)$$

where e_n is the nth element of sequence e. The conditions for a valid moment set (Eqs. 2a and 2b) are easily reformulated in terms of the alphas. Considering the order in which the determinants appear, it follows that the non-negativity conditions:

$$\alpha_n \geq 0 \qquad (n = 1, 2, \cdots) \tag{3.4}$$

are equivalent to the determinant inequalities (2.2a and 2.2b). Thus a moment sequence is valid if and only if inequalities (3.4) hold. To show this by induction, assume non-negativity through α_{n-1}. Eq. 3.3 shows that α_n will be non-negative if and only if the determinant e_n is non-negative – so the equivalence between inequalities 2.2a and 2.2b, and 3.4 holds also at the level of α_n. Inspection of the determinant sequence shows that e_n is the first determinant to enlist the nth moment, μ_{n-1}. Consider the case that μ_{n-1} is the first sufficiently corrupted moment that the sequence $\{1, \mu_1, \mu_2, \cdots, \mu_{n-1}\}$ is invalid (the previous sequence $\{1, \mu_1, \mu_2, \cdots, \mu_{n-2}\}$ is valid by assumption). In this case both e_n and α_n will be negative; the former because the determinant inequality on e_n is violated for the invalid sequence and the latter by Eq. 3.3 that defines the alpha set. The method of PASE correction is to set the first negative entry, here α_n, and all higher values of alpha, $\alpha_{n+1}, \alpha_{n+2}, \cdots$ to zero. Because the modified alpha sequence now satisfies inequalities 3.4, a valid moment sequence is guaranteed. Equations 3.3 have been introduced mainly to show equivalence of the inequalities 2.2 and 3.4. For computational purposes handling products and quotients of determinants is not recommended given that a much more efficient and well-conditioned approach is available for generating the alpha sequence, quadrature points, and valid moment sets - as now described.

Inequalities 3.4 (rather than 2.2) will be used to test for corrupted moment sets. To obtain the alpha sequence the widely available Numerical Recipes subroutine ORTHOG is used to first obtain a tridiagonal Jacobi matrix from the moment sequence. For six moments $\mu_0 = 1$ through μ_5 this has the form:

$$J_3 = \begin{pmatrix} a_1 & \sqrt{b_2} & 0 \\ \sqrt{b_2} & a_2 & \sqrt{b_3} \\ 0 & \sqrt{b_3} & a_3 \end{pmatrix} \tag{3.5}$$

where the matrix elements are computed from the moments using ORTHOG. A similar construction applies for $2n$ moments and J_n. Note that ORTHOG works with modified moments, which are better conditioned than powers of r, but require that the weight function $f(r)$ be known. ORTHOG also works with the ordinary moments defined by Eq. 2.1. For this purpose the coefficients of the modified moment recurrence relations used in ORTHOG need to be set to zero. Because the ordinary moments are not as well conditioned, only lower-order moments (powers up to about r^9) should be used. Fortunately, even fewer moments have proven adequate for most applications requiring both reliable dynamics and accurate estimation of aerosol physical and optical properties from moments [McGraw et al. 1995; Wright et al., 2002].

The Jacobi matrix elements are expressed in terms of the alpha sequence as follows [Gordon, 1968]. Note a switch in notation from Gordon's b_i^2 to b_i in the equations below. This is consistent with the off-diagonal elements used in defining J_n (Eq. 3.5) and Numerical Recipes [Press et al., 1992].

$$\begin{aligned} a_1 &= \alpha_2 & b_2 &= \alpha_2 \alpha_3 \\ a_2 &= \alpha_3 + \alpha_4 & b_3 &= \alpha_4 \alpha_5 \\ a_3 &= \alpha_5 + \alpha_6 & & \\ &\ \vdots & &\ \vdots \end{aligned} \tag{3.6}$$

Inverting Eq. 3.6 gives a continued fraction expansion for the alphas in terms of the Jacobi matrix elements:

$$\begin{aligned} \alpha_1 &= 1 & \alpha_5 &= \cfrac{b_3}{a_2 - \cfrac{b_2}{a_1}} \\ \alpha_2 &= a_1 & & \\ \alpha_3 &= \frac{b_2}{a_1} & \alpha_6 &= a_3 - \cfrac{b_3}{a_2 - \cfrac{b_2}{a_1}} \\ \alpha_4 &= a_2 - \frac{b_2}{a_1} & &\ \vdots \end{aligned} \tag{3.7}$$

This concludes generation of the alpha sequence. Quadrature abscissas and weights are obtained by solving the eigenvalue problem associated with the Jacobi matrix. The abscissas $\{r_i\}$ are just the eigenvalues of this matrix. The corresponding weights, $\{w_i\}$, are given by squares of the first components of the corresponding eigenvectors. Thus if w_i is the weight corresponding to abscissa/eigenvalue r_i, v_i the corresponding eigenvector, and $v_{i,1}$ its first component, then $w_i = \mu_0 (v_{i,1})^2 = v_{i,1}^2$, where the second equality applies for normalized moments [Press et al., 1992]. As with $f(r)$ (Eq. 2.1), each of the abscissas and weights must be non-negative for a valid moment set. Indeed, one can just as well examine the quadrature points obtained form the Jacobi matrix to determine validity of a moment set - especially if the QMOM is already being used. Most moment sequences will pass this inversion test; in case an invalid set is detected, e.g., by the appearance of a negative eigenvalue, an alpha sequence can be generated from the matrix elements using Eq. 3.7 and the PASE correction applied.

The quadrature points, obtained through ORTHOG and matrix diagonalization, give approximations to integrals of the form:

$$I = \int_0^\infty \sigma(r)f(r)dr \approx \sum_{i=1}^n \sigma(r_i)w_i \qquad (3.8)$$

for known kernel functions, including an n-point quadrature estimate for μ_k in the special case that kernel is of the form $\sigma(r) = r^k$:

$$\mu_k \approx \sum_{i=1}^n r_i^k w_i . \qquad (3.9)$$

The approximate equality of Eq. 3.9 is exact for moments μ_0 through μ_{2n-1}. For $\sigma(r) = \exp(-sr)$, in which case Eq. 3.8 defines the Laplace transform, $I(s)$, of $f(r)$, nested and rapidity convergent pairs of upper and lower bounds to $I(s)$ are obtained form the alpha sequence [Gordon, 1968; McGraw, 2001]. Recent calculations using a model coagulation kernel show similar behavior, with rapid and nested convergence to benchmark numerical results from particle-resolved simulation [McGraw et al., 2008]. More work needs to be done to explore the mathematical basis for this result.

Two test cases showing implementation of the PASE method are given in Tables 1 and 2. Table 1 illustrates the processing of a valid moment set. Here the alphas are non-negative and J_3 gives a valid set of quadrature abscissas and weights. Table 2 shows moments from a log-normal distribution, except that μ_3 is corrupted. Here some of the alphas are negative, an eigenvalue of J_3 (not shown) is negative, and PASE (corrected alpha sequence column) applied. Elimination of higher-order alpha values has reduce the order of the Jacobi matrix to J_2 and for this smaller matrix a valid set of 2 quadrature points and a valid moment sequence, with exact recovery of the first 3 moments, are obtained. Table 2 shows a case where an odd-number of moments (here μ_0 through μ_2) passes the test. With an odd number of moments Eq. 3.8 describes Gauss-Radau quadrature, which places an abscissa at one of the boundaries of the domain of $f(r)$ - in this case at $r = 0$ [McGraw et al., 2008]. For certain kernels division by zero is a problem, e.g., vapor condensation by the continuum particle growth law has a kernel of the form $\sigma(r) \propto 1/r$. So it is sometimes safer to work with even numbers of moments. Even in the most unfavorable situation two physically valid moments, e.g. number and radius or number and mass, always result when using a positive advection scheme. Transport errors in either of both of these are possible, even likely, but such errors will cause no violation of the moment inequalities, which only require that the first two moments be positive or, in the trivial case, zero (see first two inequalities of Eq. 2.2).

moments	J_3 elements ORTHOG	alpha sequence Eq. 3.7	quadrature from J_3
$\mu_0 = 1$	$a_1 = 5.0$	$\alpha_1 = 1$	$r_1 = 15.2853$
$\mu_1 = 5.0$	$\sqrt{b_2} = 2.88675$	$\alpha_2 = 5.0$	$w_1 = 0.0283736$
$\mu_2 = 33.3333$	$a_2 = 8.33343$	$\alpha_3 = 1.66666$	$r_2 = 7.18626$
$\mu_3 = 277.778$	$\sqrt{b_3} = 4.7139$	$\alpha_4 = 6.66677$	$w_2 = 0.452838$
$\mu_4 = 2777.78$	$a_3 = 11.6673$	$\alpha_5 = 3.33308$	$r_3 = 2.52914$
$\mu_5 = 32407.4$		$\alpha_6 = 8.33423$	$w_3 = 0.518788$

Table 1. Valid moment set from a model cloud droplet size distribution with particle radii in micron.

moments	J_3 elements ORTHOG	alpha sequence Eq. 3.7	corrected alpha sequence	quadrature from J_2
$\ln \mu_0 = 0$	$a_1 = 2.71828$	$\alpha_1 = 1$	$\alpha_1 = 1$	$r_1 = 20.0855$
$\ln \mu_1 = 1.0$	$\sqrt{b_2} = 6.87089$	$\alpha_2 = 2.71828$	$\alpha_2 = 2.71828$	$w_1 = 0.135335$
$\ln \mu_2 = 4.0$	$a_2 = 2.68355$	$\alpha_3 = 17.3673$	$\alpha_3 = 17.3673$	$r_2 = 0.0$
$\ln \mu_3 = 6.0$	$\sqrt{b_3} = 433.747$	$\alpha_4 = -14.6837$	$\alpha_4 = 0$	$w_2 = 0.864665$
$\ln \mu_4 = 16.0$	$a_3 = 8096.25$	$\alpha_5 = -12812.6$	$\alpha_5 = 0$	
$\ln \mu_5 = 25.0$		$\alpha_6 = 20908.9$	$\alpha_6 = 0$	

Table 2. Moment set from a log-normal distribution. The third radial moment μ_3 is corrupted. The natural logarithms of the corrected moments (Eq. 3.9 using the abscissas and weights from J_2) are {0, 1, 4, 7, 10, 13} for moments 0 through 5, respectively. Moments of lower order than the corrupted moment are reproduced exactly from J_2.

4. Correcting invalid moment sequences by the filter method

Because corruption of moment sequences through advective transport tends to be infrequent, it is likely to result from improper assignment of one, or at most a few of the moments in the sequence. Accordingly, we would like to adjust only *those* moments.

For a sufficiently long sequence the filter method is a way to achieve this goal.

Difference tables: The construction of difference tables is especially useful for spotting errors in an ordered sequence of data [Lanczos, 1988]. The construction is simple and evident from the tables to follow. Tables 1and 2 illustrate difference tables for two sequences of six moments. The first column gives the 'data' to be evaluated, a sequence of values of $\ln \mu_k$. The ith-order difference column is labeled d_i. Column 2 contains the first-order differences, d_1, which are differences of the data entries in column 1. Column 3 contains the second-order differences, d_2, which are just the first-order differences of the entries in column 2, etc.

A necessary but not sufficient criterion for a valid moment sequence is that $\ln \mu_k$ be a convex function of index k [Feller, 1971]. This requires that the second-order differences be non-negative. In Table 1 the $\ln \mu_k$ were assigned as a quadratic function of k, as is characteristic of a log-normal distribution [Hinds, 1982]. In this case the second-order differences are constant and higher-order differences vanish. In general we cannot expect that $\ln \mu_k$ will have quadratic form, however it is reasonable to expect a smooth function of index k and moment interpolation methods have been developed that exploit smoothness in $\ln \mu_k$ [Frenklach, 2002]. In Table 4 the third moment has been corrupted and the modified sequence violates convexity as is evident from the appearance of negative elements in the column of second-order differences, d_2. Note how the error propagates with amplified oscillation in sign through the higher-order differences. Here one sees the useful property of a difference table for spotlighting errors in a data sequence through inspection of higher-order differences [Lanczos, 1988].

log moments	d_1	d_2	d_3	d_4	d_5
$\ln \mu_0 = 0$	1	2	0	0	0
$\ln \mu_1 = 1$	3	2	0	0	n
$\ln \mu_2 = 4$	5	2	0	n	n
$\ln \mu_3 = 9$	7	2	n	n	n
$\ln \mu_4 = 16$	9	n	n	n	n
$\ln \mu_5 = 25$	n	n	n	n	n

Table 3. Moment sequence and first to fifth-order differences. In this case the moments are from a lognormal distribution and $\ln \mu_k$ is quadratic in index k. 'n' means no entry.

log moments	d_1	d_2	d_3	d_4	d_5
$\ln \mu_0 = 0$	1	2	-3	12	-30
$\ln \mu_1 = 1$	3	-1	9	-18	n
$\ln \mu_2 = 4$	2	8	-9	n	n
$\ln \mu_3 = 6$	10	-1	n	n	n
$\ln \mu_4 = 16$	9	n	n	n	n
$\ln \mu_5 = 25$	n	n	n	n	n

Table 4. Similar pattern as Table 3 using the moments from Table 2 with the same corrupted value of μ_3.

For sequences of six or more moments, the third-order differences can be used to both attribute the error (i.e. identify the index of the miss-assigned moment) and provide an optimal correction in the sense of minimizing the sum of the squared differences of the elements in column d_2 so as to restore smoothness. Note that the sum of squared differences of elements in column d_2 is just the squared magnitude of the vector $\mathbf{a} = \{-3, 9, -9\}$ containing the third-order differences listed in column d_3 of Table 4. Our strategy will be to minimize the magnitude of the vector of these third-order differences, which vanishes for the special case of quadratic sequence, i.e., $|\mathbf{a}|^2 = 0$ in Table 3.

Description of the algorithm: The following minimum square gradient algorithm restores a valid moment sequence by adjusting that moment, μ_{k*}, which after adjustment maximizes smoothness through minimization of $|\mathbf{a}|^2$. To illustrate the method, we begin by first determining the response of $|\mathbf{a}|^2$ to change in an arbitrary moment, μ_k, and next determine $k*$. (In actual calculations these steps are reversed as described below.) Consider a change in the kth moment from an initial value $\mu_k(0)$ to a final value $\mu_k(1)$ and denote the ratio $\mu_k(1)/\mu_k(0)$ by c_k. Note by inspection of the difference table that if $\mu_k(1) = c_k \mu_k(0)$ or, equivalently, $\ln \mu_k(1) = \ln c_k + \ln \mu_k(0)$, then $\mathbf{a}_1 - \mathbf{a}_0 = (\ln c_k)\mathbf{b}_k$ where $\mathbf{a}_0$ and $\mathbf{a}_1$ are, respectively, the initial and final vectors of third-order differences and the "response vectors" $\mathbf{b}_k$ give the change in the vector of third-order differences to a unit increment in $\ln \mu_k$. The latter are as follows:

$$\mathbf{b}_0 = \{-1,0,0\}; \mathbf{b}_1 = \{3,-1,0\}; \mathbf{b}_2 = \{-3,3,-1\}; \mathbf{b}_3 = \{1,-3,3\}; \mathbf{b}_4 = \{0,1,-3\}; \mathbf{b}_5 = \{0,0,1\}, \quad (4.1)$$

which are related to the entries in the Pascal triangle except for oscillations in sign [Lanczos, 1988]. Next consider the value of c_k (actually $\ln c_k$) for which $|\mathbf{a}_1|^2 = |\mathbf{a}_0 + (\ln c_k)\mathbf{b}_k|^2$ is minimized. Inspection of Fig. 1 shows that minimization is achieved for the condition that $\mathbf{a}_0 + (\ln c_k)\mathbf{b}_k$ is orthogonal to $\mathbf{b}_k$. The value of c_k that satisfied this condition is:

$$\ln c_k = -\cos(\mathbf{a}_0, \mathbf{b}_k)\frac{|\mathbf{a}_0|}{|\mathbf{b}_k|} = -\frac{(\mathbf{a}_0 \cdot \mathbf{b}_k)}{|\mathbf{b}_k|^2}. \quad (4.2)$$

The last equality follows form the law of cosines, $\cos(\mathbf{a}_0, \mathbf{b}_k) = (\mathbf{a}_0 \cdot \mathbf{b}_k)/(|\mathbf{a}_0||\mathbf{b}_k|)$ where $(\mathbf{a}_0, \mathbf{b}_k)$ is the angle between vectors $\mathbf{a}_0$ and $\mathbf{b}_k$. The resulting minimum squared amplitude satisfies:

$$|\mathbf{a}_1|^2 = |\mathbf{a}_0 + \ln c_k \mathbf{b}_k|^2 = |\mathbf{a}_0|^2 \left[1 - \cos^2(\mathbf{a}_0, \mathbf{b}_k)\right] \quad (4.3)$$

which is the largest reduction in magnitude of the vector of third-order differences achievable by changing μ_k alone (Fig. 1).

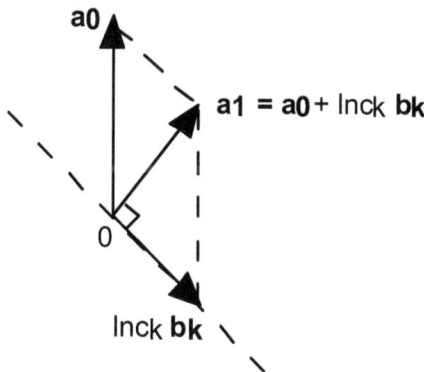

Fig. 1. Disposition of the third order difference vectors before and after correction, $\mathbf{a}_0$ and $\mathbf{a}_1 = \mathbf{a}_0 + \ln c_k \mathbf{b}_k$ respectively.

Equation 4.3 shows that maximal smoothness is achieved by adjusting the moment, μ_{k*}, corresponding to that basis vector $\mathbf{b}_{k*}$ which gives the largest $\cos^2(\mathbf{a}_0, \mathbf{b}_k)$ for any moment index k. Thus by determining which k gives the maximum value of $\cos^2(\mathbf{a}_0, \mathbf{b}_k)$, we obtain the index of the suspect moment, $k*$. That moment alone is corrected, using the factor c_{k*} from Eq. 4.2, yielding an updated moment sequence. Recalling that $\mathbf{b}_k$ gives the third-order difference response to a unit change in $\ln \mu_k$, the actual change in the moment for $k = k*$ is:

$$\ln \mu_{k*}(1) = \ln \mu_{k*}(0) + \ln c_{k*} = \ln \mu_{k*}(0) - \frac{(\mathbf{a}_0 \cdot \mathbf{b}_{k*})}{|\mathbf{b}_{k*}|^2}. \tag{4.4}$$

The other moments having $k \neq k*$ are unchanged. The new moment sequence gives the third-order difference vector $\mathbf{a}_1$ whose magnitude is in agreement with Eq. 4.3. The new moment sequence is in turn tested to insure that negative second-order differences have been removed. If not, the process is repeated, replacing $\mathbf{a}_0$ by $\mathbf{a}_1$, and obtaining $\mathbf{a}_2$, etc. Equation 4.3 assures a reduction in the amplitude of the third order difference vector on each iteration. Thus the amplitude approaches zero after many iterations, and $\ln \mu_k$ approaches a quadratic function of index k. Typically just one or two passes through the algorithm suffice to obtain a valid moment sequence.

Examples: For our first example we begin with the moments of Table 4 and show that a single pass through the filter restores the moment sequence of Table 3. Note that the third-order difference vector in Table 4 satisfies $\mathbf{a}_0 = \{-3, 9, -9\} = -3\mathbf{b}_3$. The multiplier here is understandable because $\mathbf{b}_3$ gives the response to a unit change in $\ln \mu_3$ and in passing from Table 3 to Table 4 this quantity was changed by -3. Note also that the angle $(\mathbf{a}_0, \mathbf{b}_3)$ is π and thus the maximum value of $\cos^2(\mathbf{a}_0, \mathbf{b}_k)$, which occurs here for $k* = 3$ is unity. So $|\mathbf{a}_1| = 0$ for this case, which is the reason why a single pass through the filter restores a quadratic sequence in $\ln \mu_k$. To correct the third moment, we evaluate the right hand side of Eq. 4.2 to obtain

$$\ln c_3 = -(\mathbf{a}_0 \cdot \mathbf{b}_3) / |\mathbf{b}_3|^2 = -(-3\mathbf{b}_3 \cdot \mathbf{b}_3) / |\mathbf{b}_3|^2 = 3.$$

Finally from Eq. 4.4 we obtain $\ln \mu_3(1) = \ln \mu_3(0) + \ln c_3 = 6 + 3 = 9$ showing restoration of the moment sequence of Table 3.

For our second example, consider the moment sequence $\ln \mu_k = \{0, 1, 4, 6, 16, 22\}$ for moments μ_0 though μ_5, respectively, where moments μ_3 and μ_5 both differ from the original sequence of Table 3. A check of the second-order differences shows that convexity is not satisfied. After one pass through the filter μ_3 is changed yielding the log moment sequence $\{0, 1, 4, 9.47368, 16, 22\}$. There is still a failure of convexity, although the sequence is smoother than before. After a second pass μ_5 has changed, yielding the log moment sequence $\{0, 1, 4, 9.47368, 16, 23.5789\}$ and convexity is satisfied - for this and all subsequent iterations. The last (two-pass) moment sequence not only satisfies convexity, it passes the moment inversion test to yield a valid set of three quadrature abscissas and weights.

For our final example, consider the log-moment sequence $\ln \mu_k = \{0, 1, 3, 6, 9.1, 15\}$. Here is a case that satisfies convexity (the second-order differences are positive) but is still unphysical and fails the moment inversion test. Moments 0-3 are fine but the next moment (μ_4) fails. (The PASE test of Sec. 3 also shows failure of μ_4 because α_5 is negative.) The filter method needs to include this possibility, which can be done using the computational flow scheme of Fig. 2. A single pass though the algorithm changes the failed moment from 9.1 to 10 and a valid sequence is obtained.

In the hypothetical extreme case that the moment sequence is so corrupted that $\ln \mu_k$ versus k is predominantly concave, the filter can converge to a smooth function with negative curvature and never pass the inversion test. Then, as in the PASE method, one is forced to work with two moments. Since a plot of $\ln \mu_k$ versus k defined by just two points is a straight line, the corresponding size distribution, $f(r)$, is monodisperse.

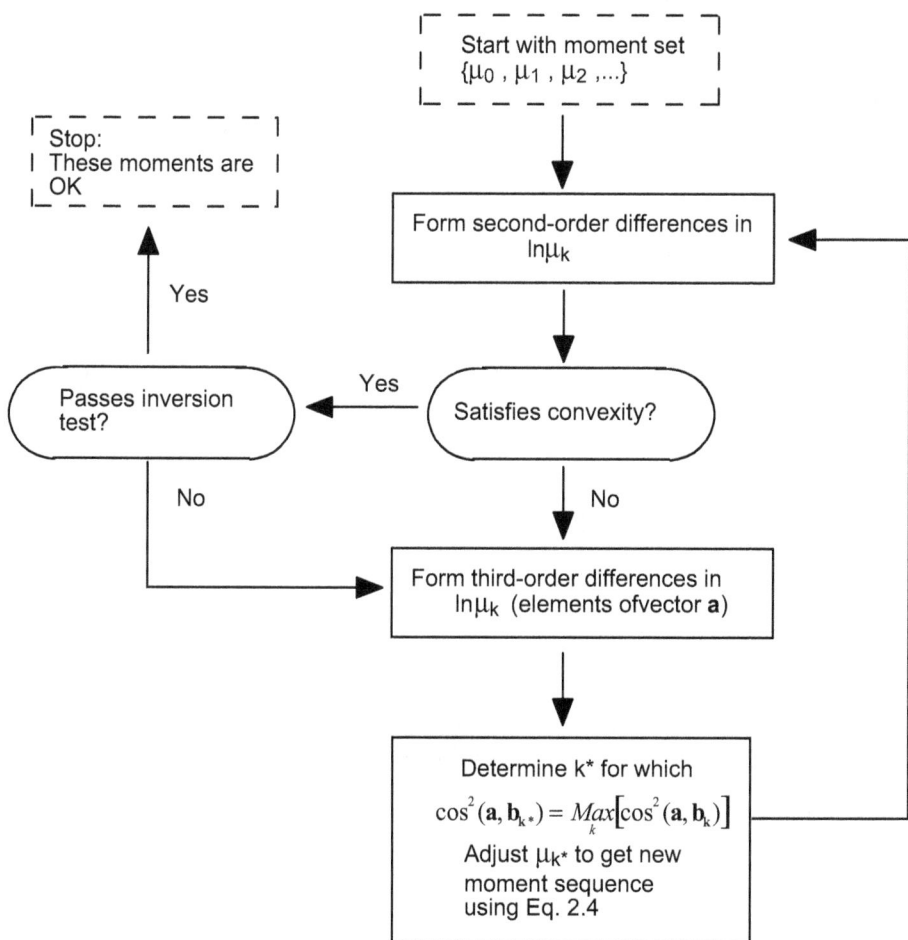

Fig. 2. Flowchart of the computations used to generate a valid moment set in the filter method.

5. Summary

This paper has introduced two independent methods for testing and correcting moment sequences undergoing advective transport in atmospheric models. Each method has been design to operate on single grid cells and require no modification to the transport scheme or storage of information from previous time steps or neighboring grid cells. With these features in place, either method, incorporated as part of an aerosol or cloud module, should enable that module to be compatible with any transport scheme.

The PASE method is applicable to sequences of as few as three moments - the minimum number for which inconsistencies of the kind described here can arise - and a corrected moment sequence is achieved in a single step. The filter method has the advantage of identifying specific corrupted moments and only correcting these, whereas the PASE method recomputed all moments of index greater than or equal to that of the corrupted moment - thus retaining less of the raw information supplied by the transport scheme. On the other hand the filter method may require multiple passes (if more than a single moment is corrupted) and needs six or more moments to operate effectively.

Both the PASE and filter schemes appear well suited for immediate use in moment-base cloud simulation. Here the particles are of uniform composition (liquid water or ice) and thus (ignoring ice crystal shape) describable by univariate moments - ideally with a separate moment sequence for each phase. Aerosols, on the other hand, are complex not only with respect to size and shape, but also mixing state. Recent multivariate extensions of the QMOM have been developed that enable such complexities to be handled through the tracking of multivariate mixed moments [Yoon and McGraw, 2004a; 2004b]. For the purpose of assigning quadrature points in higher dimension the multivariate distribution function is treated as factorizable in the principal coordinates frame, which is continuously updated in time through tracking of first and second-order mixed moments. This reduction to a direct product of univariate distributions implies that either the PASE method or filter method can still be used.

Historically, most nonlinear advection schemes in current use derive one way or another from the need to advect individual tracers in the presence of sharp gradients (e.g. fluid density in a shock front) and have not been adequately tested for transport of multiple correlated tracers. Moments, because they are so strongly and nonlinearly correlated, provide a excellent indicator of correlation failure - they serve as the "canary in the mine", so to speak. Other correlated tracers such as composition of aerosol mixtures and hydrometeor phase will also be affected and need to be considered - even if the loss of correlation is less obvious for these quantities. Quantitative metrics, which apply beyond moments to encompass these other kinds of correlated tracers, need to be developed for evaluating advective transport schemes if future climate models are to achieve optimum balance between the need to reduce numerical diffusion and the need for correlation preservation.

6. Acknowledgements

The author is grateful to Drs. Wuyin Lin and Kwinten Van Weverberg for sharing their insights into the current state of moment tracking in cloud models. This research was supported by the DOE SciDAC and ASR programs.

7. References

Feller, W. (1971), An Introduction to Probability Theory and its Applications (Vol. II) (Wiley, New York) pg 155.

Frenklach, M. (2002), "Method of moments with interpolative closure", Chem. Eng. Sci. 57, 2229-2239.

Hinds, W., C. (1982), Aerosol Technology (Wiley, New York) pg. 85.

Gordon, R. G. (1968), " Error bounds in equilibrium statistical mechanics", J. Math. Phys. 9, 655-663.

Lanczos, C. (1988), Applied Analysis (Dover, New York) Chapt. 5.

McGraw, R., Huang, P. I., and Schwartz, S. E. (1995), "Optical properties of atmospheric aerosols from moments of the particle size distribution" Geophys. Res. Letts. 22, 2929-2932.

McGraw, R. (1997), "Description of aerosol dynamics by the quadrature method of moments", Aerosol Sci. Technol. 27, 255-265.

McGraw, R. (2001), "Dynamics of barrier crossing in classical nucleation theory", J. Phys. Chem. B105, 11838-11848.

McGraw, R. (2007), " Numerical advection of correlated tracers: Preserving particle size/composition moment sequences during transport of aerosol mixtures", J. Phys.: Conf. Series 78, 012045, doi: 10.1088/1742-6596/78/1/012045.

McGraw, R., Leng, L., Zhu, W., Riemer, N., and West, M. (2008), " Aerosol dynamics using the quadrature method of moments: Comparing several quadrature schemes with particle-resolved simulation " J. Phys.: Conf. Series 125, 012020, doi: 10.1088/1742-6596/125/1/012020.

Milbrandt, J. A. and McTaggart-Cowan, R. (2010), "Sedimentation-induced errors in bulk microphysics schemes" J. Atmos. Sci. 67, 3931-3948.

Press, W. H., Teukolsky, S. A., Vetterling, W. T. and Flannery, B. P. (1992) Numerical Recipes in FORTRAN (Cambridge University Press, Cambridge).

Shohat, J. D. and Tamarkin, J. D. (1943), The problem of moments (American Mathematical Soc., New York) pg. viii.

Van Weverberg, K., Vogelmann, A. M., Morrison, H. and Milbrandt, J. (2011), "Sensitivity of idealized squall line simulations to the level of complexity used in two two-moment bulk microphysical schemes". Submitted for publication [BNL-95332-2011-JA]

Wright, D. L., Yu, S., Kasibhatla, P. S., McGraw, R., Schwartz, S. E., Saxena, V. K., and Yue, G. K. (2002), "Retrieval of aerosol properties from moments of the particle size distribution for kernels involving the step function". J. Aerosol Sci. 33, 319-337.

Wright, D. L., McGraw, R., Benkovitz, C. M. and Schwartz, S. E. (2000), "Six moment representation of multiple aerosol populations in a sub-hemispheric chemical transportation model", Geophys. Res. Lett. 27, 967-970.

Wright, D. L. (2007), " Numerical advection of moments of the particle size distribution in Eulerian models", J. Aerosol Sci. 38, 352-369.

Yoon, C. and McGraw, R. (2004a) " Representation of generally-mixed multivariate aerosols by the quadrature method of moments: I. Statistical foundation". J. Aerosol Sci. 35, 561-576.

Yoon, C. and McGraw, R. (2004b), " Representation of generally-mixed multivariate aerosols by the quadrature method of moments: II. Aerosol dynamics", J. Aerosol Sci. 35, 577-598.

13

Aerosol Radiative Forcing:
AERONET-Based Estimates

O.E. García[1], J.P. Díaz[2], F.J. Expósito[2], A.M. Díaz[2],
O. Dubovik[3] and Y. Derimian[3]
[1]*Centro de Investigación Atmosférica de Izaña, Agencia Estatal de Meteorología (AEMET)*
[2]*Grupo de Observación de la Tierra y la Atmósfera, Universidad de La Laguna*
[3]*Laboratoire d'Optique Amosphérique, Université Lille1,*
[1,2]*Spain*
[3]*France*

1. Introduction

Solar radiation is the main source of energy for the Earth-atmosphere system, and directly or indirectly, is responsible for all phenomena affecting the meteorology and climatology of this system. The variation of any component of this system affects the radiative equilibrium, resulting in, for example, changes in the temperature of the system and/or in the configuration of the atmospheric circulation. In order to quantify these effects the concept of radiative forcing was introduced in the literature to show the magnitude of radiation variation due to changes in a specific atmospheric component, for instance, clouds, gases or atmospheric aerosols.

One of the main reasons for the changes in the energy balance of the Earth-atmosphere system is the variation in the concentration of greenhouse gases, GHGs, which contribute to an increase in the system temperature. On a global scale, the total annual anthropogenic emissions of GHGs have increased 70% between 1970 and 2004 with respect to the pre-industrial era. The associated radiative forcing of this increase is expected to be +2.3 Wm^{-2}, with an uncertainty of ±0.2 Wm^{-2} (IPCC, 2007).

On the other hand, another decisive factor is atmospheric aerosol, both from natural and anthropogenic origin. These atmospheric constituents, directly and indirectly, modify the energy balance of the Earth-atmosphere system: directly through the absorption and dispersion of solar radiation in the atmosphere and, indirectly, by acting as nuclei of cloud condensation, and modifying their own properties (albedo, reflectivity, lifetime, precipitation efficiency,...) (Charlson et al., 1987; Lohmann & Fiechter, 2005; Twomey, 1977). Finally, there are semi-direct effects associated with the absorption of solar radiation by aerosols. These produce an unequal warming of the atmosphere, favouring unstable conditions and also contributing to the modification of cloud characteristics (Ackerman et al., 2000; Koren et al., 2004).

Even though the impact of atmospheric aerosols on climate (and climatic change) is undeniable, the knowledge of their radiative forcing (direct, semi-direct and indirect effects) is still not well understood (IPCC, 2007). The radiative effects of these atmospheric constituents depend largely on their size distributions and their chemical composition. Thus, on a global scale, the direct radiative forcing of sulphates is estimated to be -0.4±0.2 Wm-2, of organic carbon -0.05±0.05 Wm-2, for soot, +0.20±0.15 Wm-2, while aerosol radiative forcing from natural origins such as biomass burning and mineral dust are +0.03±0.12 Wm-2 and -0.1±0.2 Wm-2 respectively (IPCC, 2007). Although the local effects can be far greater and reach hundreds of Wm-2, mainly in those regions close to the emission sources (García et al., 2011a, 2011b; Haywood et al., 2003). These values clearly indicate significant uncertainties in the aerosol radiative forcing estimates and confirm the need to study their radiative properties and to quantify their effects on radiative balance.

According to the recent report of Intergovernmental Panel on Climate Change (2007), the direct radiative forcing of the individual aerosol species is less certain than the total direct radiative forcing by all aerosols. Likewise, recent studies show that most climate models underestimate the negative forcing by anthropogenic aerosols (Hansen et al., 2011). In this context, the long-term monitoring of different aerosol types is crucial to improve our knowledge of the radiative forcing and climate parameters. Nowadays, one of the most useful tools supporting this aim is the AErosol RObotic NETwork (AERONET, http://aeronet.gsfc.nasa.gov), which provides enough information globally to establish a ground-based aerosol climatology. An extended set of physical and optical aerosol properties, averaged in the atmospheric column (Dubovik et al., 2002a, 2006) and given at more than 180 worldwide locations, have enabled verification of global aerosol models and satellite retrievals.

Thus, this chapter describes an approach to evaluate the direct radiative effect of atmospheric aerosols from AERONET observations. The radiative net effect of key aerosol types is discussed with a homogeneous methodology, which allows a direct intercomparison and establishment of a climatology, for example, to feed into climate models or for technical proposals. To that end, the AERONET network (database, radiative transfer model and its validation) is presented in section 2, while section 3 shows a climatology of the direct radiative forcing of key aerosol types at the bottom and at the top of atmosphere. Finally, the main remarks are summarized in the section 4.

2. AErosol RObotic NETwork (AERONET)

2.1 AERONET Database

AERONET is one of the most useful global networks for monitoring atmospheric aerosols. It collects near real time observations of spectral and columnar integrated aerosol optical properties. These data are collected by automatic sun and sky scanning spectral radiometers manufactured by CIMEL Electronique and they are distributed to worldwide locations. The direct sun measurements are performed by the CIMEL radiometers on all or some of the following channels: 0.34, 0.38, 0.44, 0.50, 0.67, 0.87, 0.94, 1.02 and 1.64 μm (nominal wavelengths) with a field of view of 1.2 degrees. Then, the aerosol optical depth (AOD) is retrieved at all these wavelengths except at 0.94 μm, which is used to estimate total precipitable water content. In addition to the direct solar radiance measurements, these instruments measure the sky radiance in four spectral bands (0.44, 0.67, 0.87 and 1.02 μm) along both the solar principal plane and the solar almucantar.

Solar aureole/sky radiance together with sun measurements are used to retrieve aerosol volume size distributions ($dV(r)/d\ln(r)$ from 0.05 to 15 μm), spectral complex refractive index ($m(\lambda)$-$ik(\lambda)$) and single scattering albedo ($\omega(\lambda)$) at low solar elevations (solar zenith angles, sza, between 50° and 80°), following a flexible inversion algorithm developed by Dubovik & King (2000) (Version 1.0 inversion products). This algorithm uses models of homogeneous spheres and randomly oriented spheroids (Dubovik et al., 2002a). Recently a new version of this inversion algorithm, Version 2.0, has been developed. The most significant improvement is the use of a spheroid mixture as a generalized aerosol model (representing spherical, non-spherical and mixed aerosols, with shape parameter between 0.3 and 3.0) (Dubovik et al., 2006), and replacing the spherical and spheroid models used separately up to now. In this vein, the Version 2.0 provides a parametrization of the degree of non-sphericity (sphericity parameter), as well as the same set of retrieved aerosol parameters given in the Version 1.0. The AERONET inversion scheme is clearly summarized in paper by Dubovik et al., (2011).

The appropriate characterization of the surface albedo is a critical issue to estimate aerosol radiative effect (Myhre et al., 2003) as well as an important error source in the retrieval of aerosol properties (Dubovik et al., 2000; Sinyuk et al., 2007). For that reason, one of the most important improvements in the Version 2.0 is the assumption of a dynamic spectral and spatial satellite and model estimation of the surface reflectivity (SR), including the bidirectional reflectance distribution function (BRDF), in the place of an assumed surface reflectivity (Dubovik et al., 2000). Thus, the BRDF Cox-Munk model over water was used, which takes the wind effect over water into account by using the wind speed data from NCEP/NCAR database (NOAA National Weather Service NOMADS NCEP server). For land surface covers, the Lie-Ross model was adopted, where the BRDF parameters are taken from the MODIS Ecotype generic BRDF models for vegetation, snow and ice (http://aeronet.gsfc.nasa.gov). Finally, another important addition in the AERONET inversion products Version 2.0 is that a set of radiative magnitudes are estimated at any AERONET station: spectral and broadband fluxes, besides aerosol radiative forcing and aerosol radiative forcing efficiency at the bottom of the atmosphere (BOA) and at the top of atmosphere (TOA), which allow to study the radiative effects under different aerosol regimes. The AERONET radiative transfer model used to estimate them is explained in detail in the subsection 2.2, while the validation of AERONET solar flux estimates is shown in the subsection 2.3.

The level of accuracy in the CIMEL measurements is a critical issue in the inversion process, since the retrieval algorithm is set to fit the data to the level of AERONET measurement uncertainty, i.e., the nominal error in AOD is assumed 0.015 cos (sza), while for sky radiance measurements the error is ±5%. These values are determined by calibration conditions. For the direct measurements, the calibration of field instruments is performed by a transfer of the calibration from reference CIMELs, which are calibrated by the Langley plot technique at Mauna Loa Observatory (Hawaii). Typically, the total uncertainty in spectral aerosol optical depth for the field instruments ranges from 0.01 to 0.02 under cloud-free conditions for air mass equal one (Eck et al., 1999), with the highest errors (0.02) associated with the ultraviolet wavelengths. For the sky radiance measurements, the calibration is performed by comparing to a reference integrating sphere with an accuracy of ±5% or better at the NASA

Goddard Calibration Facility (Holben et al., 1998). Regarding the long-term stability of the calibration coefficients, the optical interference filters are the main limiting factors. On average, a decrease from 0 to ~5% per year is expected, depending largely on material deposition on the optics.

The AERONET aerosol products are computed for three data quality levels: Level 1.0 (unscreened), Level 1.5 (cloud-screened), and Level 2.0 (cloud-screened and quality-assured) (http://aeronet.gsfc.nasa.gov). The AERONET data used in this study are taken from the highest quality version under clear-sky conditions, i.e., Version 2.0 at level 2.0.

2.2 AERONET radiative transfer approach

As part of the efforts to enhance the value of retrieval products, a new radiative transfer module has been integrated into operational AERONET inversion code. This module uses the detailed size distribution, complex refractive index and fraction of spherical particles retrieved by AERONET (Dubovik and King, 2000, Dubovik et al. 2006) and provides the fluxes and aerosol radiative forcing values as part of AERONET operational product. Similar to the AERONET retrieval, the AERONET computations of solar fluxes account for the absorption and multiple scattering effects using the Discrete Ordinates DISORT approach (Nakajima & Tanaka, 1988; Stamnes et al., 1988). The solar broadband fluxes are calculated by spectral integration from 0.2 to 4.0 μm, using more than 200 size sub-intervals. In each of these sub-intervals the extinction, single scattering albedo and phase function are calculated using the retrieved size distribution in the exact same manner as in the AERONET retrieval scheme. The values of $m(\lambda)$ and $k(\lambda)$ are linearly interpolated and extrapolated from the values $m(\lambda)$ and $k(\lambda)$ retrieved at the AERONET wavelengths. Likewise, the spectral dependence of surface reflectance is linearly interpolated and extrapolated from surface albedo values assumed in the retrieval of the sun/sky-radiometer measurements. Note that the AERONET solar fluxes are only evaluated for solar zenith angles between 50° and 80°, where the solar geometry conditions are the most appropriate for retrieving the aerosol properties (Dubovik et al., 2002b, 2000).

The integration of atmospheric aerosol scattering and absorption, gaseous absorption, molecular scattering and underlying surface reflection effects are conducted using developments employed in the GAME (Global Atmospheric ModEl) code (Dubuisson et al. 1996; Roger et al., 2006). In the GAME code, gaseous absorption (mainly H_2O, CO_2 and O_3), is calculated from the correlated k-distribution. The correlated k-distribution allows for the interactions between gaseous absorption and multiple scattering to be accounted for with manageable computational time. Coefficients of the correlated k-distribution were estimated from reference calculations using a line-by line code (Dubuisson et al., 2004). Regarding the gaseous content in the atmospheric column, the GAME model accounts for spectral gaseous absorption: ozone in the ultraviolet-visible spectral range (0.2-0.35 μm and 0.5-0.7 μm) and water vapour in the shortwave infrared spectrum (0.8-3 μm). The instantaneous water vapour content retrieved by AERONET (Smirnov et al. 2000) is employed, while the total ozone content is taken from monthly climatology values (1978-2004) based on the NASA Total Ozone Mapping Spectrometer (TOMS) measurements (http://jwocky.gsfc.nasa.gov/). The atmospheric gaseous profile, US standard 1976 atmosphere model, was scaled to match with the gaseous concentrations in column. The GAME code has a fixed spectral resolution of 100 cm^{-1} from 2500 to 17700 cm^{-1} (4 to 0.6 µm) and 400 cm^{-1} from 17700 cm^{-1} to 50000 cm^{-1} (0.6 to 0.2 µm).

The AERONET calculation of broadband radiation is focused on accurate accounting for the spectral dependence of the aerosol optical properties and surface reflectivity used as inputs. Recent studies showed that both to neglect completely the spectral dependence and to consider only specific spectral range of aerosol properties are important error sources. These studies found uncertainties up to 30% for total aerosol radiative effect, combining the spectral influence and solar zenith angle variation (Myhre et al., 2003; Zhou et al., 2005). Also, it should be noted that in contrast with simplified approaches (accounting only for asymmetry of phase function) flux calculations employed in AERONET processing use the detailed phase function (12 moments).

As aforementioned, the radiative forcing is introduced in literature to account for changes in the solar radiation levels due to changes in the atmospheric constituents. Hence, the direct radiative forcing of atmospheric aerosols, denoted as ΔF, is defined as the difference in the energy levels between a situation where aerosols are present, F^A, and a situation where these atmospheric particles are absent, F^C. AERONET estimates these values at the BOA and the TOA. Thus, radiative forcing can be defined at these two levels as:

$$\Delta F_{BOA} = F_{BOA}{}^{\downarrow A} - F_{BOA}{}^{\downarrow C} \tag{1}$$

$$\Delta F_{TOA} = F_{TOA}{}^{\uparrow C} - F_{TOA}{}^{\uparrow A} \tag{2}$$

where the arrows indicate the direction of the solar global fluxes: $\downarrow \equiv$ downward flux and $\uparrow \equiv$ upward flux, which are computed as aforementioned. This sign criterion implies that negative values of ΔF at the BOA and at the TOA are associated with an aerosol cooling effect, while a warming effect is due to positive values of ΔF at the BOA and at the TOA. In climate applications it is also often used the net radiative forcing, which is defined at the BOA as:

$$\Delta F_{BOA\ net} = F_{BOA}{}^{A} - F_{BOA}{}^{C} \tag{3}$$

where $F_{BOA}{}^{C}$ and $F_{BOA}{}^{A}$ are the net (downwelling minus upwelling) fluxes in aerosol absent and aerosol present situations, respectively. Therefore, the net BOA forcing accounts for interaction with the surface and depends on the surface reflectance. At the TOA the values of the net forcing are identical to what can be found from the equation (2) because at the TOA the downwelling (extraterrestrial) flux is the same either for aerosol absent or aerosol present conditions.

The aerosol radiative forcing strongly depends on solar geometry, so the ΔF results presented in this work (section 3) have been limited to a narrow interval of solar zenith angles (60±5°), where the maximal number of AERONET retrievals are assembled. The whole AERONET scheme, including the inversion process and the simulation of solar fluxes, is summarized in the Fig. 1.

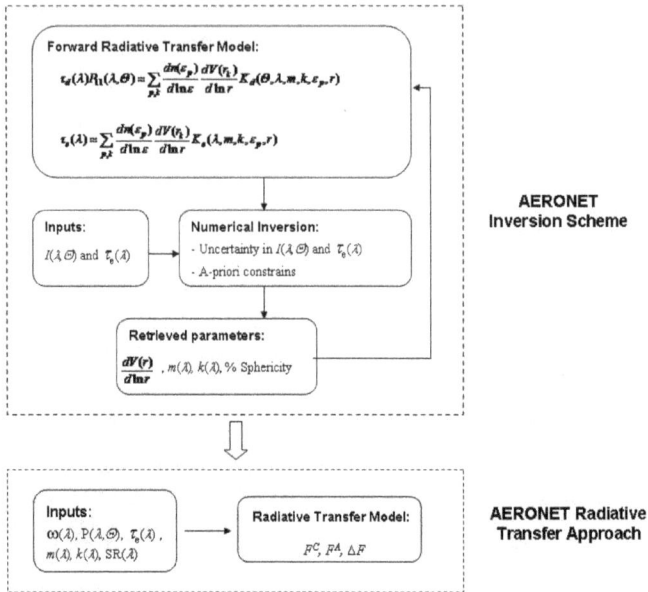

Fig. 1. AERONET scheme to compute solar broadband fluxes: AERONET inversion scheme (Dubovik et al., 2006) and AERONET Radiative Transfer approach.

2.3 Validation of AERONET solar flux estimates

High quality solar flux estimates, and hence aerosol radiative forcing values, are crucial to properly analyze annual cycles and possible trends. Furthermore, such high quality reference data are needed in order to document the quality and consistency of data obtained from space-based sensors or climate model simulations. Therefore, the intercomparison with observed solar measurements is mandatory. Thus, the AERONET broadband flux estimates have been validated in recent works by using ground-based solar measurements from global networks and during intensive field campaigns (Derimian et al., 2008; García et al., 2008). As a result, AERONET solar fluxes agree with solar observations within 10% or better.

García et al., (2008), using ground-based solar measurements under different aerosol environments, found that the straight forward comparison between the observed and modeled solar irradiance shows an excellent agreement (slope of 0.98±0.00 and bias of -5.32±1.00 Wm-2), with a correlation of 99% (see Fig. 2). As a result, in global terms, a small overestimation of +9±12 Wm-2 is found on the observed solar radiation at surface, which means a relative error of +2.1±3.0%. For each aerosol type, these differences range from -2±8 to -3±7 Wm-2 under continental background, from +8±9 to +16±10 Wm-2 for stations with mineral dust conditions, +14±10 Wm-2 and +6±13 Wm-2 for urban-industrial and biomass burning aerosols respectively, lower than +13±10 Wm-2 under oceanic environments and +2±10 Wm-2 in the free troposphere conditions observed in the Mauna Loa Observatory. These errors are expected to be of the same magnitude at the TOA, since the same methodology is used at both levels (gaseous and aerosol distribution, radiative model, etc).

These differences could be attributed, partly, to the combination of instrumental errors such as cosine response, calibration, linearity, etc., in the pyranometer measurements. All of these error sources usually give uncertainties of less than 3%-5% of the instantaneous instrument signal (Dutton et al., 2001). Given the solar zenith angle range used to AERONET sky measurements (50°-80°), one of the most influencing factor is the non-ideal angular response of this type of instruments, which limits their accuracy to about 3%, or 20-30 Wm^{-2}, for instantaneous clear-sky measurements (Michalsky et al., 1999). This effect would result in measured flux that is biased low relative to true flux, resulting thus the overestimation observed in the AERONET solar broadband fluxes.

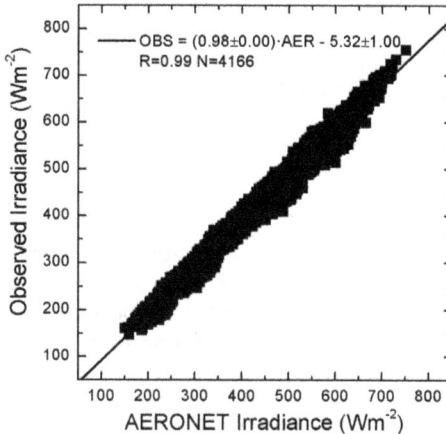

Fig. 2. Comparison between the solar irradiance (Wm^{-2}) observed by ground-based pyranometer and modelled with AERONET inputs at surface (García et al., 2008). The legend shows the least square fit parameters.

In addition to the instrumental uncertainties, the uncertainty in surface albedo and BRDF, assumed by AERONET, must also be considered as a potential error source in the modeled downwelling fluxes. This is especially true for the locations dominated by mineral dust, where the surface albedo is usually quite bright. In such conditions, the values managed in AERONET may be higher than real ones, providing an overestimation of the observed irradiance as was shown previously. This artefact will be more pronounced for high aerosol load due to the multiple scattering processes. Besides the uncertainties in the surface albedo for large solar zenith angles must be taken into account in the error interpretation.

Model assumptions as the homogeneous aerosol vertical structure considered in AERONET retrievals can be considered also as sources of uncertainties; however the effect of these errors is minor in most cases (Dubovik et al., 2000). The flux calculations are performed for multi-layered atmosphere with US standard atmosphere model for gaseous distributions and single fixed aerosol vertical distribution (exponential with aerosol height of 1 km). The deviations of these assumptions from reality are also potential source of errors, although, our tests did not show any significant sensitivity of flux estimates to these assumptions. Differences less than 1 Wm^{-2} due to different vertical profiles were observed on the downward solar flux at the bottom of the atmosphere (García et al., 2008). These differences are negligible (~0.2-3%) compared to instantaneous aerosol radiative forcing.

Regarding the aerosol shape, the AERONET solar flux estimates account for particle non-sphericity, a critical issue especially for mineral dust aerosols. Calculation of broadband radiative flux often does not consider particle non-sphericity. It is because the possible angular and spectral differences are expected to be canceled once all contributions of scattered light summed up into the total energy flux. Mishchenko et al. (1997) found that the Mie theory can be used for aerosol radiative effect calculations with adequate accuracy even for nonspherical aerosol particles. It was noted that using asymmetry parameter, which is an integrated value, the contrasting nonspherical-spherical differences in the aerosol phase function are averaged out during the integration. However, in AERONET flux estimates computations are conducted using not a limited integrated value of asymmetry parameter, but using twelve moments of phase function expansion that account for the details of the aerosol phase function. In addition, recalculating phase function at each step of spectral integration includes the spectral variations of the phase function. Using this type of computations, in a study by Derimian et al. (2008) it was shown that neglecting particle non-sphericity in the simulations of broadband fluxes in the presence of mineral dust results in an overestimation of aersosol radiative forcing by up to 10%.

In many studies in the literature, the aerosol radiative forcing simulations are carried out assuming no errors in the solar fluxes without aerosols or they are negligible with regard to the uncertainty of the observed measurements. Nonetheless, as found for the Mauna Loa Observatory, which is quite representative of a clear atmosphere, we have documented that its uncertainties are in the same range that those found under higher aerosol load. Hence, it is necessary to give special attention to this assumption because it seems to be not completely valid.

3. AERONET radiative forcing of key aerosol types

3.1 Description of key aerosol types

Atmospheric aerosols have a high temporal and spatial variability due to their different residence times in the atmosphere and the geographical distribution of their emission sources. This heterogeneity leads atmospheric aerosols to have differentiated size distributions and chemical compositions and, thus, different impacts on the radiative balance of the Earth-atmosphere system. These properties directly determine how atmospheric aerosols interact with the solar radiation and the relative importance between scattering and absorption processes. In order to account for this heterogeneity of aerosol radiative properties, this study examines the radiative effect of the individual aerosol species, emitted into the atmosphere from different source areas. Thus, the key aerosol types considered here have been shown to have an important influence on the radiative balance, according to their rates of emission into the atmosphere and to their radiative properties (Dubovik et al., 2002b; IPCC, 2007): desert mineral dust introduced into the atmosphere by erosive processes; biomass burning produced by farming activities and forest fires; urban-industrial, continental background, oceanic aerosols and those that are found in free troposphere conditions. Fig. 3 shows the scanning electron microscopy (SEM) observations of some of these atmospheric aerosols observed during MINATROC (MINeral dust Aerosol and TROpospheric Chemistry) campaign in 2002 in Tenerife (Canary Islands) (Alastuey et al., 2005): (a) spongy carbonaceous particles from urban-industrial combustion; (b) sodium chloride crystals from marine aerosols; (c) aggregates of micro-crystals of clay minerals typical of mineral dust environments and (d) silica skeletons of fresh water diatom present in lakes or ponds from Northern Africa.

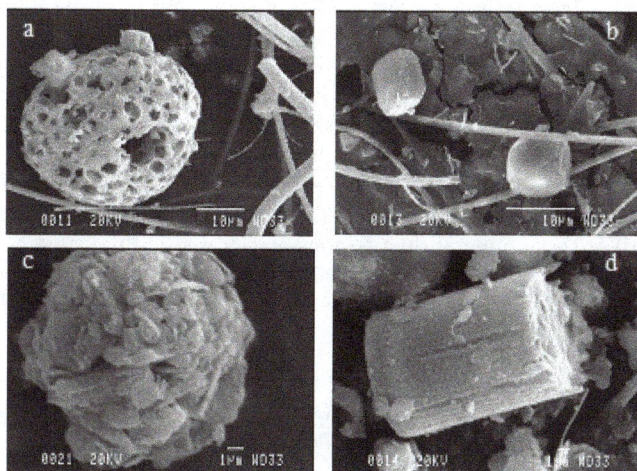

Fig. 3. SEM photographs of particle matter observed over Tenerife (Canary Islands) during the MINATROC campaign in 2002.

Following García et al. (2011b), in order to analyze the radiative effect of the aforementioned atmospheric aerosols a set of forty AERONET stations were selected (Fig. 4), which have been grouped into 14 regions to account for key aerosol types from different aerosol sources.

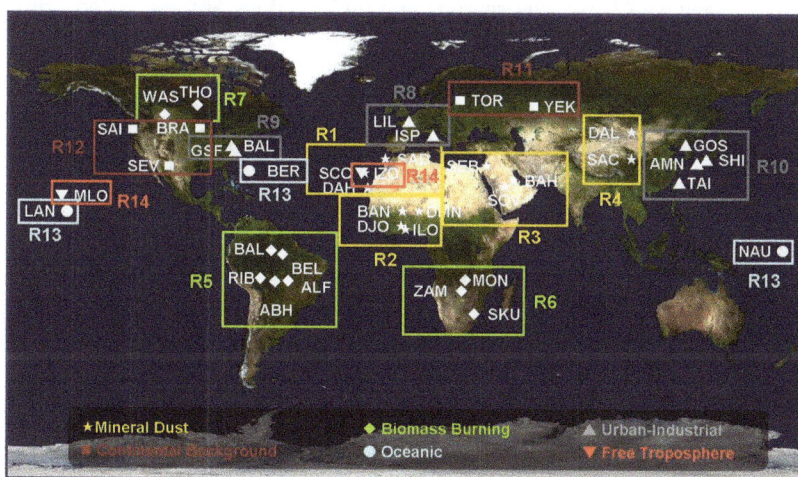

Fig. 4. Geographical distribution of the AERONET stations used. These stations were grouped into 14 regions (from R1 to R14), according to the key aerosol types and their geographic location (García et al., 2011b). Each aerosol type is labelled with a different color: mineral dust as yellow, biomass burning as green, urban-industrial as gray, background continental as brown, oceanic as blue and free troposphere as red.

3.1.1 Desert mineral dust

Desert mineral dust is one of the main aerosols emitted into the atmosphere on a global scale, and it therefore plays a decisive role in the radiative balance of the climatic system. In general, mineral dust particles show a high capacity for attenuating solar radiation (high AOD) and are mainly made up of large particles, which normally form aggregates of micro-crystals. The chemical and mineralogical composition of these particles strongly depends on the characteristics and composition of soils that generate them. For these reasons, this discussion is dedicated to mineral dust from the main desert dust areas of the world: the Sahara-Sahel desert area (distinguishing two regions: North, R1, and Central Africa, R2), the Middle East (Arabian Peninsula, R3) and the Taklamakan and the Gobi deserts (China and Mongolia, R4). All of these mineral dust regions are located within "dust belt" (5°N-40°N), where the mineral dust concentrations are maximal on a global scale (Prospero et al., 2002). Please refer to García et al. (2011b) for a detailed description of the AERONET stations used and the regions.

The mineral dust properties are clearly observed in the annual cycle of the inter-annual and regional average of the aerosol optical depth (AOD) and the single scattering albedo (ω) at 0.55 μm and the effective radius of the aerosol size distribution for all regions (Fig. 5, Fig. 6 and Fig. 7 respectively). The ω is a measure of the effectiveness of scattering relative to total extinction for the light encountering the atmospheric aerosol particles (the ratio of scattering optical depth to the total optical depth), while the effective radius is an area weighted mean radius of the aerosol particles. The combination of AOD, ω and effective radius information allows us to describe and discriminate aerosols with similar solar extinction but different sizes, such as mineral dust and biomass burning particles. An example is the contribution of biomass burning particles due to farming activities during the dry season (winter) in the Central Africa region (R2). In this season high AOD values are associated with a clear decrease of the ω and the effective radius in contrast to spring, where the mineral dust is the predominant aerosol and, thus, larger particles with a lower absorption at 0.55 μm are observed.

In order to guarantee the quality of the AERONET ω estimations this magnitude is only retrieved for high aerosol load (AOD (0.44 μm)≥0.4) (Dubovik et al., 2002b; 2000).

3.1.2 Smoke from vegetation fires

Biomass burning aerosols are dominated by fine particles of black carbon and organic material from forest fires of local and regional farming activities, mainly from the tropical regions of Amazonian forest (R5) and African savannah (R6) (25°S-25°N) during the dry season. Their radiative properties are mainly given by the different material composition and combustion processes associated with their production (with or without flames). Thus, in the African savannah, most biomass burning particles are produced by means of flaming combustion, while in deforestation fires, typical of Amazon forest, ~50% or less of the biomass burning is produced by this phase of combustion. These different production mechanisms lead to different concentrations of black carbon: 15-20% of the aerosols generated during the flaming combustion phase correspond to black carbon while less than 3% are generated during smouldering combustion (García et al., 2011b and references therein) and, thus, significant differences occur for the single scattering albedo (Fig. 6.b).

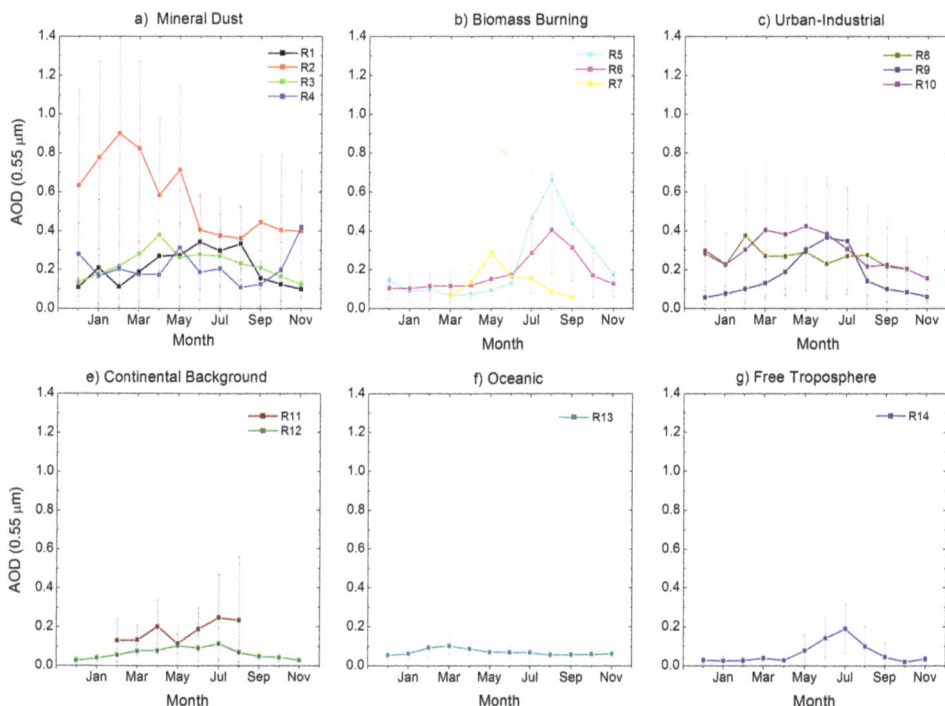

Fig. 5. Annual cycle of the inter-annual average of the aerosol optical depth (AOD) at 0.55 μm for the key aerosol types grouped into 14 regions. Error bars indicate 1σ (standard deviation).

Together with the production of biomass burning due to agriculture, this aerosol is also produced, but to a lesser extent, in fortuitous or provoked forest fires over the continental platforms of North America and Eurasia (R7). These fires occur almost exclusively during summer, when meteorological conditions are more favourable. In these cases the smouldering combustion aerosols are majority (Dubovik et al., 2002b).

Besides the aerosol size, a clear difference between mineral dust and other aerosols, such as biomass burning or urban-industrial ones aerosols is the spectral dependence of the absorption properties. Mineral dust show the highest absorption at wavelengths close to the ultraviolet solar range in contrast to biomass burning and urban-industrial aerosols, where the absorption increases in the near infrared region (Dubovik et al., 2002b, Hess et al., 1998). Also, it should be noted that, while mineral dust was previously considered as strongly absorbing (e.g. Hess et al., 1998), Kaufman et al. (2001) and (Dubovik et al., 2002b) demonstrated using AERONET and satellite observations that mineral dust is practically does not absorb in visible and near infrared.

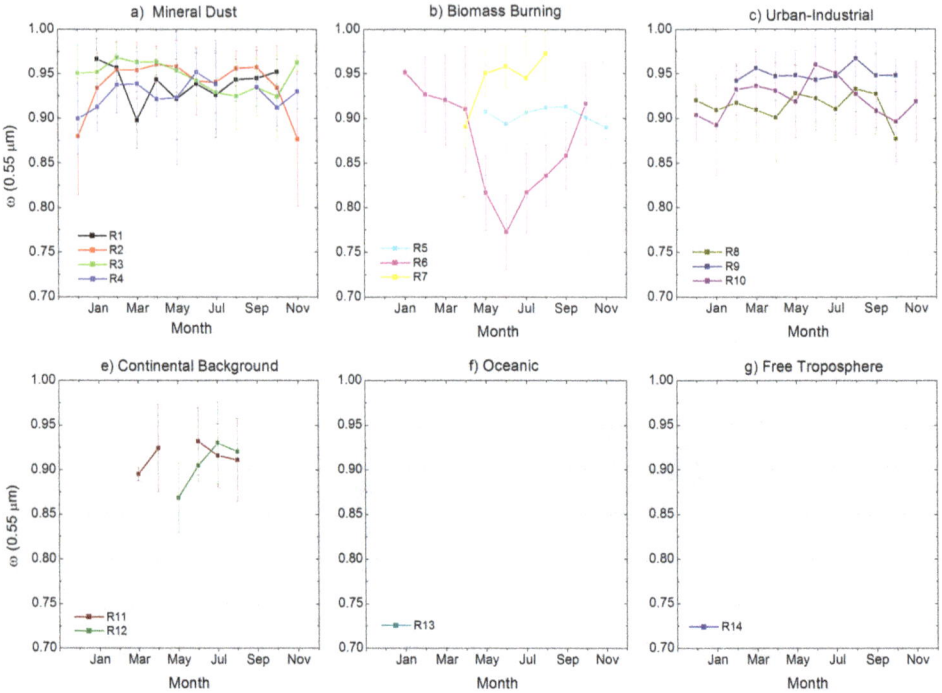

Fig. 6. Annual cycle of the inter-annual average of the aerosol single scattering albedo (ω) at 0.55 μm for the key aerosol types grouped into 14 regions. Error bars indicate 1σ (standard deviation).

3.1.3 Urban and industrial pollution

Urban-industrial aerosols are very small particles so the fine mode is predominant in the size distribution. The main sources of these aerosols are found in the most heavily populated and industrialized areas of the world, such as Europe (R8), the East and West coast of North America (R9) and Southeast Asia (R10). The industrial combustion processes, the population or even the regional meteorological conditions are responsible for the differences in their radiative properties. For example, although road traffic is a common source in all regions, and in some cases the main source of contamination, Europe has a higher proportion of diesel automobiles than United States (Dubovik et al., 2002b). Also, new chemical measurements show that downwind of the eastern United States the contribution of carbonaceous material to AOD (30%) is double that of sulphates (16%),with water intake (48%) and black carbon (6%) accounting for the rest (Kaufman & Koren, 2006). Hence, the American region exhibits a lower absorption capacity (see Fig. 6.c).

Note that in the Asiatic region there are notable episodes of mineral dust transport from the Gobi-Taklamakan deserts during spring and at the beginning of summer, which are responsible for the spring peak observed in the AOD and the effective radius annual cycles (Fig. 5.c and Fig. 7.c respectively).

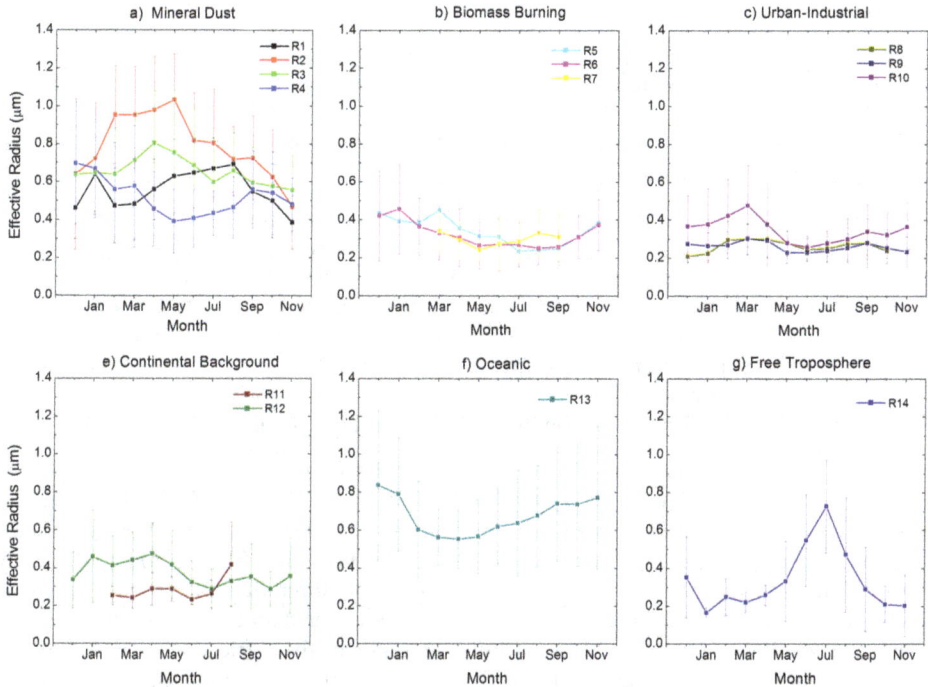

Fig. 7. Annual cycle of the inter-annual average of the effective radius (μm) for the key aerosol types grouped into 14 regions. Error bars indicate 1σ (standard deviation).

3.1.4 Background continental

Background continental aerosols are typical from non-industrial areas located over eastern European (R11) and American (R12) prairies, where the main aerosol type generally derives from continental sources, such as biogenic emissions from forests, re-suspension of particulate matter from grass and soil, etc. Normally, the presence of local anthropogenic pollution sources is not significant in these regions, although they may be affected by regional transport from nearby urban-industrial areas or forest fires. Therefore, they may show similar properties to urban-industrial (R8 and R9) and biomass burning aerosols from boreal forest fires (R7): exhibiting small sizes and a significant absorption (Fig. 7.d and Fig. 6.d respectively).

3.1.5 Oceanic aerosol

Sea spray generated from the surface of the seas and oceans is the most common component globally (R13). Nonetheless, they have a minor role in the solar radiative balance of the Earth-atmosphere system due to their low capacity for extinguishing solar radiation (low and stable AOD, Fig. 5.e). These particles are also characterized by a pronounced mode of coarse particles with effective radius of the order of mineral dust, except that oceanic aerosols have a spherical shape because of their hygroscopic character. These particles are normally the result of two physical processes: stirring the sea surface by wind and the rupture of air bubbles during the formation of sea foam.

3.1.6 Free troposphere

The atmospheric conditions for monitoring atmospheric aerosols in the free troposphere are very special. These measurements should be representative of natural background conditions. This difficult requirement is only met in a few high mountain observatories, which are exposed to a natural undisturbed atmosphere with minimal influences from natural ecosystem or human activities.

In order to evaluate the aerosol properties of the free troposphere (R14) two GAW (Global Atmospheric Watch) stations were considered: Mauna Loa and Izaña Observatories. Both places are located at a high altitude (2,367 and 3,397 m a.s.l. respectively) and above a thermal inversion layer, thus avoiding the contributions of natural and anthropogenic aerosols. The exceptions are episodes of African desert mineral dust measured at the Izaña station during the summer, months of greatest activity in the North Sahara-Sahel region. Thus, these events are responsible for the marked increase of AOD and effective radius observed in the free troposphere region during summer (Fig. 5.f and Fig. 7.f respectively).

In order to estimate a budget of aerosol emissions, the frequency of occurrence of those situations with a high extinction of solar radiation has been computed for each region and aerosol type (Fig. 8). For that, we have only considered cases with AOD (0.44 μm)$\geq$0.4 according to AERONET's high quality threshold. The Central Africa region (R2) clearly shows the highest number of cases that exceeds the AOD limit, 32% of those analyzed, in contrast to 15% in the Arabian mineral dust or values lower than 5% in the other regions under mineral dust. Regarding biomass burning regions, the African savannah has the majority impact on the AOD levels (~15%) and doubles the cases observed in the Amazonian region. This latter region and the urban-industrial areas show similar percentages, ~8%, while for background continental regions they represent less than 1% of the total. Finally, in the oceanic and the free troposphere regions there are no situations that fulfilled the AOD threshold.

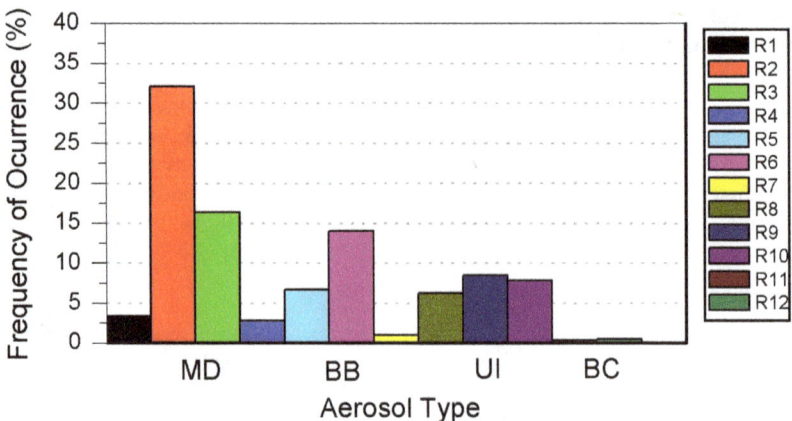

Fig. 8. Frequency of Occurrence (%) of cases with AOD (0.44 μm)$\geq$0.4 for the key aerosol types grouped into 14 regions. The key aerosol types are labelled as: MD$\equiv$Mineral Dust; BB$\equiv$Biomass Burning; UI$\equiv$Urban-Industrial; BC$\equiv$Background Continental; O$\equiv$Oceanic and FT$\equiv$Free Troposphere.

3.2 Radiative Forcing at the BOA and at the TOA

As aforementioned, AERONET computes the direct radiative forcing at the BOA and at the TOA in order to study the effect of atmospheric aerosols on the climate system. The high density of the AERONET stations allows for a detailed analysis of ΔF and thus the estimates of the total radiative effect of key aerosol types.

The aerosol radiative forcing strongly depends on the total aerosol extinction (AOD), the solar geometry and the surface type. This latter parameter is especially critical in estimating the ΔF at the TOA, since a clear decrease of the ΔF absolute values has been documented as surface reflectivity increases, even changing the sign of the radiative forcing. Recent studies found that over dark surfaces (oceanic and vegetative covers, i.e., surface reflectivity, SR, lower than 30%) atmospheric aerosols always cool the Earth-atmosphere system, regardless of aerosol type (García et al., 2011a, 2011b). Nonetheless, over the brightest surfaces, the total radiative effect depends on the aerosol absorption properties and on the SR values. Therefore, in order to make a consistent comparison of the net aerosol effect, the analysis has been partitioned into two ranges of surface reflectivity: SR≤30% and SR>30%. Besides, the data shown are limited at sza of 60±5° to take the nearly same geometry into account.

The annual evolution of the ΔF is mainly determined by the annual variations of total aerosol extinction (AOD), as shown in Fig. 9 and 10 (ΔF values at the BOA, eq. 1, and at the TOA, eq. 2, respectively for SR≤30%). Thus, the ΔF annual cycle of each key aerosol type, both at the BOA and the TOA, perfectly follows its respective AOD annual evolution (see Fig. 5). The highest annual ΔF amplitude (in absolute value) is observed for the mineral dust regions, while the oceanic aerosols present the flattest ΔF annual cycle. The lowest ΔF is observed for the free tropospheric region, however its amplitude variations present a significant peak value in summer, associated with the arrival of mineral dust from the Saharan desert to the Izaña station. The biomass burning aerosols also show a clear seasonality, increasing the ΔF values (in absolute value) during the biomass burning season (from July to December). Finally, the urban-industrial regions have a wide ΔF amplitude mainly from January to August. It is important to mention that the ΔF inter-annual averages were negative for all regions both at the BOA and the TOA, denoting a global cooling effect by atmospheric aerosols.

The maximal values of ΔF at the BOA and at the TOA are found in the Central Africa region during winter, associated with the mixture of mineral dust from the Sahara-Sahel desert and biomass burning aerosols, around -180 Wm^{-2} at the BOA and higher than -40 Wm^{-2} at the TOA. During the summer period (wet season) the ΔF values are comparable to other mineral dust regions or other aerosol species (like biomass burning or urban-industrial particles). In this region, we also found the highest AOD levels among those analyzed. Furthermore, it is remarkable the peak observed in the biomass burning regions from South America and South Africa during the fall, a period of maximal biomass burning activity, with ΔF values higher than -80 Wm^{-2} at the BOA and around -40 Wm^{-2} at the TOA. These values are even more important than Saharan (R1), Arabian (R3) and Asian (R4) mineral dust and of the order of those observed under urban-industrial environments. These latter regions correspond to the most industrialized areas in the world (Southeast Asia, Europe and the East coast of North America), where their radiative forcing values range between 0 and -80 Wm^{-2} at the BOA and from 0 to -40 Wm^{-2} at the

TOA, during the first half of the year, diminishing slightly in the second half. Regarding the areas dominated by background continental aerosols, the results show that the forcing values in the European region (R11) are double those observed in the American region (R12). Finally, the lowest values of forcing are obtained in the oceanic region and under free troposphere conditions, with values lower than -10 Wm^{-2} both at the BOA and at the TOA.

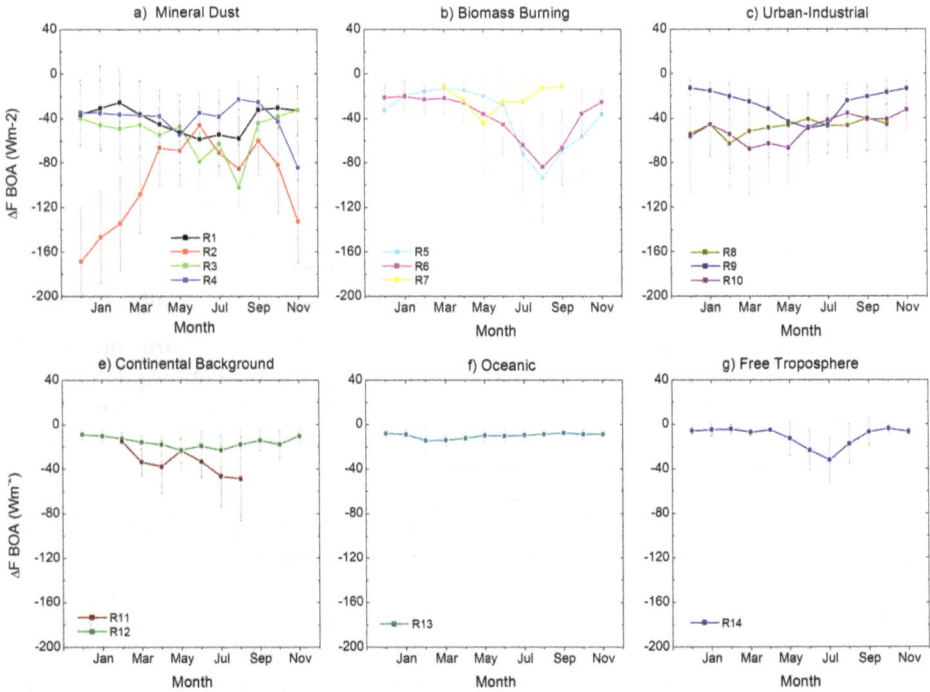

Fig. 9. Annual cycle of the inter-annual average of the aerosol radiative forcing (ΔF, Wm^{-2}) at the BOA for the key aerosol types grouped into 14 regions and for surface reflectivity SR$\leq$30%. Error bars indicate 1σ (standard deviation)

The BOA radiative forcing provides decrease of the solar radiation at the surface due to scattering of radiation back to space and due to trapping of radiation in the atmospheric layer. Therefore, decrease of the solar radiation at the surface in conjunction with low aerosol scattering effectiveness (low single scattering albedo) indicates increase of radiation trapped in the atmospheric layer due to aerosol absorption, which is directly related to the aerosol semi-direct effect. This absorption produces an atmospheric temperature gradient contributing to the modification of atmospheric dynamic and cloud characteristics. As can be seen from Fig. 9 the most significant BOA forcing occurs in the Amazonian, South Africa and Central Africa regions. The Amazonian and South Africa are known as biomass burning regions, while the Central Africa in winter time is characterised by the mixture of mineral dust and biomass burnings particles, in all these cases the single scattering albedo (Fig. 6) is low presenting important aerosol absorption.

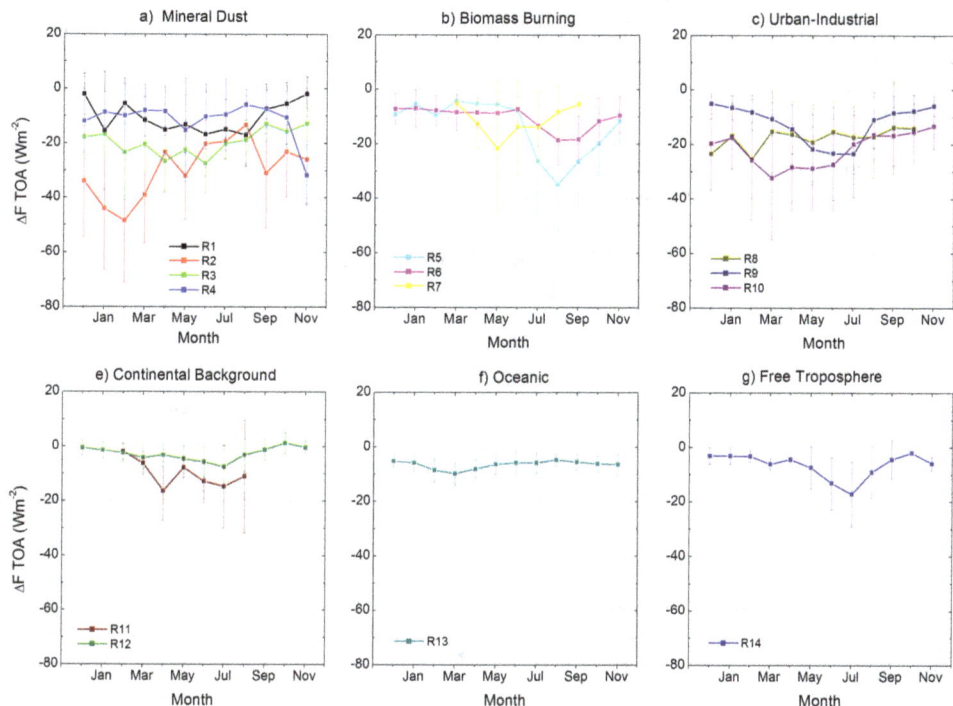

Fig. 10. Annual cycle of the inter-annual average of the aerosol radiative forcing (ΔF, Wm^{-2}) at the TOA for the key aerosol types grouped into 14 regions and for surface reflectivity SR≤30%. Error bars indicate 1σ (standard deviation).

How atmospheric aerosols affect cloud cover (increase or decrease) is one of the largest uncertainties in climate models (IPCC, 2007). In fact, recent studies show that heavy smoke over the Amazon forest and pollution over China decrease the cloud cover by heating the atmosphere and cooling the surface and these effects may balance some of the large negative aerosol forcing (Kaufman & Koren, 2006, and references therein). Likewise, Satheesh & Ramanathan (2000) have shown that, over the tropical Indian Ocean during winter months, absorbing aerosols produce important atmospheric heating, which is translated into a diurnal mean heating rate perturbation. Therefore, the assessment of the seasonal variations in the ΔF values at the BOA and at the TOA presented here is fundamental to document properly these aerosol effects in climate modelling both at the regional and global scale. Furthermore, the cloud properties, one of the most crucial climatic elements, strongly depend on these seasonal behaviours.

On the other hand, a very interesting issue is to observe whether there are differences in radiative forcing when the different aerosol types under study are grouped according to their origin. Thus, the regions R1, R2, R3, R4, R7, R13 and R14 were considered as natural, and the regions R5, R6, R8, R9 and R10 as anthropogenic. With this selection we determine that the natural aerosols carry the most weight in the radiative balance. The ΔF annual average at BOA for anthropogenic aerosols is -88±36 Wm^{-2} (AOD=0.54±0.28 at 0.55 µm) while it is -121±47 Wm^{-2} (AOD=0.77±0.43) for natural aerosols. The ΔF at the TOA for anthropogenic aerosols is -31±15 Wm^{-2} versus -36±20 Wm^{-2} observed for natural aerosols.

The influence of the surface reflectivity on ΔF is clearly observed, for example, for the mineral dust regions (Fig. 11), especially at the TOA. Thus, reductions of the absolute values of the ΔF are documented for surfaces with SR higher than 30%, reaching even positive mean values. In these latter cases, atmospheric aerosols lead to a warming of the Earth-atmosphere system, contributing to the GHGs effect.

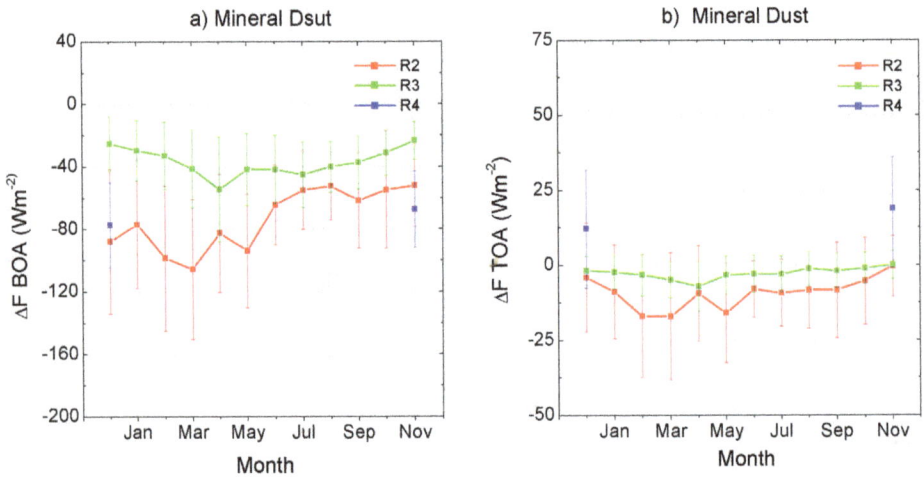

Fig. 11. Annual cycle of the inter-annual average of the aerosol radiative forcing (ΔF, Wm^{-2}) at the BOA (a) and at the TOA (b) for surface reflectivity SR>30% for the mineral dust regions. Error bars indicate 1σ (standard deviation).

From a global perspective, table 1 summarizes the annual and regional radiative forcing averages at the BOA, TOA and the AOD at 0.55 µm, considering together the whole range of surface reflectivity. Note that in order to compute mean values representative of the free troposphere on a global scale the period with possible mineral dust events at the Izaña Observatory were ruled out here. The results allow us to quantify, in clear sky conditions, the relative influence of each key aerosol type as well as the importance of the different mechanisms for modifying the energy balance of the Earth-atmosphere system (directly or by storing energy in the system). These mean values may be considered as a first reference at solar zenith angles of 60±5°, although it would be necessary to estimate annual average values from the daily radiative forcing values in order to make more solid conclusions.

Aerosol Type	Region	ΔF_{BOA} (Wm^{-2})	ΔF_{TOA} (Wm^{-2})	AOD (0.55 μm)
MD	R1	-44±28	-11±12	0.23±0.20
MD	R2	-89±51	-16±22	0.58±0.46
MD	R3	-40±23	-4±9	0.23±0.17
MD	R4	-41±32	-8±15	0.20±0.20
BB	R5	-54±45	-19±18	0.33±0.37
BB	R6	-46±37	-11±8	0.21±0.18
BB	R7	-24±26	-13±15	0.14±0.20
UI	R8	-48±30	-18±15	0.27±0.24
UI	R9	-25±20	-11±11	0.15±0.17
UI	R10	-49±34	-20±15	0.28±0.24
BC	R11	-37±22	-7±14	0.17±0.16
BC	R12	-17±12	-3±6	0.17±0.16
O	R13	-10±6	-7±4	0.07±0.04
FT	R14	-6±3	-5±3	0.03±0.01

Table 1. Annual and regional average and standard deviation of ΔF_{BOA} (Wm^{-2}), ΔF_{TOA} (Wm^{-2}) and AOD at 0.55 μm for the key aerosol types grouped into 14 regions. The key aerosol types are labelled as: MD≡Mineral Dust; BB≡Biomass Burning; UI≡Urban-Industrial; BC≡Background Continental; O≡Oceanic and FT≡Free Troposphere.

4. Conclusions

Atmospheric aerosols play an essential role in the configuration of the radiative balance of the Earth-atmosphere system. The spatial and temporal heterogeneity of their physical-chemical properties makes global networks fundamental tools for their complete characterization. Likewise, atmospheric aerosols are a decisive factor for modifying cloud properties, since they serve as cloud condensation nuclei, and hence affect the hydrological cycle as well. Moreover, those species with important absorption properties may lead to an unequal warming of the atmosphere and may also modify cloud formation and cloud cover. Therefore, improving our knowledge about the direct radiative effects of atmospheric aerosols is a necessary step in order to address the indirect and semi-direct aerosol effects.

In this context, AERONET is today the most important network for monitoring atmospheric aerosols. AERONET aerosol products are widely used by the international research community to compute aerosol climatologies, to provide data for climate models and to validate space-based sensors. Thus, this chapter presents valuable information, describing and summarizing an approach for estimating broadband solar fluxes and hence the direct radiative forcing of atmospheric aerosols from AERONET retrievals. A detailed description of the radiative transfer model and its inputs has been shown as well as the uncertainties to be expected by comparing with observations of insolation. Finally, the direct radiative forcing of key aerosol types has been analyzed: mineral dust, biomass burning, urban-industrial, background continental, oceanic and in free troposphere. These magnitudes have been evaluated at the surface and at the top of the atmosphere, allowing us to document the actual and net radiative balance of these aerosol types.

A next important step for this work would be to exploit the cloud products recently released by AERONET in order to make a more complete description of the relationships between aerosols, clouds and the energy budget in the Earth-atmosphere system. These new products, together with the information obtained from space-based and ground-based sensors, mainly the A-Train satellite constellation, will promote a deeper knowledge of the processes affecting cloud properties.

5. Acknowledgements

We acknowledge to the MEC (Ministry of Education and Science, Spain) for the following support: projects CGL2008-04740/CLI and CGL2010-21366-C04-01. We gratefully acknowledge the data provided by the AERONET network and we wish to express our appreciation to the operators of the stations for their help in operating the instruments.

6. References

Ackerman, A. S., Toon, O. B., Stevens, D. E., Heymsfield, A. J., Ramanathan, V. &Welton, E. J. (2000). *Reduction of tropical cloudiness by soot*, Science 288: 1,042–1,047.

Alastuey, A., Querol, X., Castillo, S., Escudero, M., Ávila, A., Cuevas, E., Torres, C., Romero, P. M., Expósito, F. J., García, O. E., Díaz, J. P., Dingenen, R. V. & Putaud, J. P. (2005). *Characterisation of TSP and PM2.5 at Izaña and Santa Cruz de Tenerife (Canary Island, Spain) during a Saharan dust episode (July 2002)*, Atmos. Environ. 39: 4,715–4,728.

Charlson, R. J., Lovelock, J. E., Andreae, M. O. &Warren, S. G. (1987). *Oceanic phytoplankton, atmospheric sulphur, cloud albedo and climate*, Nature 326: 655–661.

Derimian, Y., León, J.-F., Dubovik, O., Chiapello, I., Tanré, D., Sinyuk, A., Auriol, F., Podvin, T., Brogniez, G. & Holben, B. N. (2008). *Radiative properties of aerosol mixture observed during the dry season 2006 over m0bour, Senegal (African monsoon multidisciplinary analysis campaign)*, J. Geophys. Res. 113.

Dubovik, O. & King, M. D. (2000). *A flexible inversion algorithm for retrieval of aerosol optical properties from sun and sky radiance measurements*, J. Geophys. Res. 105: 20,673–20,696.

Dubovik, O., Smirnov, A., Holben, B. N., King, M. D., Kaufman, Y. J., Eck, T. F. & Slutsker, I. (2000). *Accuracy assessment of aerosol optical properties retrieval from AERONET sun and sky radiance measurements*, J. Geophys. Res. 105: 9791–9806.

Dubovik, O., Holben, B. N., Lapyonok, T., Sinyuk, A., Mishenko, M. I. & Slustker, I. (2002a). *Non-spherical aerosol retrieval method employing light scattering by spheroids*, Geophys. Res. Lett. 29: 54-1 – 54-4.

Dubovik, O., Holben, B. N., Eck, T. F., Smirnov, A., Kaufman, Y. J., King, M. D., Tanré, D. & Sluster, I. (2002b). *Variability of absorption and optical properties of key aerosol types observed in worlwide locations*, J. Atmos. Sci. 59: 590–608.

Dubovik, O., Sinyuk, A., Lapyonok, T., Holben, B. N., Mishchenko, M., Yang, P., Eck, T. F., Volten, H., Munoz, O., Veihelmann, B., van der Zande, W. J., Leon, J.-F., Sorokin, M. & Slutsker, I. (2006). *Application of light scattering by spheroids for accounting for particle non-sphericity in remote sensing of desert dust*, J. Geophys. Res. 111.

Dubovik, O., Herman, M., Holdak, A., Lapyonok, T., Tanré, D., Deuzé, J. L., Ducos, F., Sinyuk, A. & Lopatin, A. (2011). *Statistically optimized inversion algorithm for enhanced retrieval of aerosol properties from spectral multi-angle polarimetric satellite observations,* Atmos. Meas. Tech. 4: 975–1,018.

Dubuisson, P., Buriez, J. C. & Fouquart, Y. (1996). *High spectral resolution solar radiative transfer in absorbing and scattering media, Appliation to the sattelite simulations,* J. Quant. Spectrosc. Radiat. Transfer 55: 103–126.

Dubuisson, P., Dessailly, D., Vesperini, M. & Frouin, R. (2004). *Water vapor retrieval over ocean using near-infrared imagery,* J. Geophys. Res. 109.

Dutton, E. G., Michalsky, J. J., Stoffel, T., Forgan, B.W., Hickey, J., Nelson, D.W., Alberta, T. L. & Reda, I. (2001). *Measurements of broadband diffuse solar irradiance using current commercial instrumentation with a correction of thermal offset errors,* J. Atmos. Oceanic Technol. 18: 297–314.

Eck, T. F., Holben, B. N., Reid, J. S., Dubovik, O., Smirnov, A., ON´ eill, N. T., Slustker, I. & Kinne, S. (1999). *Wavelength dependence of the optical depth of biomass burning, urban, and desert dust aerosols,* J. Geophys. Res. 104.

García, O. E., Díaz, A. M., Expósito, F. J., Díaz, J. P., Dubovik, O., Dubuisson, P., Roger, J.-C., Eck, T. F., Sinyuk, A., Derimian, Y., Dutton, E. G., Schafer, J. S., Holben, B. & García, C. A. (2008). *Validation of AERONET estimates of atmospheric solar fluxes and aerosol radiative forcing by ground-based broadband measurements,* J. Geophys. Res. 113.

García, O. E., Expósito, F. J., Díaz, J. P. & Díaz, A. M. (2011a). *Radiative forcing under aerosol mixed conditions,* J. Geophys. Res. 116.

García, O. E., Expósito, F. J., Díaz, J. P., Díaz, A. M., Dubovik, O., Derimian, Y., Dubuisson, P. & Roger, J.-C. (2011b). *Shortwave radiative forcing and efficiency of key aerosol types using AERONET data,* Atmos. Chem. Phys. Discuss. 11, 1–38, 2011

Hansen, J., Sato, M., Kharecha, P. & von Schuckmann, K. (2011). *Earth's energy imbalance and implications,* Atmos. Chem. Phys. Discuss. 11: 27,031–27,105.

Haywood, J., Francis, P., Osborne, S., Glew, M., Loeb, N., Highwood, E., Tanré, D., Myhre, G., Formenti, P. & Hirst, E. (2003). *Radiative properties and direct radiative effect of saharan dust measured by the c-130 aircraft during shade: 1. solar spectrum,* J. Geophys. Res. 108.

Hess, M., Koepke, P. & Schult, I. (1998). *Optical properties of aerosols and clouds: the software pagckage OPAC,* Bull. Am. Meteor. Soc. 79: 831–844.

Holben, B. N., Eck, T. F., Slutsker, I., Tanré, D., Buis, J. P., Setzer, A., Vermote, E., Reagan, J. A., Kaufman, Y., Nakajima, T., Lavenu, F., Jankowiak, I. & Smirnov, A. (1998). *AERONET - a federated instrument network and data archive for aerosol characterization,* Remote Sens. Environ. 66: 1–16.

IPCC (2007). *The Physical Science Basis, Fourth Assessment Report Summary,* Intergovernmental Panel on Climate Change, Cambridge University Press, New York.

Kaufman, J. Y., Tanré, D., Dubovik, O., Karnieli, A. & Remer, L.A. (2001) *Absorption of sunlight by dust as inferred from satellite and ground-based remote* sensing, Geophys. Res. Lett. 28:1,479-1,483.

Kaufman, Y. & Koren, I. (2006). *Smoke and pollution aerosol effect on cloud cover,* Science 313.

Koren, I., Kaufman, Y.J., Remer, L.A. & Martins, J.V.(2004). *Measurement of the effect of Amazon smoke on inhibition of cloud formation*, Science 303:1,342-1,345.

Lohmann, U. & Feichter, J. (2005). *Global indirect aersools effects: a review*, Atmos. Chem. Pys. 5: 715-737.

Michalsky, J., Dutton, E., Rubes, M., Nelson, D., Stoffel, T., Wesley, M., Splitt, M. & DeLuisi, J. (1999). *Optimal measurements of surface shortwave irradiance using current instrumentation*, J. Atmos. Oceanic Technol. 16: 55-69.

Mishchenko, M. I., Travis, L. D., Kahn, R. A. & West, R. A. (1997). *Modeling phase functions for dustlike tropospheric aerosols using a shape mixture of randomly oriented polydisperse spheroids*, J. Geophys. Res. 102: 16,831-16,847.

Myhre, G., Berntsen, T. K., Haywood, J. M., Sundet, J. K., Holben, B. N., Johnsrud, M. & Stordal, F. (2003). *Modeling the solar radiative impact of aerosols from biomass burning during the southern african reginal science iniative (safari-2000) experiment*, J. Geophys. Res. 108.

Nakajima, T. & Tanaka, M. (1988). *Algorithms for radiative intensity calculations in moderately thick atmospheres using a truncation approximation*, J. Quant. Spectrosc. Radiat. Transfer 40: 51-69.

Prospero, J. M., Ginoux, P., Torres, O., Nicholson, S. E. & Grill, T. E. (2002). *Environmental characterization of global sources of atmospheric soil dust identified with the NIMBUS 7 Total Ozone Mapping Spectrometer (TOMS) absorbing aerosol product*, Rev. Geophys. 40.

Roger, J.-C., Mallet, M., Dubuisson, P., Cachier, H., Vermote, E., Dubovik, O. & Despiau, S. (2006). *A synergetic approach for estimating the local direct aerosol forcing: Applications to an urban zone during the ESCOMPTE experiment*, J. Geophys. Res. 111.

Satheesh, S. K. & Ramanathan, V. (2000). *Large differences in tropical aerosol forcing at the top of the atmosphere and earth's surface*, Nature 405: 60-62.

Sinyuk, A., Dubovik, O., Holben, B., Eck, T. F., Breon, F.-M., Martonchik, J., Kahn, R., Diner, D. J., Vermote, E. F., Roger, J.-C., Lapyonok, T. & Slutsker, I. (2007). *Simultaneous retrieval of aerosol and surface properties from a combination of AERONET and satellite data*, Rem. Sens. Environ. 107: 90-108.

Smirnov, A., Holben, B.N., Eck, T.F., Dubovik, O. & Slutsker, I. (2000) *Cloud screening and quality control algorithms for the AERONET data base*, Rem. Sens. Environ. 73:337-349.

Stamnes, K., Tsay, S. C., Wiscombe, W. & Jayaweera, K. (1988). *Numerically stable algorithm for discrete-ordinate-method radiative transfer in multiple scattering and emitting layered media*, Appl. Opt. 27: 2,502-2,509.

Twomey, S. (1977). *Atmospheric aerosols*, New York, USA.

Zhou, M., Y., H., Dickinson, R., Dubovik, O. & Holben, B. N. (2005). *A normalized description of the direct effect of key aerosol types on solar radiation as estimated from Aerosol Robotic Network aerosols and Moderate Resolution Imaging Spectroradiometer albedos*, J. Geophys. Res. 110.

Part 5

Modeling Climate Impacts

14

The Effects of Climate Change on Orangutans: A Time Budget Model

Charlotte Carne, Stuart Semple and Julia Lehmann
University of Roehampton
England

1. Introduction

The investigation of the potential effects of climate change on species distributions is a major focus of conservation biology (Guisan & Thuiller, 2005; Sinclair et al., 2010). Numerous predictive models have been developed, the majority of which utilise correlations between the observed distribution of a species and climate variables to produce a 'climate envelope' in which the species is predicted to live (Pearson & Dawson, 2003). However, these models provide little insight into the mechanisms that determine distributions or the effect that climate change will have on behaviour (Lehmann et al., 2010). Time budget models have been developed that incorporate behaviour as an intermediate link between climate and survival, allowing for a more in depth analysis of the factors that limit distribution patterns. These models can identify probable future distributions, and predict the potential effects of climate change on behaviour and sociality (see Dunbar et al., 2009 for a review). The basic assumption of these models is that time is an important constraint that can affect a species' ability to survive in a given habitat. Theoretically it is always possible for an individual to meet its nutritional requirements regardless of food quality, provided that there is sufficient time available to find, ingest and digest an adequate amount of forage (Dunbar et al., 2009). Thus, it is essentially time that constrains the animal's ability to survive. For an animal to survive in a given location it is therefore vital that it can perform all of its essential activities within the time available (Dunbar, 1992). Time constraints are clearly important factors that are overlooked by the more conventional species distribution models.

In this chapter we use a time budget model to examine orangutan distribution patterns under future climate change scenarios. There are two species of orangutan currently recognised, *Pongo pygmaeus* and *Pongo abelii*, which inhabit the islands of Borneo and Sumatra respectively (Goossens et al., 2009). Both species of orangutan are endangered; in 2004 only 6,500 Sumatran orangutans and 54,000 Bornean orangutans were estimated to remain in the wild (Wich et al., 2008). Orangutans are large-bodied arboreal great apes, with a primarily frugivorous diet (Morrogh-Bernard et al., 2009). Although orangutans spend the majority of their time alone, they have been shown to form clusters with neighbouring individuals and can therefore be viewed as social animals (Singleton & van Schaik, 2002). Orangutans are characterised by slow life histories with long maturation rates and interbirth intervals (Delgado & van Schaik, 2000; Wich et al., 2009). This makes them particularly vulnerable to extinction (Cardillo et al., 2004), as populations can take many years to recover from loss of individuals for example due to hunting (Singleton et al., 2004). Great apes in general are also particularly threatened by habitat loss from deforestation, as a result of their large home range and dietary requirements (Campbell-Smith et al., 2011). The effects of further habitat loss due to future climatic change could therefore have a considerable impact on the survival chances of the orangutan. The investigation of the effect of climate change on the range of the orangutan could provide valuable information for the conservation of these two endangered species.

Previous work on the effects of climate change on primate distribution patterns has shown that climate change is likely to significantly reduce the range of both gelada baboons (Dunbar, 1998) and the African great apes (Lehmann et al., 2010). Because time budget models utilise existing relationships between climatic, dietary and demographic variables and their effects on behaviour, they can be used to predict the amount of time that animals would need to devote to each activity under different climate scenarios. The climate affects individual behaviour through direct effects on the animals (e.g. thermoregulation) as well as indirect effects on vegetation quality and distribution, which can in turn influence diets and the amount of time required for feeding, moving and resting. In addition, because most primates are social, the size of the social group needs to be factored in to the models. Group size can affect time budgets through food competition, leading to an increase in feeding, moving or resting time (Dunbar, 1996), as well as through the need to maintain social relationships via grooming (Lehmann et al., 2007), which in large groups can take up a substantial amount of time. Time budget models take these effects into account and allow us to determine not only the presence or absence of a species in a habitat, but also to calculate the group sizes at which it could persist in a given location. The group size at which all available time has been allocated to time budget demands is the maximum ecologically tolerable group size that can be adopted in a specific location (Dunbar, 1992).

Time budget models therefore provide advantages over the correlative approach of bioclimate envelope models. Time budget models have been shown to predict species distributions as accurately as climate envelope models (Korstjens & Dunbar, 2007; Willems & Hill, 2009), but can also be used to understand the mechanisms that determine distributions, both now and in the future, as well as potential impacts on behaviour (Dunbar et al., 2009; Lehmann et al., 2010). Time budget models have been created for three species of great apes to date: gorillas, chimpanzees and bonobos (Lehmann et al., 2008a, 2008b, 2010; Lehmann & Dunbar, 2009). In this chapter we investigate the potential effects of climatic change on orangutan distribution patterns using a time budget model. Anthropogenic effects such as deforestation and hunting have a considerable impact on the distribution of the orangutan (Rijksen & Meijaard, 1999); therefore we also incorporate land cover and human population density data into the model.

2. Methods

2.1 Overview

Regression equations were created to predict the amount of time that orangutans ought to devote to each of the time budget variables from climatic, dietary and demographic data. Using raster data in ArcGIS 9.3, time budget allocations were predicted across Borneo and Sumatra, and summed to calculate the maximum ecologically tolerable group size of the orangutan across the islands. This was combined with land cover and human population density data to create a presence/absence map for the orangutan. The model was then re-run using a number of future climate projections to predict the effect of climate change on the distribution of the orangutan.

2.2 Data collection

Data were collected from published studies to compile a dataset of orangutan time budgets (time spent feeding, moving and resting), diet (percentage of feeding time spent eating fruit and leaves) and group size. Data were collected from 13 different orangutan study sites in total, with time budget data available for 12 of the sites (Table 1).

Climate data were obtained from Worldclim (http://www.worldclim.org/) and are displayed in Table 2. Worldclim provides a set of global climate layers for 19 bioclimatic variables at a resolution of 30 arc seconds. These layers were generated through interpolation of average monthly climate data from weather stations across the world, from 1950-2000 (Hijmans et al., 2005). All 19 variables were used in the equation finding process. The percentage of forest cover data, displayed in Table 2, were obtained from the Advanced Very High Resolution Radiometer (AVHRR) satellite data (DeFries et al., 2000; http://glcf.umd.edu/data/treecover/).

2.3 Model components

The main components of the model are diet, group size and time budgets. In addition, we included anthropogenic effects as present day primate distributions are heavily affected by human activities, such as deforestation, land cover changes and human population densities, which are not reflected in the climate data.

The orangutan diet is composed mainly of fruit and leaves (Morrogh-Bernard et al., 2009; Russon et al., 2009), which make up more than 80% of their feeding time (see Table 1); thus only these two food categories were included in the model. These variables were measured as the percentage of feeding time spent consuming fruit and leaves, which were assumed to be mutually exclusive. Diet was assumed to be influenced by the climate and the percentage of forest cover.

The majority of the orangutan's time is spent in four essential behaviours, treated here as mutually exclusive: feeding, moving, resting and socialising. Resting time is included as a key time budget variable as it is assumed to represent time that the animals are forced to devote to resting, as a result of thermoregulation and digestion demands. Feeding, moving and resting time allocations were assumed to be affected by the climate, percentage forest cover, diet and group size. Although orangutans spend the majority of their time alone, they form relationships with neighbouring individuals (Singleton & van Schaik, 2002); therefore grooming time was also included in the model, as grooming is seen as an activity that bonds individuals together. Grooming time was estimated from a generic equation calculated by Lehmann et al (2007) that determines the amount of time that individuals ought to devote to grooming to maintain group cohesion in a group of a particular size. This equation is based on the observation that grooming time in primates increases as group size increases, a result of the increased time required for maintaining relationships and group cohesion (Lehmann et al., 2007)

Site	Latitude	Longitude	Island	Species/subspecies	Group Size	%Feeding	%Moving	%Resting	%Fruit	%Leaves
Danum[1]	5.02	117.75	Borneo	*P. pygmaeus morio*	-	47.2	16.9	34.4	60.9	22.2
Kinabatangan[2]	5.53	118.28	Borneo	*P. pygmaeus morio*	-	34.1	10.3	53.6	68.0	22.9
Mentoko[3]	0.40	117.27	Borneo	*P. pygmaeus morio*	1.28	45.2	10.4	43.1	53.8	29.0
Ulu Segama[4]	5.07	117.80	Borneo	*P. pygmaeus morio*	1.93	32.3	16.4	51.4	51.5	-
Sabangau[5]	-2.32	113.90	Borneo	*P. pygmaeus wurmbii*	-	61.3	15.9	19.7	73.8	5.1
Cabang Panti[6]	-1.22	110.12	Borneo	*P. pygmaeus wurmbii*	1.04	36.1	9.9	52.8	70.0	13.4
Tanjung Puting[7]	-2.75	111.95	Borneo	*P. pygmaeus wurmbii*	1.18	60.1	18.7	18.2	60.9	14.7
Tuanan[8]	-2.15	114.43	Borneo	*P. pygmaeus wurmbii*	1.13	50.6	16.8	30.9	69.8	17.2
Sungai Lading[9]	-2.25	114.37	Borneo	*P. pygmaeus wurmbii*	1.03	-	-	-	61.0	-
Batang Seragan[10]	3.73	98.19	Sumatra	*P. abelii*	-	24.0	15.0	54.0	46.0	13.0
Ketambe[11]	3.68	97.65	Sumatra	*P. abelii*	2.04	48.2	12.8	38.8	62.8	20.8
Ranun[12]	3.25	97.92	Sumatra	*P. abelii*	1.85	44.1	16.0	40.0	84.7	10.2
Suaq Balimbing[13]	3.07	97.43	Sumatra	*P. abelii*	1.90	53.1	17.3	27.2	66.2	15.5

Table 1. Orangutan study sites and their locations (latitude, longitude and island), with the species/subspecies studied and the behavioural data (group size, feeding, moving and resting time) and dietary data (% fruit and leaves in the diet) used in the equation finding process. Averages were used when data from more than one study were available for the same site, so that each site is only represented once in the dataset. Behavioural data were taken from: 1 Kanamori et al., 2010; 2 Morrogh-Bernard et al., 2009; 3 Mitani, 1989, 1990; Rodman, 1973, 1979; 4 Mackinnon, 1974; 5 Morrogh-Bernard et al., 2009; 6 Mitani, 1991; Morrogh-Bernard et al., 2009; van Schaik, 1999; 7 Galdikas, 1984, 1988; 8 Bastian et al., 2010; Morrogh-Bernard et al., 2009; 9 Bastian et al., 2010; 10 Campbell-Smith, 2010; Campbell-Smith et al., 2011; 11 Sugardjito et al., 1987; van Schaik, 1999; Wich et al., 2006; 12 Mackinnon, 1974; 13 Fox et al., 2004; van Schaik, 1999

Site	Forest (%)	Climate Variables																		
		T_a	T_d	T_i	T_s	T_x	T_m	T_r	T_{we}	T_{dr}	T_{wa}	T_c	P_a	P_w	P_d	P_s	P_{we}	P_{dr}	P_{wa}	P_c
Danum	80	26.1	7.6	8.7	36.4	30.7	22.0	8.7	25.7	25.9	26.5	25.5	2386	264	146	15	718	523	523	698
Kinabatangan	64	27.0	7.9	8.5	46.8	32.0	22.8	9.2	26.3	27.3	27.5	26.3	2711	362	131	31	977	479	479	977
Mentoko	80	26.1	7.0	8.8	23.7	30.2	22.3	7.9	26.2	25.9	26.3	25.9	1973	228	102	27	645	325	543	363
Ulu Segama	72	26.3	7.7	8.7	37.1	31.0	22.2	8.8	25.9	26.5	26.7	25.7	2416	275	143	17	742	516	516	727
Sabangau	79	26.6	8.2	8.6	36.8	31.8	22.3	9.5	26.1	26.9	27.0	26.1	2579	294	123	29	836	395	434	812
Cabang Panti	52	25.3	8.6	9.1	31.0	30.2	20.8	9.4	24.8	25.4	25.5	24.8	3258	364	134	27	1069	495	787	1069
Tanjung Puting	80	26.8	8.2	8.7	35.9	31.9	22.5	9.4	26.9	27.0	27.1	26.2	2723	295	148	22	824	471	506	783
Tuanan	80	26.6	8.1	8.7	30.5	31.6	22.3	9.3	26.2	26.8	26.9	26.2	2522	285	108	32	837	366	399	837
Sungai Lading	80	26.6	8.1	8.4	35.0	31.8	22.2	9.6	26.2	26.8	27.0	26.1	2508	285	109	33	841	364	398	838
Batang Seragan	33	26.8	9.6	8.7	43.1	32.5	21.5	11.0	26.5	26.5	27.3	26.2	2608	311	123	26	862	465	623	720
Ketambe	80	24.8	9.8	8.8	38.4	30.6	19.5	11.1	24.4	24.9	25.3	24.3	2563	298	123	27	871	440	626	763
Ranun	69	24.4	9.9	8.8	38.0	30.2	19.0	11.2	24.0	24.4	24.9	23.9	2868	331	138	26	956	500	787	852
Suaq Balimbing	80	26.9	9.5	8.7	43.7	32.6	21.8	10.8	26.5	27.0	27.5	26.3	2950	344	158	25	953	559	716	821

Ta = annual mean temperature; Td = mean temperature; Td = mean temperature diurnal range (mean of monthly max temp - min temp); Ti = isothermality ((mean diurnal range/temperature annual range) * 100); Ts = temperature seasonality (standard deviation *100); Tx = maximum temperature of warmest month; Tm = minimum temperature of coldest month; Tr = temperature annual range (maximum temperature of warmest month - minimum temperature of coldest month); Twe = mean temperature of wettest quarter; Tdr = mean temperature of driest quarter; Twa = mean temperature of warmest quarter; Tc = mean temperature of coldest quarter; Pa = annual precipitation; Pw = precipitation of wettest month; Pd = precipitation of driest month; Ps = precipitation seasonality (coefficient of variation); Pwe = precipitation of wettest quarter; Pdr = precipitation of driest quarter; Pwa = precipitation of warmest quarter; Pc = precipitation of coldest quarter

Table 2. Forest cover and 19 climate variables at each site (Temperature variables are in °C and precipitation variables are in mm).

For the other model components, bivariate Pearson correlation analyses, together with visual screening of the data, were used to identify possible linear and curvilinear relationships between the behavioural and dietary variables (Table 1) and the climate and forest cover data (Table 2). Stepwise regression analyses were then performed to obtain best-fit equations to predict the percentage of fruit in the diet, feeding, moving and resting time, using the 19 climate variables, diet and group size as possible predictor variables. Variables were only added to an equation if their inclusion could be justified using biological first principles, and if they explained a significant proportion of the variation (Dunbar, 1992). Because of the small sample size, a maximum of three predictor variables were included in the regression equations (Korstjens & Dunbar, 2007); the three variables that explained the highest proportion of the variance were selected. The predictions of the equations were constrained to within biologically realistic values, to prevent the model from producing mathematically possible, but biologically improbable results (Willems & Hill, 2009). Feeding and resting time budget allocations were constrained to a minimum value of 10% and moving time was constrained to a minimum of 5%. The percentage of time that orangutans spend feeding on fruit and leaves was constrained to values between 0 and 100%.

In addition to these core model components we also included anthropogenic effects. In the past century the total human population of Borneo and Sumatra has increased dramatically (Hirschman, 1994), leading to intense deforestation and hunting pressures (Sodhi et al., 2004). Although the percentage of forest cover, an indication of deforestation, was included in the model, the data were based on satellite imagery collected in 1992-1993 and are therefore unlikely to be representative of the current situation, especially given the extremely rapid rate of deforestation in Borneo and Sumatra - forest cover declined by 1.3% in Borneo and 2.7% in Sumatra per year between 2000 and 2010 (Miettinen et al., 2011b). In order to account for the increasing impact of anthropogenic factors on the orangutans, a recent land cover map from 2010 was incorporated into the model (Miettinen et al., 2011a; http://www.eorc.jaxa.jp/SAFE/LC_MAP/), so that areas where human induced land use changes are likely to prevent orangutans from persisting could be identified. Land cover categories were separated into those suitable for orangutans or unsuitable (Table 3).

Suitable Habitat	Unsuitable habitat
Mangrove	Plantation/regrowth
Peatswamp forest	Lowland mosaic
Lowland forest	Montane mosaic
Lower montane forest	Lowland open
Upper montane forest	Montane open
	Urban
	Large-scale palm plantation

Table 3. Land cover categories (suitable and unsuitable habitat) from the 2010 land cover map (Miettinen et al., 2011a).

Human population density (HPD) was also added to the model as it can be used as a summary measure of human impacts (Cardillo et al., 2004). HPD data were obtained from the Oak Ridge National Laboratory (UT-Battelle, LLC) LandScan 2008™ High Resolution Global Population Data Set (http://www.ornl.gov/sci/landscan/). In order to choose an appropriate threshold for the value of HPD above which orangutans were unlikely to survive, the distribution of HPD values within locations in the observed orangutan range was examined. Although some locations within the orangutan range have extremely high values of HPD, this is likely to be the result of errors in estimation of either HPD or the orangutan range, both of which were determined based on relatively crude methods. A threshold of 20 people per square kilometre was chosen as an appropriate threshold as almost 97% of locations within the orangutan range were characterised by HPD values lower than this threshold (Table 4). In addition, increasing the threshold above 20 people per square kilometre leads to relatively small increases in the percentage of locations with HPD values below the threshold, while decreasing the threshold below 20 leads to increasingly large decreases in this percentage. This indicates that orangutans may struggle to survive in locations with HPD above 20 people per square kilometre.

Threshold value of HPD (people per km²)	Proportion of range with HPD < threshold
30	97.7
25	97.3
20	96.8
15	95.2
10	92.7
5	86.3

Table 4. The proportion of the current orangutan range with human population densities (HPD) below different threshold values.

2.4 Model procedure

The equations were implemented in ArcGIS version 9.3 to predict time budget allocations of orangutans for each pixel in a raster image across Borneo and Sumatra. The maximum ecologically tolerable group size was determined by calculating the sum of the time budget allocations for each pixel. Group size was then increased algorithmically from 1, until the total time budget allocations exceeded 100%. The maximum ecologically tolerable group size is defined as the group size of the previous iteration. Although orangutans have been shown to form clusters with neighbouring individuals (Singleton et al., 2009; Singleton & van Schaik, 2002), they spend the majority of their time alone (Bastian et al., 2010; Galdikas et al., 1981; van Schaik, 1999); therefore they are assumed to be able to survive in a location if the predicted maximum group size is greater than or equal to 1. Thus, the orangutan is predicted to be able to survive in all pixels with a predicted maximum ecologically tolerable group size greater than or equal to 1, a suitable land cover category and a HPD value less than 20 people per square kilometre.

2.5 Validating the model

The model was validated in three ways. Firstly, the predicted distribution was compared against the observed range of the orangutan, to determine how accurately the model predicted the current distribution. The observed distribution was obtained from UNEP-WCMC (Meijaard et al., 2004) and was based on extensive ground and aerial surveys and forest data from both satellite imagery and the Indonesian Ministry of Forestry (Wich et al.,

2008). Secondly, predicted values for orangutan maximum ecologically tolerable group size were compared with observed values from the 9 sites for which there were data. Observed group sizes should fall equal to or below the predicted maximum values. Observed and predicted group sizes were compared using Wilcoxon signed-ranks tests. Finally, sensitivity analyses were used to evaluate the robustness of the model to errors in parameter estimates. The parameter estimates in the regression equations were each changed one at a time by +5%, -5%, +50% or -50% and the model re-run. All parameters were then changed simultaneously by +5%, -5%, +50% or -50% and the four resulting models run. The predictions of the altered models were compared to the predictions of the original model to determine the sensitivity of the model to error. Ideally, the model should be robust to small changes in parameter estimates (~5%) but sensitive to large changes (~50%). Comparing the predictions for every pixel in the raster image is problematic as the pixels are not independent; therefore a grid consisting of 374 locations across Borneo and Sumatra separated by 0.5° longitude and latitude was created. The prediction (i.e. presence or absence) at each of these locations was extracted for the original model and all sensitivity analysis models for comparison. This improves the independence of the data compared to pixel level resolution and thereby allows the predictions of the altered models to be compared with those of the original model. Chi-squared tests were used to determine if the altered equations produced significantly different results from the original model. All statistical analyses were performed in SPSS version 17.

2.6 Climate change and human population growth effects

In order to assess the effects of climate change on the biogeographical range of orangutans, the model was re-run using predicted future scenarios. All analyses were restricted to the orangutan's current distribution; therefore we did not consider range increases, as it is assumed that anthropogenic factors would prevent any future expansions. Predicted climate data for 2080 (the date furthest in the future for which data are available) were downloaded from the International Centre for Tropical Agriculture (Ramirez & Jarvis, 2008; http://ccafs-climate.org/). The data had been statistically downscaled using the delta method (Ramirez & Jarvis, 2010). Two future emissions scenarios were chosen (SRES A2a and SRES B2a), representing alternative predictions for the future. The A2a scenario is characterised by high energy requirements and continuous human population growth, while the B2a scenario is based on assumptions of lower energy requirements and slower population growth (Nakicenovic et al., 2000). Within these scenarios, data were available from four different and widely used Global Circulation Models (CCCMA-CGCM2 (hereafter CCCMA), CSIRO-MK2 (hereafter CSIRO), NIES99 and HACCPR HadCM3 (hereafter HadCM3)) for each of the two scenarios. Data were obtained for all four models in order to examine a range of future predictions (Jenkins et al., 2011). The models produce predictions at very coarse resolutions (CCCMA = 3.7° by 3.7°; CSIRO = 5.6° by 3.2°; HadCM3 = 2.5° by 3.8°; NIES99 = 5.6° by 2.8°), therefore statistically downscaled data were used to increase the resolution of the data. All models predict warming climates, while predictions for precipitation vary, with CCCMA predicting a decrease in annual rainfall across Borneo and Sumatra and the other three models predicting increases (Table 5).

	Change in mean annual temperature (°C)		Change in mean annual precipitation (mm)	
	A2a	**B2a**	**A2a**	**B2a**
CCCMA	+ 2.51	+ 1.60	- 197.12	- 143.46
CSIRO	+ 3.20	+ 1.99	+ 352.17	+ 270.65
HadCM3	+ 3.47	+ 3.04	+ 183.40	+ 176.24
NIES99	+ 3.48	+ 2.19	+ 288.23	+ 88.68

Table 5. Changes in mean annual temperature and precipitation predicted for 2080 across Borneo and Sumatra under the 8 combinations of 4 climate models and 2 emissions scenarios.

The time budget model was re-run using future climate data from all four models (CCCMA, CSIRO, NIES99 and HadCM3) under both emissions scenarios (SRES A2a and B2a). Because we know very little about the effect of climate change on tropical forests (Clark, 2004), forest cover was assumed to remain constant. Similarly, land cover data were maintained as in the original model, as it is impossible to predict these changes. The human population density in 2080 across the region was predicted by calculating the percentage increase in global population from 2008 to 2080 predicted by the two emissions scenarios. The population density of each pixel was then increased by the same percentage (an increase of 98% under the A2a scenario and 43% under the B2a scenario).

3. Results

3.1 Equations

The regression equations obtained are presented in Table 6 and summarised in Figure 1. In addition to the equations for feeding, moving and resting time, a regression equation was produced for the percentage of fruit in the diet from which the percentage of leaves in the diet can be calculated, as this variable was a significant predictor of feeding time.

Variable	Equation	R^2adj	N	F	P
Feeding	22.081 + 0.665*Forest – 0.806*Leaves – 0.032*Precipitation of Wettest Month	0.91	11	35.99	<0.001
Moving	-97.085 + 0.358*Temperature Seasonality – 0.0005*Temperature Seasonality ^{2} + 0.189*Mean Temperature of Wettest Quarter	0.56	12	5.61	<0.05
Resting	128.548 + 27.417*Group Size – 0.968*Forest – 0.176* Temperature Seasonality	0.88	8	17.58	<0.01
Fruit	174.100 – 0.488* Precipitation of Warmest Quarter + 0.0004* Precipitation of Warmest Quarter ^{2} + 0.267*Forest	0.56	13	6.06	<0.05
Leaves	100 - Fruit				
Social	1.01 + 0.23*Group Size	Generic equation†			

† Equation from Lehmann et al, 2007

Table 6. Multiple regression equations for time budget and diet variables, with the adjusted R^2, the sample size, and the significance of the equation.

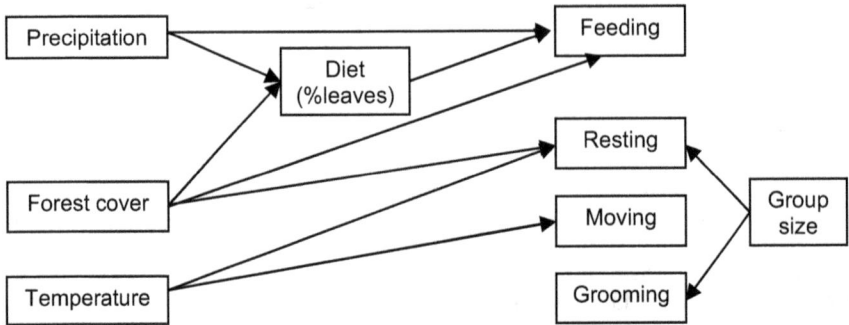

Fig. 1. Flow chart depicting relationships between climate and forest cover variables and the time budget components. The arrows indicate directional relationships used in the model.

The equations highlight the important effect of forest cover on time budget allocations. The positive relationship between forest cover and fruit could be the result of a higher availability of fruit in dense forest than in more degraded habitats. Forest cover also predicted feeding and moving time, and it is likely that these relationships reflect the effect of forest cover on diet. In densely forested areas orangutans may need to spend less time resting and more time feeding, possibly as a result of the increased availability of high quality food that requires less time for digestion but more time for extraction (which would be attributed to feeding time). In contrast, in more sparse forests, orangutans may be forced to rest more to conserve energy (and to digest the higher percentage of leaves in their diet) and feed less. Resting time was also affected by group size and temperature seasonality. As group size increases fruit sources will be depleted faster and orangutans may be forced to consume more leaves which require more digestion time. The relationship with temperature seasonality indicates that orangutans spend more time resting in areas with more consistent temperatures. This may reflect higher resting time requirements in areas closer to the equator, where temperatures are higher and less variable (Chuan, 2005). When temperatures are high animals are unable to perform many of their essential activities, and are thus forced to rest (Korstjens et al., 2010).

Feeding time was negatively related to the amount of leaves in the diet. Similarly, among the African great apes, it was found that feeding time increased as the percentage of fruit in the diet increased (Lehmann et al., 2008b). These relationships may reflect the reduced foraging and processing times required to eat leaves.

3.2 Model validation

The current range of the orangutan is displayed in Figure 2.

The model produced a presence/absence map for the orangutan. This was compared against the observed distribution (Figure 3).

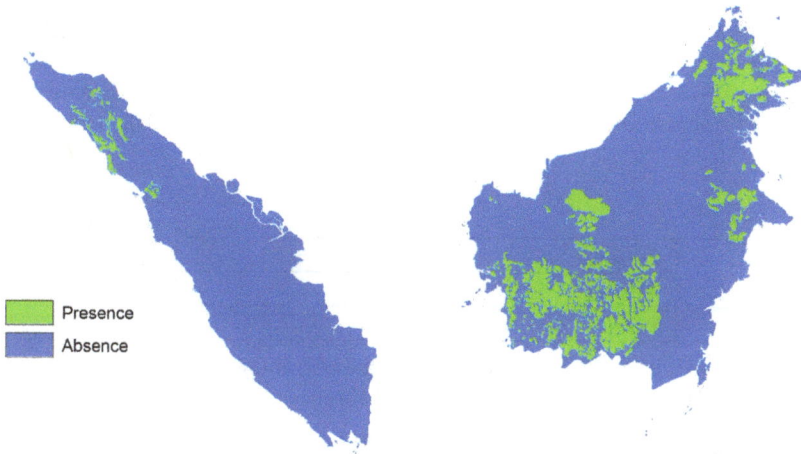

Fig. 2. The current distribution of the orangutan (Meijaard et al., 2004).

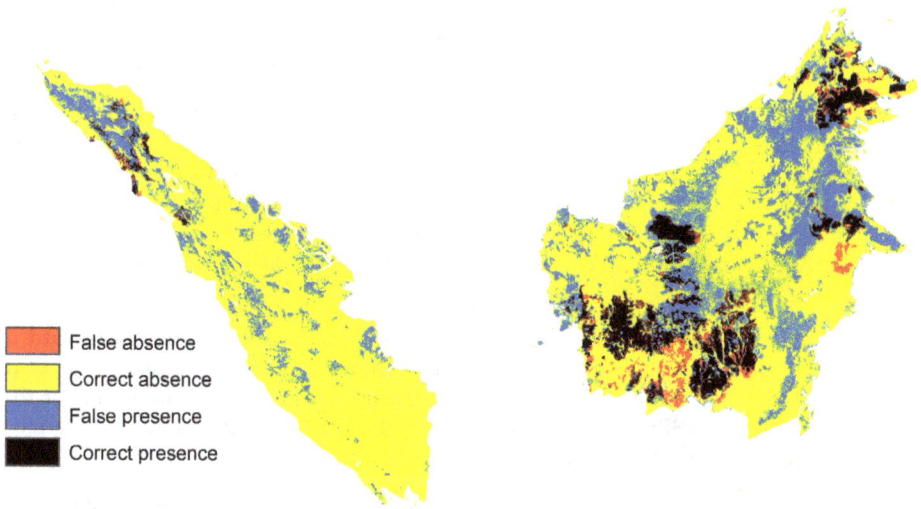

Fig. 3. Predicted distribution of the orangutan based on the model compared against the observed distribution. False absences indicate locations where orangutans are present but the model predicted absence, correct absences are locations where orangutans are absent and the model predicted absence, false presences are locations where orangutans are absent and the model predicted presence and correct presences are locations where orangutans are present and the model predicted presence.

Overall, the model correctly predicted the presence or absence of orangutans in 77.0% of raster pixels. The model correctly predicted orangutan absence from 78.1% of the areas from which they are currently thought to be absent and correctly predicted their presence in 68.3% of their current range (Table 7). The number of correct predictions (when analysed across the grid of 374 points) is significantly higher than would be expected by chance, based on the observed proportion of presences and absences ($\chi^2 = 11.87$, df = 1, N = 374, P < 0.001).

		Observed	
		Present	Absent
Predicted	**Present**	68.3	21.9
	Absent	31.7	78.1
	Total	100	100

Table 7. Table of model performance, displaying the percentage of observed presences and absences that were predicted to be presences and absences.

The model therefore incorrectly predicted the presence or absence of orangutans in just over 20% of raster pixels. The majority of these false predictions were those where the model predicted suitable habitat for orangutans but no orangutans were recorded to live there in 2007, i.e. the model overestimated the current orangutan range. To investigate this further, an orangutan distribution map was obtained for the island of Borneo from 1930 (Rijksen & Meijaard, 1999), and compared with the model predictions for Borneo. This showed that 45.2% of these false presences are in land that was previously suitable for orangutans according to the 1930 map, thereby confirming that climatically these areas may be suitable.

Similarly, 58.3% of those locations that were incorrectly identified as unsuitable for orangutans by the model were in areas that are now classified as plantations or regrowth in the 2010 land cover map, and were thus classed as unsuitable for orangutans. This conversion to plantations appears to be relatively recent, as 75.2% of the locations classified as plantations in 2010 were classed as forested in a land cover map from 2000 (Global Land Cover 2000, 2003; http://bioval.jrc.ec.europa.eu/products/glc2000/glc2000.php).

Observed group sizes were compared against the maximum ecologically tolerable group sizes predicted by the model. Observed group sizes should be less than or equal to the maximum ecologically tolerable group size; figure 4 shows that this is the case for all of the points. Predicted maximum ecologically tolerable group sizes are significantly higher than observed group sizes (WSR: Z = -2.668, N = 9, P < 0.01), and still within a realistic range of the observed grouping patterns (Figure 4).

Sensitivity analyses were performed to determine the robustness of the model to errors in parameter estimates. Changes of 5% resulted in presence/absence distributions that were significantly different from the original model in only 1 of 36 runs, while changes of 50% led to significantly different distributions in 21 of 36 runs (all $\chi^2 > 3.84$, df = 1, N = 374, P < 0.05). Thus, the model is robust to small errors and sensitive to large errors in parameter estimates. Changing all parameters at once by both ±5% and ±50% led to significantly different presence/absence distributions than the original model in all four runs (all $\chi^2 > 11$, df = 1, N = 374, P < 0.001). This indicates that it is the particular set of parameter values obtained that is important.

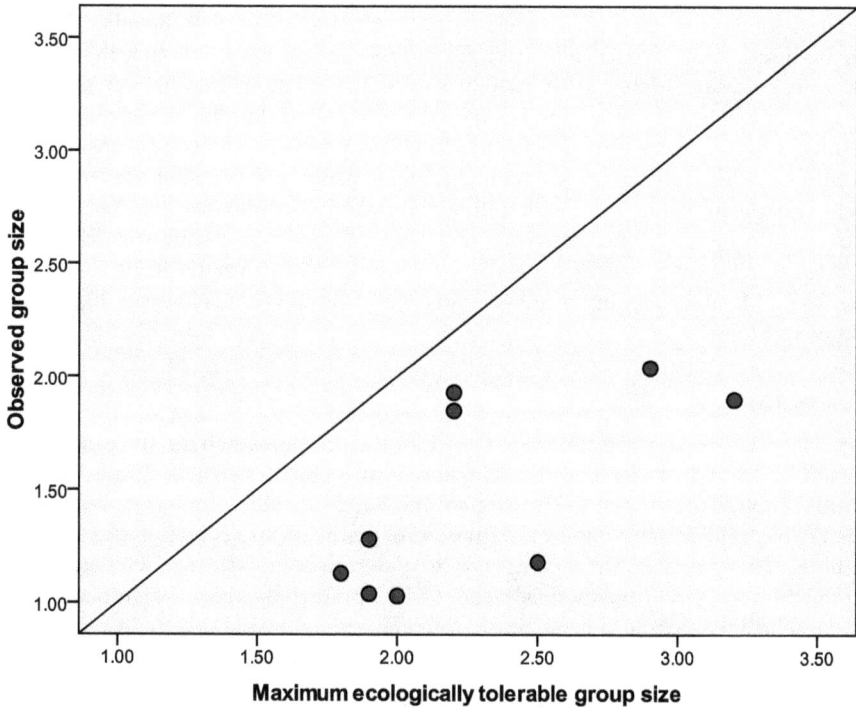

Fig. 4. Observed group sizes versus predicted maximum ecologically tolerable group sizes (the line of equality is the line on which the maximum ecologically tolerable group size equals the observed group size).

3.3 The effect of climate change and human population growth

3.3.1 Biogeography

The percentage of the orangutan's current range that is predicted to become unsuitable as a result of climate change and human population growth was calculated (the areas classed as false absences in the original model were excluded) (Table 8). The majority of the models predict that the orangutan will lose approximately 5% of its current range under scenario A2a and around 3% under scenario B2a. However, the CSIRO model predicts a larger reduction in the suitable range of the orangutan, by just over 15% under scenario A2a and 13% under scenario B2a.

	A2a	B2a
CCCMA	5.11	3.28
CSIRO	15.34	13.43
HadCM3	4.83	3.35
NIES99	6.61	3.05

Table 8. The percentage of the orangutans' current range that is predicted to be lost as a result of climate change and human population growth under the 8 combinations of 4 climate models and 2 emissions scenarios.

The CSIRO model thus produces a much larger range reduction than the other three models under both scenarios. A particular strength of time budget models is that they allow us to investigate in greater detail what exactly it is that is making this habitat unsuitable for orangutans, i.e. we can investigate the mechanisms underlying the range reduction. Mean time budget values were obtained for the locations predicted to become unsuitable under the CSIRO model, and compared with the mean value at these sites under the original model. The mean value for feeding time increased considerably by 2080 under both scenarios, with moving time increasing to a lesser degree and resting time changing very little (Mean$_{A2a}$: Feeding = 47.57 versus 30.24, Moving = 14.10 versus 10.26, Resting = 33.64 versus 33.22; Mean$_{B2a}$: Feeding = 51.84 versus 30.81, Moving = 14.44 versus 10.94, Resting = 31.77 versus 31.94).

3.3.2 Group size

Group size estimates were compared within the locations where the orangutans are predicted to survive under both current and future climates. For each model/emissions scenario combination as well as the original model, predicted values were obtained for all points on the grid in which the orangutan was predicted to survive under both climatic conditions. These values were compared to determine the effect of climate change on grouping patterns. All of the models except CSIRO predicted either no significant change or an increase in the maximum ecologically tolerable group size (Table 9). The CSIRO model predicts a decrease in the maximum ecologically tolerable group size in the future.

	A2a				B2a			
	Direction	Z	N	P	Direction	Z	N	P
CCCMA	NS	-1.000	24	0.317	NS	-1.342	24	0.180
CSIRO	Decrease	-2.000	20	< 0.05	Decrease	-2.646	21	< 0.01
HadCM3	Increase	-2.646	24	< 0.01	Increase	-2.646	24	< 0.01
NIES99	NS	-1.342	24	0.180	Increase	-2.121	24	< 0.05

Table 9. Results of Wilcoxon signed-ranks tests, indicating the direction and significance of predicted changes to the maximum ecologically tolerable group size within locations in which the orangutan is predicted to survive under both current and future conditions, for all 8 combinations of the 4 climate models and 2 emissions scenarios (NS = not significant).

3.3.3 Time budgets

Time budget estimates were also obtained for each of the locations in the grid in which the orangutan was predicted to survive under both climatic conditions, to compare current and future values. The predicted changes to time budget allocations are displayed in Table 10.

The NIES99 model and the CSIRO model predict that feeding will only increase significantly under the B2a scenario. The other model/scenario combinations predict no significant change. Moving time is predicted to increase under both scenarios by the CCCMA and CSIRO models, and under the A2a scenario by the NIES99 model. The CCCMA, HadCM3 and NIES99 models all predict a decrease in resting time under both possible futures, while the CSIRO model predicts no significant change.

		A2a				B2a			
		Direction	Z	N	P	Direction	Z	N	P
Feeding	CCCMA	NS	-0.983	24	0.326	NS	-0.522	24	0.602
	CSIRO	NS	-0.742	20	0.458	Increase	-2.185	21	< 0.05
	HadCM3	NS	-0.524	24	0.600	NS	-0.403	24	0.687
	NIES99	NS	-1.689	24	0.091	Increase	-2.142	24	< 0.05
Moving	CCCMA	Increase	-2.621	24	< 0.01	Increase	-2.560	24	< 0.05
	CSIRO	Increase	-4.054	20	< 0.001	Increase	-4.055	21	< 0.001
	HadCM3	NS	-0.017	24	0.986	NS	-0.281	24	0.779
	NIES99	Increase	-4.082	24	< 0.001	NS	-0.137	24	0.891
Resting	CCCMA	Decrease	-3.662	24	< 0.001	Decrease	-2.784	24	< 0.01
	CSIRO	NS	-0.829	20	0.407	NS	-1.269	21	0.204
	HadCM3	Decrease	-4.018	24	< 0.001	Decrease	-3.949	24	< 0.001
	NIES99	Decrease	-3.742	24	< 0.001	Decrease	-4.019	24	< 0.001

Table 10. Results of Wilcoxon signed-ranks tests, indicating the direction and significance of the predicted changes to time budget allocations within locations in which the orangutan is predicted to survive under both current and future conditions, for all 8 combinations of the 4 climate models and 2 emissions scenarios (NS = not significant).

4. Discussion

4.1 Overview

Time budget models have previously been shown to produce species distribution maps comparable in accuracy to climate envelope models (Korstjens & Dunbar, 2007; Willems & Hill, 2009). In addition, time budget models can also be used to assess the effects of changes in the climate on individual behaviour, diet and group sizes, helping us to understand what exactly is causing the changes in biogeography. In this study, we found that only one of the scenarios tested predicted a substantial change in orangutan distribution by 2080, while a relatively mild effect on the range of the orangutan was found for the remaining three models, which predicted a reduction of between 3 and 7%. However, given the additional pressures of deforestation and hunting by humans, this reduction could have a severe impact on the survival of the species.

4.2 Model performance

Although the model predicts the overall distribution of the orangutan encouragingly accurately, it tended to overestimate orangutan ranges in certain areas, i.e. the model identified habitats as climatically suitable for the orangutan where they are not currently recorded to live. A number of possible reasons may help to explain these deviations. Firstly, areas may indeed be suitable, but dispersal barriers have prevented the orangutan from colonising them. In the south-east of Borneo, for example, the Barito and Mahakam rivers may have prevented the orangutan from entering this region, as orangutans are unlikely to attempt to cross wide rivers with fast flowing water (Rijksen & Meijaard, 1999). Secondly, although we included some anthropogenic factors in the model, some effects such as hunting by small-scale societies were not included. Archaeological evidence indicates that orangutans have been hunted for thousands of years by the indigenous people of both

Borneo and Sumatra (Harrison, 2000; Rijksen & Meijaard, 1999). It has been suggested that the absence of the orangutan from areas in northern Borneo is the result of past hunting by indigenous people (Rijksen & Meijaard, 1999) while the areas where sizeable populations remain in Borneo are those inhabited by Muslim groups, who for religious reasons do not hunt the orangutan (Sugardjito, 1995). Thus, many of the locations predicted as suitable where the orangutan is not recorded to live, particularly those in northern Borneo, may be climatically suitable areas for the orangutans but represent locations where they have been extirpated by prehistoric human populations. Hunting by contemporary populations may also explain the absence of orangutans from otherwise suitable areas. Marshall et al (2006) found that the distance to the nearest village known to hunt orangutans was the most important predictor of orangutan density in East Kalimantan. Orangutans have an extremely slow reproductive rate (Knott et al., 2009; Wich et al., 2009) and therefore even relatively mild hunting pressures can have dramatic effects on orangutan population sizes (Marshall et al., 2009; Singleton et al., 2004). Thus, both past and current hunting pressures can significantly impact on the distribution of the orangutan, and may explain why orangutans were not found to live in some of the areas in which they are predicted to be able to survive. Although the model included human population density in an attempt to account for such human effects, hunting of orangutans is often performed by small groups of indigenous people that are unlikely to be characterised by high population densities.

Many of the locations in which the orangutan is absent but the model predicted them to live were those adjacent to its current range. The range of the orangutan has contracted dramatically in the past century, and has become increasingly fragmented (Husson et al., 2009). The estimated distribution of the orangutan in 1930 indicates that it once ranged much more continuously across southern Borneo and along the eastern coast (Rijksen & Meijaard, 1999). In Borneo, 45.2% of the locations incorrectly predicted as suitable habitat are within the orangutan's 1930 range. These areas may therefore be climatically suitable, but uninhabited by orangutans as a result of recent anthropogenic effects not incorporated in the model.

Another problem that causes inaccuracies in any kind of climate model is the limited availability of accurate and recent data on distribution patterns, land cover and climate. Although we were able to use a very recent land cover map in the current study, orangutan distribution patterns were based on data from 2007, while climate data were from long-term averages over a period of 50 years. Thus, changes in climate related to land cover changes are unlikely to be reflected in the climate data. Recent orangutan distribution shifts in response to land cover change are also not recorded. These effects may at least partially explain some of the inaccuracies in the model.

Finally, the model predicted orangutans to be absent from a number of locations where they were recorded to be present in 2007. These were primarily in locations adjacent to those known to contain orangutans. The distribution of orangutans is often extremely patchy, with densities varying spatially and temporally in accordance with food availability (McConkey, 2005). Thus, there are likely to be areas that do not clearly fit into a strict definition of presence or absence, for example, those used by roaming males (Husson et al., 2009). This may explain why orangutans have been observed in areas in which the model did not predict them to be able to subsist. This is supported by the fact that 58.3% of these 'false absences' were in pixels categorised as plantations or regrowth in the land cover map, and are therefore unlikely to support breeding populations, but may be used irregularly by orangutans with ranges on the borders of the forests. In addition, these areas may represent locations that have been converted to plantations since the orangutan distribution data were collected. Indeed 75.2% of these plantations were categorised as forest in 2000. Forest cover in South-East Asia declines by around 1% per year (Miettinen et al., 2011b), and much of the 2007 range may therefore now be unsuitable for the orangutan.

4.3 The effect of climate change and human population growth on the orangutan

Under most of the scenarios used climate change and human population growth are predicted to have a relatively mild effect on orangutan distribution patterns. The current range of the orangutan will most likely not contract considerably, with a reduction of between 3 and 7% predicted by three out of four climate models. The exception is the CSIRO model, which predicts a much larger range reduction under both scenarios (around 15% under A2a and 13% under B2a). This appears to be the result of an increase in feeding time caused by changes in precipitation patterns; the CSIRO models predict the largest increases in mean annual rainfall under each scenario. However, rainfall increases more in the NIES99 model under the A2a scenario than the CSIRO model under the B2a scenario, yet the range reduction is considerably less (only 6.61%). The dramatic range reduction under the CSIRO models is therefore likely to be the result of regional increases in rainfall specifically within the orangutan range. Changing precipitation patterns could have both a direct effect on feeding time allocations and an indirect effect through their influence on diet. As there is considerable uncertainty in climate predictions, the results predicted under the CSIRO model should be treated with caution; however, they do highlight the vulnerability of the orangutan to changes in feeding time, caused by changing precipitation patterns.

The models did not produce a consistent effect of climate change on group sizes within the areas where the orangutan is predicted to persist. Although under the CSIRO models group sizes were projected to decrease, the other models predicted either an increase in group size or no significant change. Group size in orangutans therefore appears to be sensitive to variations in climate predictions, but if indeed a group size of close to one individual is a viable minimum, orangutans appear to be somewhat buffered against the effects of climate change.

However, the changes in climate conditions are likely to force orangutans to shift some of their behavioural patterns. Although resting time demands may go down (with 6 out of 8 model/emissions scenario combinations predicting a decrease), moving time demands may become increasingly high (with 5 out of 8 model/emissions scenario combinations predicting an increase). Feeding time appears less likely to change in the future, with only 2 model/emissions scenario combinations predicting an increase in feeding time allocations. However, higher moving time demands will in turn lead to increased energy requirements, which will have a knock-on effect on other time budget variables. Feedback loops like this are not currently included in the model, but it is important to consider them as they may lead to exaggerated effects of climate change on orangutan populations.

It is important to note that the models may be seriously underestimating the effect of climate change on the orangutan. The possible effects of climate change on the energy content of orangutan food sources and on the percentage of forest cover were not incorporated into the model. Although the response of tropical forests to climate change is extremely uncertain (Clark, 2004), it is possible that the increased temperatures could lead to considerable forest dieback and an increase in the frequency and severity of forest fires (Allen et al., 2009; Bonan, 2008). The percentage of forest cover was found to be an important variable in the time budget model; therefore if climatic change causes strong negative effects on forest cover, this could be devastating for the remaining orangutan populations. Furthermore, the effect of future deforestation was not incorporated in the model. Future research is needed to investigate the inclusion of both the effect of climate change on forest cover as well as deforestation projections to provide more realistic predictions of the future distribution of the orangutan. Finally, the statistical downscaling procedure used to generate these data at a high resolution may lead to uncertainties in climate estimations (Ramirez & Jarvis, 2010).

Although the orangutan time budget model suggests that the effect of climate change and human population growth on the orangutan will be relatively small, commercial logging and the conversion of forests for plantations continues unabated in Borneo and Sumatra (Koh et al., 2011; Miettinen et al., 2011a), rapidly destroying and fragmenting the remaining orangutan habitat. Forest fires consume vast tracts of rainforest as well as the orangutans that inhabit them, while hunting for both bushmeat and for the pet trade reduce populations to critically low levels (Nellemann et al., 2007). Thus, orangutans are currently extremely vulnerable to extinction and even a small decline in their range due to climate change may have a large impact on the survival chances of the species.

5. Conclusion

Although the predicted effect of climate change on orangutan biogeographical ranges appears to be relatively mild, it is important to remember that there are other more imminent threats to orangutan survival. In addition, the results of this study highlight the precarious balance between climate, behaviour and biogeography – if one of the components is affected this can have knock-on effects on other variables, exaggerating the consequences for orangutan populations. Moving and feeding time demands, for example, are predicted to become more severe in the future, while resting time may become less restrictive, and these changes will have secondary effects on other variables. It is worrying that orangutans were predicted to lose up to 15% of their habitat under one of the climate change scenarios, especially considering that the effects of land and forest cover changes were not incorporated in the climate change models. However, it is important to mention that future climate predictions remain tentative, and the regression equations did not produce a perfect fit to the data leading to further uncertainty in model predictions. Overall, the results indicate that the current range of the orangutan will decline by approximately 3-7% by 2080 as a direct result of climatic change and human population growth, although this is likely to be an underestimate as land and forest cover changes were not included. However, even a range reduction of 5% may have a dramatic effect on the survival of the two orangutan species, particularly as suitable habitat becomes more fragmented through deforestation.

6. Acknowledgements

We sincerely thank Mary Mackenzie, Lionel Gunn, Frimpong Twum and Martin Evans for their invaluable technical support. We also thank Simon Blyth at the UNEP-WCMC for providing the orangutan distribution data. The University of Roehampton provided financial support to Charlotte Carne.

7. References

Allen, C.D.; Macalady, A.K.; Chenchouni, H.; Bachelet, D.; Mcdowell, N.; Vennetier, M.; Kitzberger, T.; Rigling, A.; Breshears, D.D.; Hogg, E.H.; Gonzalez, P.; Fensham, R.; Zhang, Z.; Castro, J.; Demidova, N.; Lim, J.-H.; Allard, G.; Running, S.W.; Semerci, A. & Cobb, N. (2010). A Global Overview of Drought and Heat-Induced Tree Mortality Reveals Emerging Climate Change Risks for Forests. *Forest Ecology and Management*, Vol. 259, No. 4, (February 2010), pp. 660-684, ISSN 0378-1127

Bastian, M.L.; Zweifel, N.; Vogel, E.R.; Wich, S.A. & Van Schaik, C.P. (2010). Diet Traditions in Wild Orangutans. *American Journal of Physical Anthropology*, Vol. 143, No. 2, (October 2010), pp. 175-187, ISSN 0002-9483

Bonan, G.B. (2008). Forests and Climate Change: Forcings, Feedbacks, and the Climate Benefits of Forests. *Science*, Vol. 320, No. 5882, (June 2008), pp. 1444-1449, ISSN 0036-8075

Campbell-Smith, G.A. (2010). *Bittersweet Knowledge: Can People and Orangutans Live in Harmony?* People's Trust for Endangered Species, Available from http://www.ptes.org/files/503_indonesian_orangutans.pdf

Campbell-Smith, G.; Campbell-Smith, M.; Singleton, I. & Linkie, M. (2011). Apes in Space: Saving an Imperilled Orangutan Population in Sumatra. *Plos One*, Vol. 6, No. 2, (February 2011), ISSN 1932-6203

Cardillo, M.; Purvis, A.; Sechrest, W.; Gittleman, J.L.; Bielby, J. & Mace, G.M. (2004). Human Population Density and Extinction Risk in the World's Carnivores. *Plos Biology*, Vol. 2, No. 7, (July 2004), pp. 909-914, ISSN 1544-9173

Chuan, G.K. (2005). The Climate of Southeast Asia, In: *The Physical Geography of Southeast Asia*, A. Gupta, (Ed.), 80-93, Oxford University Press, ISBN 9780199248025, New York

Clark, D.A. (2004). Sources or Sinks? The Responses of Tropical Forests to Current and Future Climate and Atmospheric Composition. *Philosophical Transactions of the Royal Society of London Series B-Biological Sciences*, Vol. 359, No. 1443, (March 2004), pp. 477-491, ISSN 0962-8436

DeFries, R.; Hansen, M.; Townshend, J.R.G.; Janetos, A.C. & Loveland T.R. (2000). 1 Kilometer Tree Cover Continuous Fields, 1.0. Department of Geography, University of Maryland, College Park, Maryland, 1992-1993, available from www.landcover.org

Delgado, R.A. & van Schaik, C.P. (2000). The Behavioral Ecology and Conservation of the Orangutan (*Pongo pygmaeus*): A Tale of Two Islands. *Evolutionary Anthropology*, Vol. 9, No. 5, (October 2000), pp. 201-218, ISSN 1060-1538

Dunbar, R.I.M. (1992). Time - A Hidden Constraint on the Behavioral Ecology of Baboons. *Behavioral Ecology and Sociobiology*, Vol. 31, No. 1, (July 1992), pp. 35-49, ISSN 0340-5443

Dunbar, R.I.M. (1996). Determinants of Group Size in Primates: A General Model. *Proceedings of the British Academy; Evolution of social behaviour patterns in primates and man*, Vol. 88, pp. 33-57

Dunbar, R.I.M. (1998). Impact of Global Warming on the Distribution and Survival of the Gelada Baboon: A Modelling Approach. *Global Change Biology*, Vol. 4, No. 3, (July 1998) pp. 293-304, ISSN 1354-1013

Dunbar, R.I.M.; Korstjens, A.H. & Lehmann, J. (2009). Time as an Ecological Constraint. *Biological Reviews*, Vol. 84, No. 3, (July 2009) pp. 413-429, ISSN 1464-7931

Fox, E.A.; Van Schaik, C.P.; Sitompul, A. & Wright, D.N. (2004). Intra- and Interpopulational Differences in Orangutan (Pongo Pygmaeus) Activity and Diet: Implications for the Invention of Tool Use. *American Journal of Physical Anthropology*, Vol. 125, No. 2, (August 2004), pp. 162-174, ISSN 0002-9483

Galdikas, B.M.F.; Teleki, G.; Coelho, A.M.; Eckhardt, R.B.; Fleagle, J.G.; Hladik, C.M.; Kelso, A.J.; McGrew, W.C.; Nash, L.T.; Nishida, T.; Paterson, J.D.; Savage-Rumbaugh, S.; Smith, E.O.; Sugiyama, Y. & Wrangham, R.W. (1981). Variations in Subsistence Activities of Female and Male Pongids: New Perspectives on the Origins of Hominid Labor Division. *Current Anthropology*, Vol. 22, No. 3, (June 1981), pp. 241-256, ISSN 0011-3204

Galdikas, B.M.F. (1984). Adult Female Sociality among Wild Orangutans at Tanjung Puting Reserve Indonesia, In: *Monographs in Primatology, Vol. 4. Female Primates: Studies by Women Primatologists*, Small, M.F. (Ed.), pp. 217-236, Alan R. Liss, Inc., New York

Galdikas, B.M.F. (1988). Orangutan Diet, Range, and Activity at Tanjung Puting, Central Borneo. *International Journal of Primatology*, Vol. 9, No. 1, (February 1988), pp. 1-35, ISSN 0164-0291

Global Land Cover 2000 database. (2003). European Commision, Joint Research Centre. Available from http://bioval.jrc.ec.europa.eu/products/glc2000/glc2000.php

Goossens, B.; Chikhi, L.; Jalil, M.F.; James, S.; Ancrenaz, M.; Lackman-Ancrenaz, I. & Bruford, M.W. (2009). Taxonomy, Geographic Variation and Population Genetics of Bornean and Sumatran Orangutans, In: *Orangutans: Geographic Variation in Behavioral Ecology and Conservation*, S.A. Wich; S.S.U. Atmoko; T.M. Setia & C.P. van Schaik, (Eds.), 77-96, Oxford University Press, ISBN 978-0199213276, New York

Guisan, A. & Thuiller, W. (2005). Predicting species distribution: offering more than simple habitat models. *Ecology Letters*, Vol.8, No.9, (September 2005), pp. 993-1009, ISSN 1461-0248

Harrison, T. (2000). Archaeological and Ecological Implications of the Primate Fauna from Prehistoric Sites in Borneo. *Indo-Pacific Prehistory Association Bulletin*, Vol. 20, pp. 133–146, ISSN 0156-1316

Hijmans, R.J.; Cameron, S.E.; Parra, J.L.; Jones, P.G. & Jarvis, A. (2005). Very high resolution interpolated climate surfaces for global land areas. *International Journal of Climatology*, Vol. 25, No. 15, (December 2005), pp. 1965-1978, ISSN 0899-8418

Hirschman, C. (1994). Population and Society in 20th-Century Southeast-Asia. *Journal of Southeast Asian Studies*, Vol. 25, No. 2, (September 1994), pp. 381-416, ISSN 0022-4634

Husson, S.J.; Wich, S.A.; Marshall, A.J.; Dennis, R.D.; Ancrenaz, M.; Brassey, R.; Gumal, M.; Hearn, A.J.; Meijaard, E.; Simorangkir, T. & Singleton, I. (2009). Orangutan Distribution, Density, Abundance and Impacts of Disturbance, In: *Orangutans: Geographic Variation in Behavioral Ecology and Conservation*, S.A. Wich; S.S.U. Atmoko; T.M. Setia & C.P. van Schaik, (Eds.), 77-96, Oxford University Press, ISBN 978-0199213276, New York

Jenkins, C.N.; Sanders, N.J.; Andersen, A.N.; Arnan, X.; Brühl, C.A.; Cerda, X.; Ellison, A.M.; Fisher, B.L.; Fitzpatrick, M.C.; Gotelli, N.J.; Gove, A.D.; Guénard, B.; Lattke, J.E.; Lessard, J.; McGlynn, T.P.; Menke, S.B.; Parr, C.L.; Philpott, S.M.; Vasconcelos, H.L.; Weiser, M.D. & Dunn, R.R. (2011). Global diversity in light of climate change: the case of ants. *Diversity and Distributions*, Vol.17, No.4, (July 2011), pp. 652–662, ISSN 1366-9516

Kanamori, T.; Kuze, N.; Bernard, H.; Malim, T. P. & Kohshima, S. (2010). Feeding Ecology of Bornean Orangutans (Pongo Pygmaeus Morio) in Danum Valley, Sabah, Malaysia: A 3-Year Record Including Two Mast Fruitings. *American Journal of Primatology*, Vol. 72, No. 9, (September 2010), pp. 820-840, ISSN 0275-2565

Knott, C.D.; Thompson, M.E. & Wich, S.A. (2009). The Ecology of Female Reproduction in Wild Orangutans, In: *Orangutans: Geographic Variation in Behavioral Ecology and Conservation*, S.A. Wich; S.S.U. Atmoko; T.M. Setia & C.P. van Schaik, (Eds.), 77-96, Oxford University Press, ISBN 978-0199213276, New York

Koh, L.P.; Miettinen, J.; Liew, S.C. & Ghazoul, J. (2011). Remotely Sensed Evidence of Tropical Peatland Conversion to Oil Palm. Proceedings of the National Academy of Sciences of the United States of America, Vol. 108, No. 12, (March 2011), pp. 5127-5132, ISSN 1091-6490

Korstjens, A.H. & Dunbar, R.I.M. (2007). Time Constraints Limit Group Sizes and Distribution in Red and Black-and-White Colobus. *International Journal of Primatology*, Vol. 28, No. 3, (June 2007), pp. 551-575, ISSN 0164-0291

Korstjens, A.H.; Lehmann, J. & Dunbar, R.I.M. (2010). Resting Time as an Ecological Constraint on Primate Biogeography. *Animal Behaviour*, Vol. 79, No. 2, (February 2010), pp. 361-374, ISSN 0003-3472

Lehmann, J.; Korstjens, A.H. & Dunbar, R.I.M. (2007). Fission-Fusion Social Systems as a Strategy for Coping with Ecological Constraints: A Primate Case. *Evolutionary Ecology*, Vol. 21, No. 5, (September 2007), pp. 613-634, ISSN 0269-7653

Lehmann, J.; Korstjens, A.H. & Dunbar, R.I.M. (2008a). Time Management in Great Apes: Implications for Gorilla Biogeography. *Evolutionary Ecology Research*, Vol. 10, No. 4, (July 2008), pp. 517-536, ISSN 1522-0613

Lehmann, J.; Korstjens, A.H. & Dunbar, R.I.M. (2008b). Time and Distribution: A Model of Ape Biogeography. *Ethology Ecology & Evolution*, Vol. 20, No. 4, (September 2008), pp. 337-359, ISSN 0394-9370

Lehmann, J. & Dunbar, R.I.M. (2009). Implications of Body Mass and Predation for Ape Social System and Biogeographical Distribution. *Oikos*, Vol. 118, No. 3, (March 2009), pp. 379-390, ISSN 0003-3472

Lehmann, J.; Korstjens, A.H. & Dunbar, R.I.M. (2010). Apes in a Changing World - the Effects of Global Warming on the Behaviour and Distribution of African Apes. *Journal of Biogeography*, Vol. 37, No. 12, (December 2010), pp. 2217-2231, ISSN 0305-0270

Mackinnon, J. (1974). Behavior and Ecology of Wild Orangutans (Pongo-Pygmaeus). *Animal Behaviour*, Vol. 22, (February 1974), pp. 3-74, ISSN 0003-3472

Marshall, A.J.; Nardiyono; Engstrom, L.M.; Pamungkas, B.; Palapa, J.; Meijaard, E. & Stanley, S.A. (2006). The Blowgun Is Mightier Than the Chainsaw in Determining Population Density of Bornean Orangutans (Pongo Pygmaeus Morio) in the Forests of East Kalimantan. *Biological Conservation*, Vol. 129, No. 4, (May 2006), pp. 566-578, ISSN 0006-3207

Marshall, A.J.; Lacy, R.; Ancrenaz, M.; Byers, O.; Husson, S.J.; Leighton, M.; Meijaard, E.; Rosen, N.; Singleton, I.; Stephens, S.; Traylor-Holzer, K.; Atmoko, S.S.U.; van Schaik, C.P. & Wich, S.A. (2009). Orangutan Population Biology, Life History, and Conservation, In: *Orangutans: Geographic Variation in Behavioral Ecology and Conservation*, S.A. Wich; S.S.U. Atmoko; T.M. Setia & C.P. van Schaik, (Eds.), 77-96, Oxford University Press, ISBN 978-0199213276, New York

McConkey, K. (2005). Sumatran Orangutan (*Pongo abelii*), In: *The World Atlas of Great Apes & their Conservation*, J. Caldecott, & L. Miles (Eds.), pp. 185-204, University of California Press, ISBN 978-0520246331, Berkeley

Meijaard. E.; Dennis, R. & Singleton, I. (2004). Borneo Orangutan PHVA Habitats Units: Composite dataset developed by Meijaard & Dennis (2003) and amended by delegates at the Orangutan PHVA Workshop, Jakarta, January 15-18, 2004. Subsequently further updated by Erik Meijaard, Serge Wich and Rona Dennis. (2007). For further information and full report please see: http://www.grida.no/publications/rr/orangutan/

Miettinen, J.; Shi, C.; Tan, W.J. & Liew, S.C. (2011a). 2010 Land Cover Map of Insular Southeast Asia in 250-m Spatial Resolution. Remote Sensing Letters, Vol. 3, No. 1, (March 2011), pp. 11-20, ISSN 2150-7058 available from http://www.eorc.jaxa.jp/SAFE/LC_MAP/

Miettinen, J.; Shi, C. & Liew, S.C. (2011b). Deforestation rates in insular Southeast Asia between 2000 and 2010. *Global Change Biology*, Vol. 17, No. 7, (July 2011), pp. 2261-2270, ISSN 1365-2486

Mitani, J.C. (1989). Orangutan Activity Budgets - Monthly Variations and the Effects of Body Size, Parturition, and Sociality. *American Journal of Primatology*, Vol. 18, No. 2, pp. 87-100, ISSN 0275-2565

Mitani, J.C. (1990). Experimental field studies of Asian ape social systems. *International Journal of Primatology*, Vol. 11, No. 2, (April 1990), pp. 103-126, ISSN 0164-0291

Mitani, J.C.; Grether, G.F.; Rodman, P.S. & Priatna, D. (1991). Associations among Wild Orangutans - Sociality, Passive Aggregations or Chance. *Animal Behaviour*, Vol. 42, No. 1, (July 1991), pp. 33-46, ISSN 0003-3472

Morrogh-Bernard, H.C.; Husson, S.J.; Knott, C.D.; Wich, S.A.; van Schaik, C.P.; van Noordwijk, M.A.; Lackman-Ancrenaz, I.; Marshall, A.J.; Kanamori, T.; Kuze, N. & bin Sakong, R. (2009) Orangutan Activity Budgets and Diet: A Comparison between Species, Populations and Habitats, In: *Orangutans: Geographic Variation in Behavioral Ecology and Conservation*, S.A. Wich; S.S.U. Atmoko; T.M. Setia & C.P. van Schaik, (Eds.), 77-96, Oxford University Press, ISBN 978-0199213276, New York

Nakicenovic, N.; Davidson,O.;Davis,G.; Grübler, A.; Kram,T.; La Rovere, E.L.; Metz, B.; Morita, T.; Pepper, W.; Pitcher, H.; Sankovski, A.; Shukla, P.; Swart, R.; Watson, R. & Dadi, Z. (2000). *Special Report on Emissions Scenarios: A Special Report of Working Group III of the Intergovernmental Panel on Climate Change*, Cambridge University Press, Retrieved from http://www.grida.no/climate/ipcc/emission/index.htm

Nellemann, C.; Miles, L.; Kaltenborn, B.P.; Virtue, M. & Ahlenius, H. (Eds.). 2007. The last stand of the orangutan – State of emergency: Illegal logging, fire and palm oil in Indonesia's national parks. United Nations Environment Programme, GRID-Arendal, Norway, Available from www.grida.no

Pearson, R.G. & Dawson, T.P. (2003). Predicting the Impacts of Climate Change on the Distribution of Species: are Bioclimate Envelope Models Useful? *Global Ecology and Biogeography*, Vol. 12, No. 5, (September 2003), pp. 361-371, ISSN 0960-7447

Ramirez, J. & Jarvis, A. (2008). High Resolution Statistically Downscaled Future Climate Surfaces, Available from http://ccafs-climate.org/

Ramirez, J. & Jarvis, A. (2010). Downscaling Global Circulation Model Outputs: The Delta Method Decision and Policy Analysis Working Paper No. 1, Available from http://ccafs-climate.org/docs/Downscaling-WP-01.pdf

Rijksen, H.D. & Meijaard, E. (1999). *Our vanishing relative*, Kluwer Academic Publishers, ISBN 978-0792357544, Dordrecht

Rodman, P.S. (1973). Population composition and adaptive organization among orangutans, In: *Comparative Ecology and Behavior of Primates*, J. Crook & R. Michael, (Eds.), 171-209, Academic Press, ISBN 0124934501, London

Rodman, P.S. (1979). Individual activity patterns and the solitary nature of orangutans, In: *The great apes*, Vol. 5, D. L. Hamburg & E. R. McCown, (Eds.), 234–255, W. A. Benjamin, ISBN 9780805336696, London

Russon, A.E.; Wich, S.A.; Ancrenaz, M.; Kanamori, T.; Knott, C.D.; Kuze, N.; Morrogh-Bernard, H.C.; Pratje, P.; Ramlee, H.; Rodman, P.; Sawang, A.; Sidiyasa, K.; Singleton, I. & van Schaik, C.P. (2009). Geographic Variation in Orangutan Diets, In: *Orangutans: Geographic Variation in Behavioral Ecology and Conservation*, S.A. Wich; S.S.U. Atmoko; T.M. Setia & C.P. van Schaik, (Eds.), 77-96, Oxford University Press, ISBN 978-0199213276, New York

Sinclair, S. J.; White, M. D. & Newell, G. R. (2010). How Useful Are Species Distribution Models for Managing Biodiversity under Future Climates? *Ecology and Society*, Vol. 15, No. 1, ISSN 1708-3087

Singleton, I. & van Schaik, C.P. (2002). The Social Organisation of a Population of Sumatran Orang-utans. Folia Primatologica, Vol. 73, No. 1, (January 2002), pp. 1-20, ISSN 0015-5713

Singleton, I.; Wich, S.; Husson, S.; Stephens, S.; Utami Atmoko, S.; Leighton, M.; Rosen, N.; Traylor-Holzer, K.; Lacy, R. & Byers, O. (Eds.). (2004). *Orangutan Population and Habitat Viability Assessment: Final Report.* IUCN/SSC Conservation Breeding Specialist Group, Apple Valley, MN, Available from http://www.cbsg.org/cbsg/workshopreports/23/orangutanphva04_final_report. pdf

Singleton, I.; Knott, C.D.; Morrogh-Bernard, H.C.; Wich, S.A. & van Schaik, C.P. (2009). Ranging Behaviour of Orangutan Females and Social Organisation, In: *Orangutans: Geographic Variation in Behavioral Ecology and Conservation,* S.A. Wich; S.S.U. Atmoko; T.M. Setia & C.P. van Schaik, (Eds.), 77-96, Oxford University Press, ISBN 978-0199213276, New York

Sodhi, N.S.; Koh, L.P.; Brook, B.W. & NG, P.K.L. (2004). Southeast Asian Biodiversity: An Impending Disaster. *Trends in Ecology & Evolution,* Vol. 19, No. 12, (December 2004), pp.654-660, ISSN 0169-5347

Sugardjito, J.; Teboekhorst, I.J.A. & Van Hooff, J. (1987). Ecological Constraints on the Grouping of Wild Orangutans (*Pongo-pygmaeus*) in the Gunung-Leuser-National-Park, Sumatra, Indonesia. *International Journal of Primatology,* Vol. 8, No. 1, (February 1987), pp. 17-41, ISSN 0164-0291

Sugardjito, J. (1995). Conservation of Orangutans - Threats and Prospects, In: *The Neglected Ape,* R.D. Nadler; B.F.M. Galdikas; L.K. Sheeran & N. Rosen (Eds.), pp. 45-49, Plenum Press, ISBN 0306-45213-8, New York

Van Schaik, C. P. (1999). The Socioecology of Fission-Fusion Sociality in Orangutans. *Primates,* Vol. 40, No. 1, (January 1999), pp. 69-86, ISSN 0032-8332

Wich, S.A.; Utami-Atmoko, S.S.; Setia, T.M.; Djoyosudharmo, S. & Geurts, M.L. (2006). Dietary and Energetic Responses of Pongo Abelii to Fruit Availability Fluctuations. *International Journal of Primatology,* Vol. 27, No. 6, (December 2006), pp. 1535-1550, ISSN 0164-0291

Wich, S.A.; Meijaard, E.; Marshall, A.J.; Husson, S.; Ancrenaz, M.; Lacy, R.C.; Van Schaik, C. P.; Sugardjito, J.; Simorangkir, T.; Traylor-Holzer, K.; Doughty, M.; Supriatna, J.; Dennis, R.; Gumal, M.; Knott, C.D. & Singleton, I. (2008). Distribution and Conservation Status of the Orang-Utan (Pongo Spp.) on Borneo and Sumatra: How Many Remain? *Oryx,* Vol. 42, No. 3, (July 2008), pp. 329-339, ISSN 0030-6053

Wich, S.A.; de Vries, H.; Ancrenaz, M.; Perkins, L.; Shumaker, R.W.; Suzuki, A. & van Schaik, C.P. (2009). Orangutan Life History Variation, In: *Orangutans: Geographic Variation in Behavioral Ecology and Conservation,* S.A. Wich; S.S.U. Atmoko; T.M. Setia & C.P. van Schaik, (Eds.), 77-96, Oxford University Press, ISBN 978-0199213276, New York

Willems, E.P. & Hill, R.A. (2009). A Critical Assessment of Two Species Distribution Models: A Case Study of the Vervet Monkey (Cercopithecus Aethiops). *Journal of Biogeography,* Vol. 36, No. 12, (December 2009), pp. 2300-2312, ISSN 0305-0270

Permissions

The contributors of this book come from diverse backgrounds, making this book a truly international effort. This book will bring forth new frontiers with its revolutionizing research information and detailed analysis of the nascent developments around the world.

We would like to thank Leonard M. Druyan, for lending his expertise to make the book truly unique. He has played a crucial role in the development of this book. Without his invaluable contribution this book wouldn't have been possible. He has made vital efforts to compile up to date information on the varied aspects of this subject to make this book a valuable addition to the collection of many professionals and students.

This book was conceptualized with the vision of imparting up-to-date information and advanced data in this field. To ensure the same, a matchless editorial board was set up. Every individual on the board went through rigorous rounds of assessment to prove their worth. After which they invested a large part of their time researching and compiling the most relevant data for our readers. Conferences and sessions were held from time to time between the editorial board and the contributing authors to present the data in the most comprehensible form. The editorial team has worked tirelessly to provide valuable and valid information to help people across the globe.

Every chapter published in this book has been scrutinized by our experts. Their significance has been extensively debated. The topics covered herein carry significant findings which will fuel the growth of the discipline. They may even be implemented as practical applications or may be referred to as a beginning point for another development. Chapters in this book were first published by InTech; hereby published with permission under the Creative Commons Attribution License or equivalent.

The editorial board has been involved in producing this book since its inception. They have spent rigorous hours researching and exploring the diverse topics which have resulted in the successful publishing of this book. They have passed on their knowledge of decades through this book. To expedite this challenging task, the publisher supported the team at every step. A small team of assistant editors was also appointed to further simplify the editing procedure and attain best results for the readers.

Our editorial team has been hand-picked from every corner of the world. Their multi-ethnicity adds dynamic inputs to the discussions which result in innovative outcomes. These outcomes are then further discussed with the researchers and contributors who give their valuable feedback and opinion regarding the same. The feedback is then collaborated with the researches and they are edited in a comprehensive manner to aid the understanding of the subject.

Apart from the editorial board, the designing team has also invested a significant amount of their time in understanding the subject and creating the most relevant covers. They scrutinized every image to scout for the most suitable representation of the subject and create an appropriate cover for the book.

The publishing team has been involved in this book since its early stages. They were actively engaged in every process, be it collecting the data, connecting with the contributors or procuring relevant information. The team has been an ardent support to the editorial, designing and production team. Their endless efforts to recruit the best for this project, has resulted in the accomplishment of this book. They are a veteran in the field of academics and their pool of knowledge is as vast as their experience in printing. Their expertise and guidance has proved useful at every step. Their uncompromising quality standards have made this book an exceptional effort. Their encouragement from time to time has been an inspiration for everyone.

The publisher and the editorial board hope that this book will prove to be a valuable piece of knowledge for researchers, students, practitioners and scholars across the globe.

List of Contributors

Pallav Ray and Tim Li
International Pacific Research Center (IPRC), University of Hawaii, USA

Chidong Zhang
Rosenstiel School of Marine and Atmospheric Science (RSMAS), University of Miami, USA

Jim Dudhia and Mitchell W. Moncrieff
National Center for Atmospheric Research (NCAR), USA

Leonard M. Druyan and Matthew Fulakeza
Center for Climate Systems Research, Columbia University, NASA/Goddard Institute for Space Studies, New York, USA

June-Yi Lee and Bin Wang
Department of Meteorology and International Pacific Research Center, University of Hawaii, Honolulu, HI, USA

Fu Qin and Hao Zhen-chun
State Key Laboratory of Hydrology-Water Resources and Hydraulic Engineering, Hohai University, Nanjing, China

Ou Geng-xin
School of Natural Resources, University of Nebraska–Lincoln, Lincoln, USA

Wang Lu
Delft University of Technology, Delft, The Netherlands

Zhu Chang-jun
College of Urban Construction, Hebei University of Engineering, Handan, China

Sonia R. Gámiz-Fortis, María Jesús Esteban-Parra and Yolanda Castro-Díez
University of Granada, Spain

Paul Nyeko-Ogiramoi
Ministry of Water and Environment, Directorate of Water Development, Rural Water and Sanitation Department, Kampala, Uganda

Katholieke Universiteit Leuven, Civil Engineering Department, Hydraulics Laboratory, Leuven, Belgium
Makerere University, Civil Engineering Department, Kampala, Uganda

Gaddi Ngirane-Katashaya
Makerere University, Civil Engineering Department, Kampala, Uganda

Patrick Willems and Victor Ntegeka
Katholieke Universiteit Leuven, Civil Engineering Department, Hydraulics Laboratory, Leuven, Belgium

O. Rafael García-Cueto and Néstor Santillán-Soto
Universidad Autónoma de Baja California, Instituto de Ingeniería, México

Kazuyuki Saito
University of Alaska Fairbanks, USA
Japan Agency for Marine-Earth Science and Technology, Japan

M.N. Lorenzo, J.J. Taboada and I. Iglesias
University of Vigo / Ephyslab, Spain

Kazuo Mabuchi
Meteorological Research Institute, Japan

Abhilash S. Panicker and Dong-In Lee
Department of Environmental Atmospheric Sciences, Pukyong National University, Busan, South Korea

Robert McGraw
Atmospheric Sciences Division, Environmental Sciences Department, Brookhaven National Laboratory, Upton, NY, USA

O.E. García
Centro de Investigación Atmosférica de Izaña, Agencia Estatal de Meteorología (AEMET), Spain

J.P. Díaz, F.J. Expósito and A.M. Díaz
Grupo de Observación de la Tierra y la Atmósfera, Universidad de La Laguna, Spain

O. Dubovik and Y. Derimian
Laboratoire d'Optique Amosphérique, Université Lille1, France

Charlotte Carne, Stuart Semple and Julia Lehmann
University of Roehampton, England